Nonthermal Pulsed Electric Field Technology for the Food Industry

Innovations in Agricultural and Biological Engineering

Nonthermal Pulsed Electric Field Technology for the Food Industry

Preeti Birwal, PhD

Megh R. Goyal, PhD, PE

Baldev Singh Kalsi, PhD

Ankita Kataria, PhD

Editors

AAP | APPLE ACADEMIC PRESS

First edition published 2026

Apple Academic Press Inc.
1265 Goldenrod Circle, NE,
Palm Bay, FL 32905 USA

760 Laurentian Drive, Unit 19,
Burlington, ON L7N 0A4, Canada

CRC Press
2385 NW Executive Center Drive,
Suite 320, Boca Raton FL 33431

4 Park Square, Milton Park,
Abingdon, Oxon, OX14 4RN, UK

Library and Archives Canada Cataloguing in Publication

Title: Nonthermal pulsed electric field technology for the food industry / Preeti Birwal, PhD, Megh R. Goyal, PhD, PE, Baldev Singh Kalsi, PhD, Ankita Kataria, PhD, editors.
Names: Birwal, Preeti, editor | Goyal, Megh R., editor | Kalsi, Baldev Singh, editor. | Kataria, Ankita, editor.
Series: Innovations in agricultural and biological engineering.
Description: First edition. | Series statement: Innovations in agricultural and biological engineering | Includes bibliographical references and index.
Identifiers: Canadiana (print) 20250146355 | Canadiana (ebook) 20250146363 | ISBN 9781774918869 (hardcover) | ISBN 9781003605904 (ebook)
Subjects: LCSH: Food—Preservation—Technological innovations. | LCSH: Electric fields—Industrial applications. | LCSH: Food industry and trade.
Classification: LCC TP371.7 .N66 2025 | DDC 664/.028—dc23

Library of Congress Cataloging-in-Publication Data

Names: Birwal, Preeti editor http://id.loc.gov/rwo/agents/no2021018820 | Goyal, Megh R. editor http://id.loc.gov/rwo/agents/n82033477 | Kalsi, Baldev Singh editor http://id.loc.gov/rwo/agents/no2025015820 | Kataria, Ankita editor http://id.loc.gov/rwo/agents/no2025015824

Title: Nonthermal pulsed electric field technology for the food industry / Preeti Birwal, Megh R. Goyal, Baldev Singh Kalsi, Ankita Kataria, editors.

Description: First edition. | Palm Bay, FL : Apple Academic Press, 2025. | Series: Innovations in agricultural and biological engineering | Includes bibliographical references and index. | Summary: "While food preservation has been undertaken since time immemorial through techniques such as heating, cooking, drying, concentration, freezing, and cooling, a big challenge today is saving food from deterioration in terms of loss of nutrients, texture, and sensorial characteristics. This new book, Nonthermal Pulsed Electric Field Technology for the Food Industry, illustrates the contemporary pulsed electric field processing techniques for food preservation, an emerging technique that has certain advantages over conventional thermal food processing techniques. Nonthermal PEF produces fresh-like food without affecting the sensory and nutritional properties of food. This volume explores and provides in-depth knowledge and understanding of the pulsed electric field and its applications. It introduces the role of nonthermal technology in novel food product development for ensuring the quality of the product. This book is divided into three parts: Part I: Introduction and Design of Pulsed Electric Field provides an introduction to pulsed electric field, pulse generation technologies, PEF processing system, and safety and critical parameters. Part II: Advances and Applications of Pulsed Electric Field addresses the influence of PEF on microorganisms and enzymes, presents an analysis of kinetics governing the inactivation of enzymes as well as microorganisms under PEF treatment, and covers the application of PEF in hurdle technology. Part III: Role of Pulse Electric Field in Processing and Extraction of Byproducts covers applications and future aspects of the technology in fruits, vegetables, processing of alcoholic beverages; egg, meat, and poultry products; milk products; and cereals, legumes, and oilseeds; as well as in byproducts technology and its usage as a sustainable technique of extracting phytochemicals. This book will be a helpful resource for academicians and professionals in food science, technology, and engineering around the world. It will serve as a reference for researchers, students, scholars, industries, universities, and research centers as well"-- Provided by publisher.

Identifiers: LCCN 2025005531 (print) | LCCN 2025005532 (ebook) | ISBN 9781774918869 hardcover | ISBN 9781003605904 ebook

Subjects: LCSH: Food--Preservation--Case studies | Electric fields--Industrial applications v Case studies | Food industry and trade v Case studies | LCGFT: Case studies. https://id.loc.gov/authorities/genreForms/gf2017026140

Classification: LCC TP371.7 .N66 2025 (print) | LCC TP371.7 (ebook) | DDC 664/.0285028--dc23/eng/20250411

LC record available at https://lccn.loc.gov/2025005531

LC ebook record available at https://lccn.loc.gov/2025005532

ISBN: 978-1-774918-86-9 (hbk)
ISBN: 978-1-003605-90-4 (ebk)
DOI: 10.1201/9781003605904

ABOUT THE BOOK SERIES: INNOVATIONS IN AGRICULTURAL AND BIOLOGICAL ENGINEERING

Under this book series, Apple Academic Press Inc. is publishing book volumes over a span of 8–10 years in the specialty areas defined by the American Society of Agricultural and Biological Engineers (asabe.org). Apple Academic Press Inc. is a principal source of books in agricultural and biological engineering. We welcome book proposals from readers in areas of their expertise.

The mission of this series is to provide knowledge and techniques for agricultural and biological engineers (ABEs). The book series offers high-quality reference and academic content on agricultural and biological engineering (ABE) that is accessible to academicians, researchers, scientists, university faculty and university-level students, and professionals around the world.

Agricultural and biological engineers ensure that the world has the necessities of life, including safe and plentiful food, clean air and water, renewable fuel and energy, safe working conditions, and a healthy environment by employing knowledge and expertise of the sciences, both pure and applied, and engineering principles. Biological engineering applies engineering practices to problems and opportunities presented by living things and the natural environment in agriculture.

ABE embraces a variety of the following specialty areas (asabe.org): aquaculture engineering, biological engineering, energy, farm machinery and power engineering, food and process engineering, forest engineering, information and electrical technologies, soil and water conservation engineering, natural resources engineering, nursery and greenhouse engineering, safety and health, and structures and environment.

For this book series, we welcome chapters on the following specialty areas (but not limited to):

- Academia-to-industry to end-user loop in agricultural engineering
- Agricultural mechanization
- Aquaculture engineering
- Biological engineering in agriculture

- Biotechnology applications in agricultural engineering
- Energy source engineering
- Food and bioprocess engineering
- Forest engineering
- Hill land agriculture
- Human factors in engineering
- Information and electrical technologies
- Irrigation and drainage engineering
- Nanotechnology applications in agricultural engineering
- Natural resources engineering
- Nursery and greenhouse engineering
- Potential of phytochemicals from agricultural and wild plants for human health
- Power systems and machinery design
- GPS and remote sensing potential in agricultural engineering
- Robot engineering and drones in agriculture
- Simulation and computer modeling
- Smart engineering applications in agriculture
- Soil and water engineering
- Micro irrigation engineering
- Structures and environment engineering
- Waste management and recycling
- Rural electrification.
- Sanitary engineering
- Farm to fork technologies in agriculture
- Impact of global warming and climatic change on agriculture economy
- Any other focus areas.

For more information on this series, readers may contact:

Megh R. Goyal, PhD, PE
Book Series Senior Editor-in-Chief: Innovations in Agricultural and Biological Engineering

OTHER BOOKS ON AGRICULTURAL AND BIOLOGICAL ENGINEERING BY APPLE ACADEMIC PRESS, INC.

Management of Drip/Trickle or Micro Irrigation
Megh R. Goyal, PhD, PE, Senior Editor-in-Chief
Evapotranspiration: Principles and Applications for Water Management
Megh R. Goyal, PhD, PE and Eric W. Harmsen, PhD Editors

Book Series: Research Advances in Sustainable Micro Irrigation
Senior Editor-in-Chief: Megh R. Goyal, PhD, PE

- **Volume 1:** Sustainable Micro Irrigation: Principles and Practices
- **Volume 2:** Sustainable Practices in Surface and Subsurface Micro Irrigation
- **Volume 3:** Sustainable Micro Irrigation Management for Trees and Vines
- **Volume 4:** Management, Performance, and Applications of Micro Irrigation Systems
- **Volume 5:** Applications of Furrow and Micro Irrigation in Arid and Semi-Arid Regions
- **Volume 6:** Best Management Practices for Drip Irrigated Crops
- **Volume 7:** Closed Circuit Micro Irrigation Design: Theory and Applications
- **Volume 8:** Wastewater Management for Irrigation: Principles and Practices
- **Volume 9:** Water and Fertigation Management in Micro Irrigation
- **Volume 10:** Innovation in Micro Irrigation Technology

Book Series: Innovations and Challenges in Micro Irrigation
Senior Editor-in-Chief: Megh R. Goyal, PhD, PE

- Engineering Interventions in Sustainable Trickle Irrigation: Water Requirements, Uniformity, Fertigation, and Crop Performance
- Management Strategies for Water Use Efficiency and Micro Irrigated Crops: Principles, Practices, and Performance

- Micro-Irrigation Engineering for Horticultural Crops: Policy Options, Scheduling, and Design
- Micro-Irrigation Management: Technological Advances and Their Applications
- Micro-Irrigation Scheduling and Practices
- Performance Evaluation of Micro-Irrigation Management: Principles and Practices
- Potential of Solar Energy and Emerging Technologies in Sustainable Micro-Irrigation
- Principles and Management of Clogging in Micro-Irrigation
- Sustainable Micro-Irrigation Design Systems for Agricultural Crops: Methods and Practices

Book Series: Innovations in Agricultural and Biological Engineering
Senior Editor-in-Chief: Megh R. Goyal, PhD, PE

- Advanced Research Methods in Food Processing Technologies
- Advances in Food Process Engineering: Novel Processing, Preservation and Decontamination of Foods
- Advances in Green and Sustainable Nanomaterials: Applications in Energy, Biomedicine, Agriculture, and Environmental Science
- Advances in Sustainable Food Packaging Technology
- Analytical Methods for Milk and Milk Products, 3-volume set:
    - Volume 1: Sampling Methods and Chemical and Compositional Analysis
    - Volume 2: Physicochemical Analysis of Concentrated, Coagulated and Fermented Products
    - Volume 3: Microbiological Analysis, Isolation, and Characterization
- Applications of Supercritical Fluid Extraction in Food Processing
- Biological and Chemical Hazards in Food and Food Products: Prevention, Practices, and Management
- Bioremediation and Phytoremediation Technologies in Sustainable Soil Management, 4-volume set:
    - Volume 1: Fundamental Aspects and Contaminated Sites
    - Volume 2: Microbial Approaches and Recent Trends
    - Volume 3: Inventive Techniques, Research Methods, and Case Studies
    - Volume 4: Degradation of Pesticides and Polychlorinated Biphenyls

- Dairy Engineering: Advanced Technologies and Their Applications
- Developing Technologies in Food Science: Status, Applications, and Challenges
- Emerging Technologies in Agricultural Engineering
- Engineering Interventions in Agricultural Processing
- Engineering Interventions in Foods and Plants
- Engineering Practices for Agricultural Production and Water Conservation: An Interdisciplinary Approach
- Engineering Practices for Management of Soil Salinity: Agricultural, Physiological, and Adaptive Approaches
- Engineering Practices for Milk Products: Dairyceuticals, Novel Technologies, and Quality
- Enzyme Inactivation in Food Processing: Technologies, Materials, and Applications
- Field Practices for Wastewater Use in Agriculture: Future Trends and Use of Biological Systems
- Flood Assessment: Modeling and Parameterization
- Food Engineering: Emerging Issues, Modeling, and Applications
- Food Process Engineering: Emerging Trends in Research and Their Applications
- Food Processing and Preservation Technology: Advances, Methods, and Applications
- Food Technology: Applied Research and Production Techniques
- Functional Dairy Ingredients and Nutraceuticals: Physicochemical, Technological, and Therapeutic Aspects
- Handbook of Research on Food Processing and Preservation Technologies, 5-volume set:
  - Volume 1: Nonthermal and Innovative Food Processing Methods
  - Volume 2: Nonthermal Food Preservation and Novel Processing Strategies
  - Volume 3: Computer-Aided Food Processing and Quality Evaluation Techniques
  - Volume 4: Design and Development of Specific Foods, Packaging Systems, and Food Safety / Volume 5: Emerging Techniques for Food Processing, Quality, and Safety Assurance
- Modeling Methods and Practices in Soil and Water Engineering
- Nanotechnology and Nanomaterial Applications in Food, Health, and Biomedical Sciences

- Nanotechnology Applications in Agricultural and Bioprocess Engineering: Farm to Table
- Nanotechnology Applications in Dairy Science: Packaging, Processing, and Preservation
- Nanotechnology Horizons in Food Process Engineering, 3-volume set:
  - Volume 1: Food Preservation, Food Packaging and Sustainable Agriculture
  - Volume 2: Scope, Biomaterials, and Human Health
  - Volume 3: Trends, Nanomaterials, and Food Delivery
- Nonthermal Plasma Technology for Food and Food Products
- Nonthermal Pulsed Electric Field Technology for the Food Industry
- Novel and Alternative Methods in Food Processing: Biotechnological, Physicochemical, and Mathematical Approaches
- Novel Dairy Processing Technologies: Techniques, Management, and Energy Conservation
- Novel Processing Methods for Plant-Based Health Foods: Extraction, Encapsulation and Health Benefits of Bioactive Compounds
- Novel Strategies to Improve Shelf-Life and Quality of Foods: Quality, Safety, and Health Aspects
- Ohmic Heating Technology for Processing of Foods and Food Products: Design, Food Quality, and Applications
- Phytochemicals and Medicinal Plants in Food Design: Strategies and Technologies for Improved Healthcare
- Processing of Fruits and Vegetables: From Farm to Fork
- Processing Technologies for Milk and Milk Products: Methods, Applications, and Energy Usage
- Quality Control in Fruit and Vegetable Processing: Methods and Strategies
- Scientific and Technical Terms in Bioengineering and Biological Engineering
- Soil and Water Engineering: Principles and Applications of Modeling
- Soil Salinity Management in Agriculture: Technological Advances and Applications
- State-of-the-Art Technologies in Food Science: Human Health, Emerging Issues and Specialty Topics
- Sustainable and Functional Foods from Plants: Health Impact, Bioactive Compounds, and Production Technologies
- Sustainable Biological Systems for Agriculture: Emerging Issues in Nanotechnology, Biofertilizers, Wastewater, and Farm Machines

ABOUT THE EDITORS

Preeti Birwal, PhD
Scientist, Department of Processing and Food Engineering, College of Agricultural Engineering and Technology, Punjab Agricultural University, Ludhiana, Punjab, India

Preeti Birwal, PhD, is working as a Scientist (processing and food engineering) in the Department of Processing and Food Engineering, College of Agricultural Engineering and Technology, Punjab Agricultural University, Ludhiana, Punjab, India.

She is currently working in the area of nonthermal food preservation, fermented beverages, food packaging, and technology of millet-based beer. She has served at Jain Deemed to be University, Bangalore, as a member of the board of examiners and placements. She also advises several MTech. scholars in Food Technology. She has participated in several national and international conferences and seminars and has delivered lectures as a resource person on doubling farmers' income through dairy technology in a training sponsored by the Directorate of Extension, Ministry of Agriculture and Farmers Welfare, Government of India.

She has published research papers, edited books, book chapters, popular articles, conference papers, abstracts, and editorial opinions. She has been named outstanding reviewer of the month by the journal *Current Research in Nutrition and Food Science*. She also serves as an editor and reviewer for several journals.

Dr. Birwal has successfully guided five postgraduate students for their dissertation work. She also serves as an external examiner for various Indian state agricultural universities. She is a life member of IDEA. She holds a BSc degree (2012) in Dairy Technology from ICAR-National Dairy Research Institute (NDRI), Karnal; an MSc degree (2014) in Food Process Engineering and Management from NIFTEM, Haryana; and PhD (Dairy Engineering) from ICAR-NDRI, Bangalore. She is a recipient of MHRD (2008), Nestle India (2009), GATE (2012–2014), UGC-RGN fellowships (2014–2018). She has successfully completed AUTOCAD 2D and 3D certification.

Megh R. Goyal, PhD, PE

Retired Professor in Agricultural and Biomedical Engineering, University of Puerto Rico, Mayaguez Campus; Senior Acquisitions Editor, Biomedical Engineering and Agricultural Science, Apple Academic Press, Inc.

Megh R. Goyal, PhD, PE, is currently a Retired Professor of Agricultural and Biomedical Engineering from the General Engineering Department at the College of Engineering at the University of Puerto Rico–Mayaguez Campus (UPRM); and Senior Acquisitions Editor and Senior Technical Editor-in-Chief for Agricultural and Biomedical Engineering for Apple Academic Press Inc. During his professional career, he has worked as a Soil Conservation Inspector; Research Assistant at Haryana Agricultural University and Ohio State University; Research Agricultural Engineer/Professor at the Department of Agricultural Engineering of UPRM; and Professor of Agricultural and Biomedical Engineering in the General Engineering Department of UPRM. He spent a one-year sabbatical leave in 2002–2003 at the Biomedical Engineering Department of Florida International University, Miami, USA.

Dr. Goyal was the first agricultural engineer to receive the professional license in agricultural engineering from the College of Engineers and Surveyors of Puerto Rico. In 2005, he was proclaimed the "Father of Irrigation Engineering in Puerto Rico for the Twentieth Century" by the American Society of Agricultural and Biological Engineers, Puerto Rico Section, for his pioneering work on micro irrigation, evapotranspiration, agroclimatology, and soil and water engineering.

During his professional career of 52 years, he has received many awards, including Scientist of the Year, Membership Grand Prize for the American Society of Agricultural Engineers Campaign, Felix Castro Rodriguez Academic Excellence Award, Man of Drip Irrigation by the Mayor of Municipalities of Mayaguez/Caguas/Ponce and Senate/Secretary of Agriculture of ELA, Puerto Rico, and many others. He has been recognized as one of the experts "who rendered meritorious service for the development of [the] irrigation sector in India" by the Water Technology Centre of Tamil Nadu Agricultural University in Coimbatore, India, and ASABE (American Society of Agricultural and Biological Engineers), which was bestowed on him the 2018 Netafim Microirrigation Award.

Dr. Goyal has authored more than 200 journal articles and edited more than 140 books. AAP has published many of his books, including *Management of Drip/Trickle or Micro Irrigation; Evapotranspiration: Principles and Applications for Water Management;* ten-volume set on *Research Advances in Sustainable Micro Irrigation.* He has also authored the textbooks *Elements of Agroclimatology* (Spanish) by UNISARC, Colombia, and two *Bibliographies on Drip Irrigation.* Dr. Goyal has also developed several book series with AAP, including Innovations in Agricultural & Biological Engineering (with over 60 titles in the series to date), Innovations and Challenges in Micro Irrigation; and Innovations in Plant Science for Better Health: From Soil to Fork.

Dr. Goyal received his BSc degree in Engineering from Punjab Agricultural University, Ludhiana, India, and his MSc and PhD degrees from the Ohio State University, Columbus, Ohio, USA. He also earned a Master of Divinity degree from the Puerto Rico Evangelical Seminary, Hato Rey, Puerto Rico, USA.

Baldev Singh Kalsi, PhD

Senior Research Fellow, ICAR-Central Institute of Post-Harvest Engineering & Technology (CIPHET), Ludhiana, Punjab, India

Baldev Singh Kalsi, PhD, is working as a Senior Research Fellow at ICAR-Central Institute of Post-Harvest Engineering & Technology (CIPHET), Ludhiana, Punjab, India. He is actively involved in the field of food engineering research. His master's degree focused on preservation of juices using ultrasound and in combination with thermal or/and pressure. During his PhD (completed in 2023), he worked in the area of ohmic heating-assisted extraction of valuable components from agricultural produce.. He has participated in many national and international conferences and has won a Best Oral Presentation Award at the 55th Annual Convention of the Indian Society of Agricultural Engineers (ISAE) and International Symposium on Emerging Trends in Agricultural Engineering Education, Research and Extension, Patna, Bihar; and a Best Poster Presentation Award at the VIIth International Conference on Global Research Initiatives for Sustainable Agriculture & Allied Sciences (GRISAAS–2022), Birsa Agricultural University, Jharkhand, India. He is the recipient of a University Merit Scholarship during his master's and PhD degree programs. He has two research papers in international journals, one accepted book chapter, two popular articles, and six abstracts, to his credit.

Ankita Kataria, PhD
Assistant Professor, Department of Food Science and Technology, Maharaja Ranjit Singh Punjab Technical University, Bathinda, India

Ankita Kataria, PhD, is currently working as Assistant Professor Food Technology in Maharaja Ranjit Singh Punjab Technical University (MRFSPTU), Bathinda, Punjab. She is a member of the Board of Studies, Board of Control and Departmental Doctoral Research Committee of the Department of Food Science and Technology, MRSPTU. She has been awarded a PhD Food Technology from Punjab Agricultural University, Ludhiana, Punjab, India. She is also a recipient of an ICAR-Senior Research Fellowship. She is a Silver Medalist for her MSc Food in Technology and a Gold Medalist for her BTech in Dairy Technology. She also has more than 40 publications including articles in reputed journals, book chapters, and magazines within a short career span, which also includes an editorial opinion. She has presented her work at international and national conferences and won a best presentation award. She has also been invited for television and institutional talks on topics including millets and ayurvedic food products. She has successfully organized and delivered lectures in various conferences, workshops, and hands-on training sessions. She has received a Young Scientist Award from the Agro Environmental Education & Farmer's Welfare Society (AEEFWS) of which she is also a life member. She is also a life member of the Association of Food Scientists and Technologists of India (AFSTI). She is advising several PhD, MSc, and BSc scholars in food technology and has successfully guided eight postgraduate students for their dissertation work. She is also serving as a reviewer for several journals. She has worked as R&D Manager in Britannia Industries Limited. She has also served as the first Centre Manager, Centre of Excellence for Food Fortification (CEFF) at the National Institute of Food Technology Entrepreneurship and Management (NIFTEM) and worked for setting up and inauguration of the Rice Fortification Unit under the Centre of Excellence for Food Fortification (CEFF), NIFTEM Haryana.

CONTENTS

CONTRIBUTORS

Aparna Agarwal
Department of Food Technology, Lady Irwin College, University of Delhi, New Delhi, India

V. Anushma
Department of Meat and Marine Sciences, CSIR-Central Food Technological Research Institute (CFTRI), Cheluvamba Mansion, Valmiki Road, Opposite Railway Museum, Kajjihundi, Mysuru, Karnataka, India

Ritika Arora
Department of Home Science, University of Delhi, New Delhi, India

Mandeep Singh Azad
KVK, Reasi, India

Neelu S. Babu
Department of Meat and Marine Sciences, CSIR- Central Food Technological Research Institute (CFTRI), Cheluvamba Mansion, Mysuru, Karnataka, India

Adity Bandhral
Department of Food Technology and Nutrition, School of Agriculture, Lovely Professional University, Phagwara, Punjab, India

Suradeep Basak
UIBT, Chandigarh University, Mohali, Chandigarh, India

Vimala Beera
Department of Food Safety and Quality Assurance, Dr. N. T. R. College of Food Science and Technology, Acharya N. G. Ranga Agricultural University, Bapatla, Andhra Pradesh, India

Yuvraj K. Bhosale
Department of Food Technology and Nutrition, School of Agriculture, Lovely Professional University, Phagwara, Punjab, India

Preeti Birwal
Department of Processing and Food Engineering, Punjab Agricultural University (PAU), Ludhiana, Punjab, India

Bhavesh Chavhan
Dairy Engineering, ICAR-National Dairy Research Institute, Karnal, Haryana, India

Pintu Choudhary
Department of Food Technology and Nutrition, School of Agriculture, Lovely Professional University, Phagwara, Punjab, India

Himabindu Gazula
Department of Global Food Safety Quality and Regulatory, Starbucks Coffee Company, Seattle, Washington State, USA

Gargi Ghoshal
S.S.B. UICET, Panjab University, Chandigarh, India

Megh R. Goyal
University of Puerto Rico-Mayaguez, USA

Mahendra Gunjal
Department of Food Technology and Nutrition, School of Agriculture, Lovely Professional University, Phagwara, Punjab, India

Rizwana Haleem
Department of Food Technology, Bhaskaracharya College of Applied Sciences, University of Delhi, New Delhi, India

Saumya Jain
Department of Food and Nutrition & Food Technology, Institute of Home Economics, University of Delhi, Delhi, India

Baldev Singh Kalsi
Department of Processing and Food Engineering, College of Agricultural Engineering and Technology, Punjab Agricultural University, Ludhiana, Punjab, India

Sandesh Suresh Karkal
Department of Meat and Marine Sciences, CSIR-Central Food Technological Research Institute (CFTRI), Cheluvamba Mansion, Mysuru, Karnataka, India

Ankita Kataria
Department of Food Science and Technology, Maharaja Ranjit Singh Punjab Technical University Bathinda, Punjab, India

Akashdeep Kaur
Department of Energy Research Centre, Dr. S. S. Bhatnagar University Institute of Chemical Engineering & Technology, Panjab University, Chandigarh, India

Sawinder Kaur
Department of Food Technology and Nutrition, School of Agriculture, Lovely Professional University, Phagwara, Punjab, India

Preeti Khanna
WCD Division, NITI AAYOG, New Delhi, India

Kawardeep Kour
Sher-e-Kashmir University of Animal Sciences and Technology, Jammu, India

Tanaji G. Kudre
Department of Meat and Marine Sciences, CSIR-Central Food Technological Research Institute (CFTRI), Cheluvamba Mansion, Mysuru, Karnataka

Rewa Kulshrestha
Department of Food Processing and Technology, Atal Bihari Vajpayee Vishwavidyalya, Bilaspur, Chhattisgarh, India

Sandhya Kumar
National Institute of Food Technology Entrepreneurship and Management, Thanjavur, Tamil Nadu, India

Priya Dangi Malik
Department of Food and Nutrition & Food Technology, Institute of Home Economics, University of Delhi, India

Gopinath Mummaleti
Department of Food Biotechnology, Indian Institute of Food Processing Technology, Thanjavur, Tamil Nadu, India

Sagarika Nukasani
Department of Food Process Engineering, College of Food Processing Technology & Bio-Energy, Anand Agricultural University, Anand, Gujarat, India

Krishnaveni Rajamahendravarapu
Institute of Public Enterprise, Hyderabad, Telangana, India

Reshu Rajput
Department of Food Technology, Chandigarh University, Mohali, Chandigarh, India

Nalla Bhanu Prakash Reddy
National Institute of Food Technology, Entrepreneurship and Management (NIFTEM), Thanjavur, India

Jagbir Rehal
Food Technologist (Fruit and Vegetable), Department of Food Science and Technology, Punjab Agricultural University, Ludhiana, Punjab, India

Shubham Rohilla
Department of Food Engineering and Technology, Tezpur University, Assam, India

Vandana Sablania
Department of Food and Nutrition & Food Technology, Institute of Home Economics, University of Delhi, India

Chayanika Sarma
Department of Biotechnology, Vivekanandha College of Arts and Sciences for Women (Autonomous), Elayampalayam, Ticruchengode, Tamil Nadu, India

Abhishek Dutt Tripathi
Department of Dairy Science and Food Technology, Institute of Agricultural Sciences, Banaras Hindu University, Varanasi, Uttar Pradesh, India

Srinivas Yarrakula
Centre of Excellence in Non-Thermal Processing, Planning and Monitoring Cell, National Institute of Food Technology, Entrepreneurship and Management, Thanjavur, Tamil Nadu, India

ABBREVIATIONS

AC	alternate current
ALP	alkaline phosphatase
AR	*Asparagus officinalis* root
ATP	adenosine triphosphate
CARISO	carotenoid isomerase
CFU	colony-forming unit
CLO	cereal, legumes, and oilseeds
Cm	centimeter
CP	cold plasma
CSM	control soybean milk
DC	direct current
DF	dietary fiber
DNA	deoxyribonucleic acid
DUT	deoxyuridine triphosphate pyrophosphatase
E	electric field
E. coli	*Escherichia coli*
E_a	activation energy
EFS	electric field strength
FAO	Food and Agriculture Organization
FDA	Food and Drug Administration
GTO	gate turn-off
HHP	high hydrostatic pressure
HI US	high-intensity ultrasound
HIPEF	high-intensity pulsed electric field
HPCD	high-pressure carbon dioxide
HPP	high-pressure processing
HSB	hydrated soybeans
HTST	high-temperature–short-time
Hz	Hertz
IGBT	insulated gate-commutated thyristor
IR	irradiation
IU	international unit
kV	kilovolt
LL	*Longissimus lumborum*

LOX	lipoxygenase
LPO	lacto peroxidase
MFGs	milk fat globules
MIPEF	moderate intensity pulsed electric field
mPa-s	millipascal second
NCTC	National Collection of Type Cultures
OMF	oscillating magnetic field
P	pulse width
Pe	electro-compressive force
PEF	pulsed electric field
PFGE	pulsed-field gel electrophoresis
PFN	pulse-forming network
PG	polygalacturonase
PL	pulsed light
PME	pectin methylesterase
PMF	pulse magnetic field
POD	peroxidase
PPO	polyphenol oxidase
PS	phytoene synthase
PUFA	polyunsaturated fatty acids
RNA	ribonucleic acid
SBS	soybeans slurry
SDSP-PAGE	sodium dodecyl sulfate–polyacrylamide gel electrophoresis
SGCT	symmetrical gate commutated thyristor
SM	*Semimembranosus* muscle
SMF	static magnetic field
TMF	transmembrane potential
TS	thermosonication
US	ultrasound
USFA	unsaturated fatty acids
USFDA	United States Food and Drug Administration
UV	ultraviolet
V	voltage
WPNI	whey protein nitrogen index
δ	thickness of the membrane
$\Delta\varphi_c$	critical potential difference
ε_m	dielectric constant of the empty space
ε_o	electrical conductivity of the medium
μs	microsecond

PREFACE

Food preservation has existed since our civilization began. Food with nutrition is important not only for relief from hunger but is also the demand of today's population. Food with convenience, good sensorial attributes, and longer shelf-life is today's challenge. Food preservation in the current scenario is done by heating, cooking, drying, concentration, freezing, and cooling. All these techniques have their own pros and cons. It is a challenge for food scientists and industrialists to save food from deteriorating in quality such as in loss of nutrients, texture, and sensorial characteristics while also having a longer shelf life.

This volume, *Nonthermal Pulsed Electric Field Technology for the Food Industry*, attempts to illustrate various applications of pulsed electric field processing techniques to address this challenge. Moreover, this book introduces the role of nonthermal technology in novel food product development to ensure the quality of the product. It also provides a comprehensive overview of future aspects.

This book is divided into three parts: Part 1: Introduction and designing aspects cover introduction on pulsed electric field, pulse generation technologies, pulsed electric field (PEF) processing systems, and safety and critical parameters. Part 2: This part addresses the influence of PEF on microorganisms and enzymes, provides analysis of kinetics governing the inactivation of enzymes as well as microorganisms under PEF treatment, and covers the application of PEF in hurdle technology. Part 3 covers the applications and future aspects of pulsed electric field technology on fruits, vegetables, alcoholic beverages, egg, meat, and poultry, milk products, cereals, legumes, and oilseeds. It also looks at potential applications and future potential of PEF in by-products technology, as well as its usage as a sustainable technique of extracting phytochemicals.

This book will serve academicians and professionals of food science, technology, and engineering around the world. It is a valuable reference for researchers, students, scholars, industries, universities, and research centers. This book, in short, explores and provides in-depth knowledge and understanding of the pulsed electric field and its applications.

This book has taken its present shape due to the excellent contribution by all contributing authors, who have been souls of this compendium. Their names in each chapter are included with each chapter and in the list of contributors. We are indeed indebted to them for their knowledge, dedication, and enthusiasm.

We also extend our sincere thanks to the editorial office of Apple Academic Press, Inc. for their facilitation and suggestions right through this assignment and or bringing out this quality production.

We hereby appeal to our readers for their productive suggestions that may form the basis for improving future editions.

We take this opportunity to thank (1) our families for their motivation, moral support, and blessings in counteracting every obstacle coming our way; (2) our spouses for their understanding, patience, and encouragement throughout this project; (3) our institutes: PAU, Ludhiana, Punjab, and UPRM, Mayaguez, for their support during the compilation of this publication.

—Editors

PART I

INTRODUCTION AND DESIGN OF PULSED ELECTRIC FIELD

CHAPTER 1

APPLICATIONS OF PULSED ELECTRIC FIELD IN THE FOOD INDUSTRY: AN OVERVIEW

SANDESH SURESH KARKAL, ANUSHMA V., NEELU S. BABU, and TANAJI G. KUDRE

ABSTRACT

Currently, the food industry is looking for the best replacement for the conventional methods primarily used for the preservation and processing of foods. The conventional techniques used in the food industry hold certain drawbacks, mainly concerned with the loss of quality attributes and the loss of heat-sensitive compounds. Some novel techniques have been introduced to replace the conventional methods used in the food industry. Pulse electric field is an emerging, novel green technology with the potential to substitute conventional techniques and produces foods with excellent sensory and nutritional quality and shelf life. The laboratory and pilot-scale works have shown promising results for successfully utilizing pulse electric fields for food preservation and processing. However, pitfalls such as high investment costs and elevated processing costs have limited its complete installation in the food industry. Still, research is now fully oriented to making it cost-effective and user-friendly. The studies have also focused on combining pulse electric field with other novel green techniques so that it can be used to treat almost all food products. Therefore, this chapter aims to deliver basic information on PEF for the preservation and processing of food, current status and applications, commercial implementations, pitfalls, and future prospects.

Nonthermal Pulsed Electric Field Technology for the Food Industry.
Preeti Birwal, Megh R. Goyal, Baldev Singh Kalsi, & Ankita Kataria (Eds.)

DOI: 10.1201/9781003605904-2

1.1 INTRODUCTION

Over decades, several novel techniques for the processing of food have been proposed and developed for their application in the food industries.[7] The food industries require the implementation of novel automation and procedures for the extraction of essential biomolecules, as well as manufacturing the products reliably, cheaply, and efficiently without causing damage to the environment.[87] In food industries, conventional food processing methods such as cooking, blanching, freezing, drying, and refrigeration are combined with other processing operations via various process-technology plants by utilizing novel technical prospects.[78]

Conventional food processing methods aim to inactivate pathogenic microorganisms and undesirable enzymes to provide a safe product. However, conventional food processing techniques possess some key bottlenecks despite some beneficial effects. The drawbacks include unfavorable changes in flavor, texture, color, general appearance, and nutritional values of the food products. Food scientists are therefore challenged with the task of finding alternative technologies that would preserve fresh-like characteristics of food and ensure microbial safety. In recent years, some new food processing technologies have been introduced in food industries for better food processing without affecting its quality.

These technologies are so-called "non-thermal food processing," and have been widely popularized to develop food products with extended shelf life and preserved nutritional and organoleptic characteristics. These technologies include high hydrostatic pressure (HHP), pulsed magnetic field (PMF), pulsed electric field (PEF), high power ultrasound, microwave, etc. These technologies require minimum heat, operate at low or moderate temperatures, require no or less solvent, and are eco-friendly.[7] Moreover, these techniques retain the color, flavor, taste, appearance, and nutrients of the food products apart from the inactivation of enzymes and microorganisms, which causes quality deterioration and food spoilage.

Among these novel non-thermal techniques, PEF has acquired more importance as a non-thermal and innovative technology for processing and preserving food. It has been used to treat many biological materials to inactivate microorganisms and enzymes to prolong the material's shelf life and ensure its safety without any unwanted heat or chemical effects.[78] Applying PEF technology as a non-thermal food processing and preservation technique produces high-quality food for consumers. The implementation of the PEF technique has been considerably established for the pasteurization of milk and dairy products, soup, juices, tenderization of meat, and so on.[78] PEF technology involves the release of high voltage pulses to either semi-solid or liquid material positioned in the middle of two electrodes. PEF is extensively used for inactivating enzymes and microorganisms by applying an electric field of a small pulse with 0.1–0.3 to 20–80 kV/cm range of pulse amplitude.[15,87] By applying short pulses of high voltage, electric fields in the 5–50 kV/cm range are produced between the electrodes, and this causes microbial inactivation at below thermal processing temperatures. The application of a high voltage at the range of 20–80 kV/cm for the processing of biomaterial placed between two electrodes is known as high-intensity pulsed electric field (HIPEF). Low treatment intensity allows reversible disturbance of phospholipid bilayer.[38]

High electric field intensity pulses cause perpetual lysis of cell membrane, which is the basis for microbial inactivation in non-thermal food pasteurization techniques.[36] PEF causes only minor adverse effects on food quality attributes and, therefore, are also considered superior to the traditional thermal processing method. The microbial inactivation level by PEF technology hinge on the process, medium-related, environmental (temperature), and microbial parameters. The process parameters are electric field strength, total treatment time, pulse width, and pulse waveform. Conductivity, composition, and pH are medium-related parameters. Microbial parameters include microbial strain and the growth phase.[17,70,90]

Besides food processing and preservation, various studies have been performed to determine the application of PEF. PEF has also been used to extract biomolecules from food materials or waste. The conventional techniques used to extract biomolecules are maceration, soxhlet extraction, grinding, boiling, soaking, heat reflux, water percolation, and so on. These techniques hold drawbacks such as reduced extraction yields, high consumption of solvents, long processing duration, and poor extraction efficiency.[87]

The application of PEF in the extraction of biomolecules alongside conventional extraction techniques for the bioactive compound's extraction disclosed that the treatment of biomaterials with PEF preceding the conventional extraction significantly improved the bioactive ingredients yield.[87] Research has also been carried out in combining the PEF and other green extraction techniques to enhance the desired biomolecules' recovery. The combination of these techniques has shown significant recovery of the target compound from the source. Other applications of PEF include the treatment of wastewater, PEF-assisted cooking, improvement of mass transfer, and plant tissue softening. The advantages of this technology are that it kills microorganisms by preserving food qualities such as color, flavor, and nutrition. Further benefits include no proof of toxicity, relatively shorter treatment time, quick thawing, and no environmental hazards. PEF systems are attractive because of their lower operating costs and because they yield high-quality, minimally processed products.[41]

Some of this technology's most critical technical limitations are high initial cost, no effect on enzymes and spores, suitable for liquid particles in the liquid, electrolysis products may adversely affect food, and energy efficiency is not yet specific. Its application in food is also limited because only food that can withstand high electric fields can be used. However, combining PEF with other novel, innovative techniques would help overcome certain drawbacks of PEF. This chapter provides insights into the historical approach of the pulse electric field, its working principle, and critical factors affecting the PEF. The applications of PEF in food and other industries, its merits and demerits, and prospects are also described in this chapter.

This chapter provides an overview and applications of the PEF technique in the food industry.

1.2 HISTORICAL APPROACH

The application of electric energy for processing food products, especially the pasteurization process, has been established over the last few decades. Electric energy through microwave heating, ohmic heating, low voltage alternating current, low electric field stimulation, high voltage arc discharge, and high-intensity PEF has been used to inactivate unfavorable microorganisms and enzymes in food and food products. The exploitation of the electric field for microbial inactivation has been carried out since an electric current was made available at commercial scale. Around the turn of the 19th century, the application of electric current in the inactivation of microorganisms was investigated.

Prochownik and Spaeth[62] used 300 mA direct current to inactivate *Staphylococcus aureus* in their experimental work. They observed that the electric current failed to inactivate the bacteria but noticed variations in the acidity of treated media at different locations of the treatment chamber. The early 20th century marked the introduction of new process in Europe and the USA called "Electropure." The process was mainly developed for the pasteurization of milk. In this process, the milk was treated with alternating current with a voltage of 220–4200 V in a treatment chamber comprising carbon electrodes. This process continued until the mid-20th century until novel, cost-effective thermal technologies emerged as a substitute for the preservation of food. However, in the late 1940s, the implementation of electric fields other than the inactivation of microorganisms was explored in food processing. Currently, PEF has found its extensive application in the food industry and other industries.

In 1950, another new process was developed to treat food samples. This process was mainly based on Joule heating. In this process, the food sample was treated with a high-voltage pulse discharge of electric current applied across two electrodes. The passage of electric current results in heat generation due to the resistivity of food material. This process was known as electrohydraulic treatment. The electrohydraulic treatment method proved to be the rapid, effective, cheap non-thermal process for sterilizing sewage and water without adding chemicals or increasing temperature. Besides the sterilization of water, the potential application of this method in food technology was never realized. The contamination of food materials by the electrodes and food particles disintegration was the main reason for its failure in the food industry.[38]

Following the failures of the newly discovered non-thermal, innovative, green techniques for their application in the food industry, Heinz Doevenspeck[20] carried out pioneering research on the PEF application for the processing of food. They designed the PEF unit, which is patented in Germany. The PEF-treated food materials showed improved phase separation during the processing. Within a decade, Doevenspeck developed his PEF technology even for the inactivation of microorganisms. The beer spoilage by microorganisms was overcome with the PEF treatment.[20,38] Following Doevenspeck's work, the Unilever Research Centre in United Kingdom carried out the first systematic studies on the lethal effect of PEF on the microorganism. The studies mainly aimed at determining cell lysis, ion leakage, and loss of cytoplasmic constituents. The research confirmed that PEF successfully inactivates the microorganism.[38,70]

Further, based on Doevenspeck's work, the German company Krupp Maschinentechnik developed a technical scale PEF unit with a treatment capacity of 200 kg/h for the treatment of oilseeds, fruit mashes, palm fruit, sugar beet, and meat and fish slurry. The PEF system developed by Krupp Maschinentechnik was known as ELCRACK system. But this system was dismantled in a few months owing to the problems concerned with the stability of electrodes, post-treatment protein recovery, and solid–liquid separation. Zimmermann et al.[91] introduced the promotion of cell-to-cell fusion in vitro using PEF.

Thereafter, Hulsheger et al.[37] reported the sensitivity of various microorganisms to PEF and the inactivation kinetics of microorganisms as a function of electric field strength and treatment time. In 1986, after ELCRACK failure, another pilot-scale PEF plant was developed, comprising a high voltage pulse generator (15 kV). This pilot-scale PEF plant was called ESERIL.[80] This PEF plant was mainly designed for the inactivation of microbial cells using a high-intensity electric field and was mainly used for treating liquid food. After 1990, the application of PEF in food industries has revived. Many studies were performed to determine the potential PEF applications for microbial inactivation, improved food preservation, mass transfer, and the extraction of desired compounds. Presently, research is emphasized on engineering aspects, modeling, scale-up of PEF technology, and commercialization of PEF applications toward food and non-food entities.

1.3 PRINCIPLE OF OPERATION OF PEF

PEF is a treatment wherein the number of small pulses of direct current voltage is applied for a short time duration (microseconds to milliseconds) through food material positioned between two electrodes. The electric discharges of short pulses result in an electric field (E). The electric field intensity/strength is determined by the gap between the two electrodes and the applied voltage. Generally, a field intensity less than 100 V/cm is considered a low-intensity electric field, whereas a field intensity higher than 100 V/cm is regarded as a high-intensity electric field.[66]

The electric field resulting from applying high voltage causes microorganisms' inactivation. The form of the applied electric field could be oscillatory, square wave, bipolar, or exponentially decaying at sub-ambient, ambient, and marginally beyond ambient temperature. Following the PEF treatment, the food is hygienically packaged and refrigerated. When the short pulses of electric discharges are delivered to the food material positioned between two electrodes at room temperature, the ions in the food transfer electricity, thereby exhibiting electrical conductivity to a certain degree. In the case of liquid food, the application of electric current results in the transfer of current to each point in the liquid due to the ions present.[90]

1.3.1 PEF APPARATUS

The treatment of the food material with PEF is carried out in an apparatus that comprises of following main basic components (Figure 1.1):

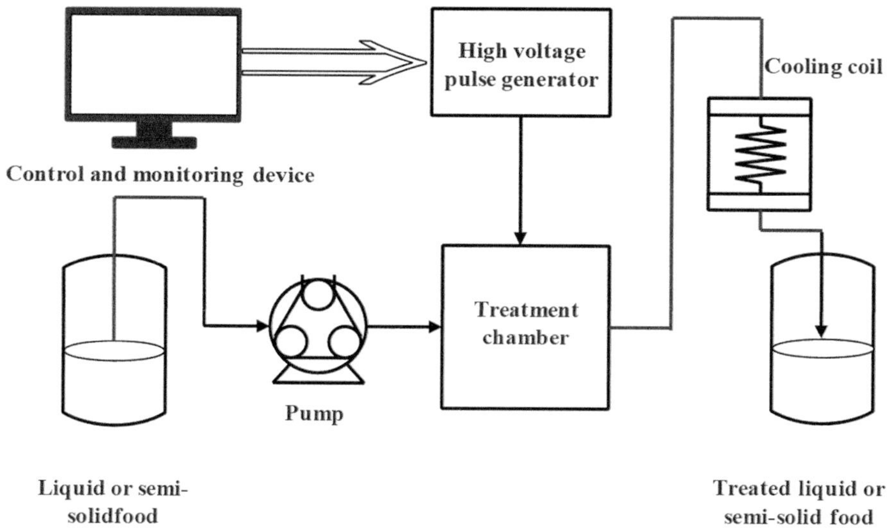

FIGURE 1.1 Schematic diagram depicting the operation of PEF for food processing.

i. **High voltage source/high voltage generator:** The high voltage source, also known as a high voltage generator, transforms alternating current to direct current and supplies a selected voltage direct current. Using the current generated, the capacitor charges and stores energy. In a given time, how often a capacitor charges or discharges limits the system's total power. The capacitor's quantity, size, and resistance of the charging resistor influence the power needed to charge the capacitor. Large capacitors require more time and power to charge compared to small capacitors. Eq. 1.1 defines the capacitance (C_0) of the capacitor:

$$C_0 = \frac{\tau}{R} = \frac{\tau\sigma A}{d} \tag{1.1}$$

where τ is the pulse duration, R is the resistance, σ is the conductivity of food, d is the treatment gap between the electrodes, and A is the electrode surface area.

The mathematical formula used to define the energy stored in the capacitor is:

$$Q = 0.5C_o V^2 \tag{1.2}$$

where Q is the energy stored in the capacitor, C_o is the capacitance, and V is the charge voltage.

ii. **Capacitors:** Either there will be a single capacitor or several parallel-connected capacitors to store electrical energy (W) temporarily. In case of insufficient voltage in the capacitor, a step-up pulse transformer is used to generate sufficient voltage for treatment. The stored energy in capacitors is utilized to produce electric and magnetic fields. The generated electric fields accelerate the charged particles, resulting in chemical, thermal, mechanical, electromagnetic wave, or breakdown effects. The magnetic field assists in extremely high pressure generation.

iii. **Voltage switch:** The voltage switch delivers the direct current voltage pulse to the electrodes and the food material. Currently, ON switches and ON/OFF switches are the two prime switches available for PEF. ON switches deliver complete discharging of the capacitor, following which this switch can be turned off. The switch is considered the essential component in the apparatus because it must be functional (able to turn on and off the circuit) at high voltage and large current in a short time frame (fraction of microseconds). Trigatron, Gas spark gap, Ignitron, and Thyratron are examples of the ON switchs used in PEF. The advantage of ON switches is that they can operate high voltages at a lower cost than ON/OFF switches, but the low rate of repetition and short life are the main drawbacks limiting their preference. ON/OFF switches were developed to control the generation of electric pulses with discharging of capacitors either entirely or partially. Symmetrical gate commutated thyristor (SGCT), insulated gate bipolar transistor (IGBT), and Gate turn off (GTO) are examples of ON/OFF switches.

iv. **Treatment chamber:** The treatment chamber is considered the most crucial and complex component in the processing system. Generally, the treatment chamber comprises two electrodes, composed of metal or carbon, parted by the food material. The primary purpose of this chamber is to keep the food material inside when the electric pulse is applied. If the applied electric field strength exceeds the electric field strength of the food product, then the food's disintegration is likely to occur. The treatment chambers are primarily organized to function in a batch or continuous mode. The PEF treatment chambers can be mainly categorized into the parallel plate and coaxial. The parallel plate chambers have been employed in batch operation, whereas coaxial chambers have been used in a continuous process. Coaxial chambers preferred mainly for continuous treatment operation are found to be advantageous since they have a higher rate of inactivation than the batch system. In continuous operation, the electric field distribution is uniform in constantly flowing media.

v. **Oscilloscope:** The oscilloscope measures the voltage across the two electrodes. The oscilloscope also displays the pulse shape. It is also essential to evaluate the temperature of the food item.

vi. Apart from significant components, additional parts are required, such as pump in a continuous PEF system, a chamber cooling system for diminishing the heating effect during treatment, and probes for measuring the voltage, current, and temperature distributed to the chamber.

1.3.2 MECHANISM OF MICROBIAL INACTIVATION BY PEF

The exact mechanism of PEF to kill or inactivate the microorganism has not been fully illustrated to date. However, many studies have postulated that the PEF-mediated microorganisms' inactivation is associated with destabilization or permeabilization of the cell membrane through electromechanical compression or dielectric rupture or electroporation, resulting in cell membrane rupture, intracellular material leaking out, and consequently cell death. Chauhan and Unni[18] postulated the complete process of inactivation of cell via PEF that applies the electric field to build-up cell's transmembrane potential (TMP). Thereafter, small metastable hydrophilic pores are generated in the cell membrane during the electric treatment and the size and population of pores increases. In the post-treatment stage, intracellular components leakage pores' shrinkage and disappearance took place. Electropermeabilization might be reversible or irreversible, depending on the degree of the structural integrity of the cell membrane. It has been demonstrated that cell membrane perforation or permeability increases with the electric field's strength. Generally, microorganism spores are more resistant to PEF treatment than the vegetative cells. Electroporation and dielectric rupture theory have been put forth to explain the microbial inactivation by PEF.

1.3.2.1 ELECTROPORATION

Electroporation is a process in which a cell membrane permeabilizes due to membrane proteins and lipid bilayer destabilization caused by exposure to high voltage electric field. In molecular biology, the electroporation technique introduces molecules such as DNA (deoxyribonucleic acid) into the cytoplasm. In food industries, the method was found to be more effective for the inactivation of microorganisms below thermal processing temperatures and increasing mass transfer.[66] Several hypothesis have been proposed to explain the initiation of electroporation. Zimmerman et al.[91] proposed an electromechanical idea in which the application of an external electric field would cause membrane compression, leading to its rupture when the electric force is more than the elastic restoring force. Another theory explains membrane permeabilization based on the denaturation of protein channels by Joule heating. The cell membrane generally comprises many protein channels, pores, and pumps. The opening and closing of protein channels are regulated by the transmembrane potential (TMP). When PEF is applied, many voltage-sensitive protein channels could open. When the applied electric current is larger than the voltage generated during normal metabolic activities, Joule heating causes the denaturation of protein channels by Joule heating.[18,81]

Further, Glaser et al.[27] explained electroporation depending on the lateral thermal fluctuations of lipid molecules. The lips in the cell membrane consist of hydrocarbon lipid tails named hydrophobic pores. The application of an electric field causes expansion of this hydrophobic pore till the pore inversion occurs. Besides these theories, the osmotic imbalance theory states that membrane pores are formed on exposure of the cell to the external electric field, resulting in the outflow of small molecules and ions and making the membrane permeable to electrochemical water breakdown and cell lysis.

1.3.2.2 DIELECTRIC RUPTURE THEORY

Regarding dielectric rupture theory, it has been postulated that a microbial cell membrane could be a dielectric material-filled capacitor having a dielectric constant of 2. The dielectric constant of the cell cytoplasm and most food is 60–80. The differences in the conductivities between cell membranes concerning cytoplasm and liquid suspension result in the accumulation of free charges at both surfaces. When the cell is exposed to an electric field of high voltage, the ions present in the cell move along the electric field until they are restrained by the membrane. This increases the cell TMP of the cell membrane as defined below:

$$U(t) = 1.5\, r\, E \cos\theta \tag{1.3}$$

where $U(t)$ is the TMP, r is cell radius, E is the strength of the electric field, and θ is the angle between a membrane site and field direction.

The opposite charges of the membrane attract each other, triggering membrane electro-compression (reduction in membrane thickness). The viscoelastic restoring forces naturally counter the electro-compressive force. At TMP of about 1V, the breakdown of the membrane occurs when the electro-compressive force increases more rapidly than the viscoelastic force. The membrane's breakdown results from the high-strength external electric field sufficient to increase the compressive force. The cells become completely inactive when TMP exceeds its critical level.[18]

1.3.3 FACTORS INFLUENCING PEF MICROBIAL INACTIVATION

Several factors influence the performance of PEF to inactivate the micro-organism in the food system. Some important factors, namely, process parameters, nature of the product, and properties of microbial cells, mostly decide the optimum performance of PEF treatment.

1.3.3.1 PEF PROCESS PARAMETERS

This includes the intensity or strength of the applied electric field, pulse length, shape, temperature, and time that affects the performance of PEF to inactivate or reduce microbial population in food.

1.3.3.1.1 Electric Field Intensity

The inactivation of the cell is very much governed by the intensity or strength of the electric field. Thus, it is considered one of the critical process parameters. For the cell inactivation, the applied electric field intensity should be greater than the cell's critical TMP. Larger cells possess higher TMP. When the applied PEF intensity is beyond the critical value, there is furthermore inactivation of the cell, i.e., with the escalation in the intensity of the electric field, cell inactivation also increases. The critical value is dependent on the pulse width.

Schoenbach et al.[72] observed that the critical value was 4.9 and 40 kV/cm for pulse width beyond 50 μs and below 2 μs, respectively. Predominantly, a higher voltage would contribute to maximizing the extraction of biomolecules.[87]

Pulsed wave shape: Electric field pulses can be used in different forms including exponential decaying, oscillatory, square, bipolar, decaying, and so on. Qin et al.[68] stated that the energy associated with square wave pulses was more than exponential decaying and oscillatory pulses and are lethally effective for inactivation of the cell. Further, the efficiency of mono-polar pulses was less than that of bipolar pulses. The advantages of the application of bipolar pulses include less energy consumption, minimized solids deposition on the surface of the electrodes, and reduced adverse electrolytic effects.[87]

1.3.3.1.2 Pulse Duration

The product of pulse number and pulse width is generally known as pulse duration. Pulse duration is known to be one of the indicators for the measurement of PEF treatment. The inactivation of the cell upsurges as the pulse width or pulse number increases. However, the higher pulse duration doesn't ensure maximum biomolecules recovery due to undesirable temperature rise. Thus, process parameters need to be optimized for enhanced cell inactivation alongside a low heating effect.[87]

1.3.3.1.3 Treatment Temperature

Very few scientific literatures are available on the effects of elevated temperatures, but room temperature was preferred for most experiments. Moreover, the degenerative temperature of the molecule limits the use of high temperatures for the experiments.[87] On another note, when the intensity of these factors escalates, the precise relationship between them and the microbial survival rate is unclear. Therefore, the optimization of these factors is essential to obtain reliable outcomes.

1.3.3.1.4 Nature of the Food Product

The physicochemical characteristics of food determine the nature of the products. PEF performance is also influenced by the nature of food products significantly. It has been stated that the efficacy of microorganisms' inactivation during PEF treatment of food depends on the media's chemical and physical properties. Physicochemical properties of food more likely influences the injured microbial cells recovery and their proliferation after the PEF exposure. Ostensibly, some food components, such as fat and protein, have preventive effects on microorganisms during PEF treatment.[36] Moreover, it has been stated that PEF treatment is ineffective for products with particles or unique structures like emulsions. Some studies have demonstrated that pH, water activity, and electrical conductivity of food products determine the competence of PEF toward the inactivation of the microorganisms. It has been noticed that if the pH of the food is ≤7.0 during the PEF treatment, greater *Salmonella* inactivation occurs.[40]

Álvarez et al.[2] reported that the *Listeria monocytogenes* inactivation was high in acidic food. The conductivity of the food medium showed an inverse relation with microbial inactivation if food has high electrical conductivity.[85] Besides, Min and Zhang[55] found that food's water activity is directly related to microbial inactivation in PEF treatment. The selection of solvents depends on several factors, such as extract solubility in the solvent, solvent polarity, and conductivity. It has been reported that the extraction rate increases with the solvent's increased conductivity due to enhanced electroporation on the cell membrane.[87] Apart from the solvent conductivity, the solubility of desired biomolecules in the solvent is also essential. If the solubility of the extract in the solvent is greater, the rate of mass transfer will be. Further, the polarity of the solvent also plays a significant role in the extraction of bioactive ingredients. The application of PEF enhances the mobility of polar molecules in the solvent by increasing the extraction rate of the molecules in the solvent-water mix.[88]

For better extraction efficiency, the ratio of solvent to raw material also plays a crucial role. He et al.[34] obtained maximum extraction of chondroitin sulfate from fish bones for a material-to-liquid ratio of 1:1.5 g/mL. The increased volume of the solvent resulted in the reduction of active ingredient concentration surrounding the sample molecules, thus causing easy dissolution of the active ingredients.[34]

1.3.3.1.5 Type of Microorganisms

Food can be *contaminated* by *different microorganisms, such as* bacteria, viruses, mold, and fungi. Microorganisms also influence the PEF treatments due to their other behavior or characteristics in the environment. Different genera, species, and stains in the same type of microorganism show different characteristics and affect the inactivation kinetics. For example, *Escherichia coli* and *Salmonella* are more prone to PEF when compared to *Listeria* and *Bacillus* sp. Moreover, the size and shape of the cell also influence the rate of inactivation. It has been claimed that Gram-positive and Gram-negative bacteria are more resilient to PEF treatment than yeast. Furthermore, bacteria and fungi spores are more resistant to PEF treatment than their vegetative cells. Therefore, the operational mechanism of PEF technology must be varied to inactivate the particular microorganism.[78]

1.4 APPLICATIONS OF PEF IN THE FOOD INDUSTRY

1.4.1 FOOD PRESERVATION

Food preservation prevents deteriorative reactions by extending the shelf life of food and assuring its safety. Some food preservation techniques include drying, canning, salting, freezing, pickling, sugaring, airtight storage, irradiation, and vacuum packaging. In some preservation cases, pre- and post-processing techniques may be essential contingents on the product and process type. Thermal processing possesses many features of an "ideal" food preservation system. In some food, the presence of high thermo-tolerant enzymes and microorganisms, primarily bacterial spores, makes the application of high temperature, which may affect food's nutritional and organoleptic properties. This makes use of an alternative non-thermal technique for the preservation of food. In non-thermal processing, the food is treated at a low temperature; therefore, only slight degradation of food quality is anticipated. Non-thermal technologies improve food quality and enhance safety levels compared to thermal aid methods.[42, 69]

In PEF food pasteurization techniques, microbial inactivation is caused by permanent cellular breakdown or cell lysis by electric pulses with high electric field intensity.[36, 68] Microbial inactivation by PEF has been explained by the "electroporation" and "dielectric rupture" theory.[18] However, Knorr et al.[44] reported the inability of high electric field pulses to inactivate bacterial spores. Therefore, the main focus of PEF applications is on inactivating microorganisms which are pathogenic and cause spoilage of food. The inactivation of microorganisms using PEF technology often relies on microbial characteristics, such as microorganism type, strains, and species.[51] PEF is one of the most favorable non-thermal methods for decontamination of microorganisms in food material.[85] With short voltage pulses (μs), electric fields (5–50 kV/cm) are generated between two electrodes. This result in the inactivation of microorganisms at a temperature below thermal methods. PEF treatment maintains the sensory and nutritional quality of food apart from extending the shelf life of food by inactivating food spoilage and pathogenic microorganisms. Therefore, PEF has more impetus over the conventional methods for food preservation.

PEF processing has promising applications in the juice, milk, and liquid egg industry, with special regard to microbial and enzymatic inactivation and prevention of off-flavoring during storage.[35] Several reports have demonstrated that the application of PEF treatments for fruit and vegetable juices, retained or enhanced the bioactive and flavor compounds. Generally, milk contaminated with *E. coli, Listeria,* and *Pseudomonas* sp. and inadequately pasteurized milk can cause several health hazards.[21] Microbial inactivation induced by PEF is thought to be an excellent method of preserving milk with no adverse effect on the quality of milk.

Sharma et al.[73] used PEF treatment for inactivating both Gram-positive as well as Gram-negative bacteria in whole milk at various dosages, time, and temperature ranges and observed a 5–6 log reduction in bacterial counts at electric field intensity of 22–28 kV/cm, 17–101 μs treatment time at 50 °C. PEF inactivation (64 pulses of 40 kV/cm) of yeast *Saccharomyces cerevisiae* in apple juice was assessed by Harrison et al.[32] by transmission electron microscopy (TEM) and observed the ultra-structural changes. These changes were similar to the proposed electroporation theory, suggesting that electroporation was the primary cause of yeast inactivation. It is well recognized that a liquid egg is a favorable growth medium for several microorganisms. *Salmonella* contamination is more prevalent in liquid egg products. Microbial inactivation in liquid egg products is generally done by conventional high-temperature short time (HTST) pasteurization which is more destructive to the structural and functional properties of egg products. Thus, PEF processing is relatively safe for retaining egg product structural and functional quality.

Monfort et al.[56] reported that 3 $\log_{10}$ cycles reduction of *Salmonella enteritidis* and *S. senftenberg* 775 W was noted at PEF intensity of 25 kV/cm for treatment duration of 105 μs (441 kJ/kg) and PEF strength of 35 kV/cm for 45 μs treatment time (369 kJ/kg), respectively. Monfort et al.[57] also found that reduction of 4 and 3 $\log_{10}$ of *S. typhimurium* and *S. aureus* populations were obtained by applying electric pulses at 45 kV/cm for 30 μs and 40 kV/cm for 15 μs, respectively. Further, the decrease in ~2.5 $\log_{10}$ cycles of *E. coli* was noticed in the liquid whole egg when 30, 68, and 11 pulses of 15, 11, and 9 kV/cm electric field strengths, respectively, were applied at 55 °C temperature.[9] Table 1.1 shows the effects of PEF in food preservation.

TABLE 1.1 Effects of PEF on Food Preservation.

Source	Experimental condition	Observations	Reference
Apple juice	64 pulses of 40 kV/cm	Inactivation of yeast *Saccharomyces cerevisiae*	Harrison et al. [32]
E. coli and *Staphylococcus aureus*	PEF of 1.6 V/xm and 60 pulses with a pulse width between 200 and 300 fis range	4–5 $\log_{10}$ cycles reduction in the *E. coli* vegetative cells population, 3–4 $\log_{10}$ cycles reduction in *S. aureus* population	Pothakamury et al. [60]
Liquid egg	Different electric field intensities: 20, 25, 30, 35, 40, 45 kV/cm, pulse width range: 3–150 µs, pulse frequency of 0.5 Hz	Inactivation at levels of 4 and 3Log_{10} cycles of *S. thyphimurium* and *S. aureus* population with 45 kV/cm, 30 µs and 40 kV/cm, 15 µs, respectively	Monfort et al. [56]
Liquid whole egg	Electric pulse strength in the range of 20–45 kV/cm, 0.5 Hz pulse frequency	Log_{10} cycles reduction of *S. enteritidis* and *S. senftenberg* at PEF intensity of 25 and 35 kV/cm for pulse width of 105 and 45 µs, respectively	Monfort et al. [57]
Liquid whole egg	Electric field intensity in 9–15 kV/cm range, 1 Hz pulse frequency, varied number of pulses up to 138, pulse duration of <138 s	*E. coli* inactivation ~2.5 log at 55 °C with electric field intensities of 15, 11, and 9 kV/cm. The corresponding pulse numbers were 30, 68, 138, respectively	Bazhal et al. [9]
Milk	PEF strength of 22–28 kV/cm, pulse width of 17–101 µs at 50 °C	Reduction in bacterial counts	Sharma et al. [73]
Orange juice	PEF at voltage intensity of 33.6–35.7 kV/cm with 35 pulses of 100 ks and a process temperature of 42–65 °C	More than five log cycles reduction (5D reduction) in the microbial count. Shelf life increased from 3 to 7 days with no significant reduction in odor and taste	Dunn and Pearlman (1987)
Pea soup	10–30 pulses with PEF of 25–33 kV/cm intensity, pulse width of 2 µs, process temperature of pea soup was <53 °C and in the range of 53–55 °C	Limited inactivation of inoculated *E. coli* and *B. subtilis* below 53 °C. Microbial reduction was approximately 4.4D when process temperature were in the range of 53–55 °C	Vega-Mercado et al. [82]
Yeast spores, *S. cerevisiae*, and *Bacillus Subtilis* bacteria	High voltage PEF of 0.54 V/ftm; 3 or 10 pulses of 90/ts with time interval of 30s	Significant reduction in the viability of vegetative cells and spores of the yeast	Yonemoto et al. [89]

1.4.2 APPLICATIONS OF PEF FOR FOOD PROCESSING

1.4.2.1 COOKING OF FOOD

The art of preparation of food for consumption using heat is cooking. Baking, roasting, frying, grilling, barbecuing, smoking, boiling, steaming, braising, and microwaving are various methods of heating used in cooking. Cooking aims to sterilize, alter moisture content, and prepare good taste and quality foods. However, heating during cooking at high temperatures has affected the basic properties of food. Therefore, fine control of the properties of food can be accomplished by PEF treatment. The fundamental changes in the basic food properties following PEF treatment, even at room temperature, are primarily related to the turgor loss and easier water and cellular components movement through the tissue.[83] PEF treatment has significant impact on the tissue microstructure, cell size, electrical conductivity, diffusion coefficient, texture, porosity, and solid-liquid expression characteristics of food tissues.[5]

PEF-induced alterations in the food properties are permitted to assist, improve, and change various existing operations in the food industry. PEF treatment at room temperature weakened the texture and increased the electrical conductivity of carrots, apples, potatoes, and other tissues.[13,46] Blahovec and Kouřím[12] reported that for PEF treatment of 600 V/cm intensity and 10 ms time interval at 25 °C, measurable improvements in the texture (50% elastic module reduction revealed by dynamic mechanical analysis) and electrical parameters (35% decrease in the actual electrical impedance portion measured by dielectric thermal analysis) for carrot tissue was observed.

It has been reported that PEF treatment causes a minimal change in texture, color, vitamin content, fatty acids, colorants, phenolic compounds, and amino acids compared to traditional heat treatments.[5] Biochemical changes such as Millard reaction, protein denaturation, and thermal degradation of compounds can be prevented by PEF treatment.[39] The PEF effect in cooking is displayed in Table 1.2. The PEF treatment of 1 kV/cm electric field intensity on Spanish dried cured sausage for 200 μs reduced the drying time up to 41–47% to obtain the same weight when it was dried without casing.[4] Results indicate that the water barrier effect produced by the casing may restrict the effectiveness of PEF in enhancing the method of producing cured meat products. Table 1.2 indicates effects of PEF on cooking.

TABLE 1.2 PEF Effect on Cooking.

Source	Experimental conditions	Observations	Reference
Carrot	PEF strength of 600 V/cm 10 ms time interval at 25 °C	Improvements in the textural and electrical parameters	Blahovec and Kouřím [12]
Spanish dried cured sausage	PEF intensity of 1 kV/cm, treatment duration of 200 μs, 28 kJ/kg	Spanish dried cured sausage	Astráin-Redín et al. [4]

1.4.2.2 PROCESSING OF MEAT AND FISH

PEF technology can be used in the processing of meat for several aspects, including enhancing permeation of cells to improve meat tenderness, attenuating microbial loads to increase shelf life, and preserving volatile meat profiles throughout storage. It has been demonstrated that PEF processing causes improvements in meat structure and function, enhances its functional properties, or helps to develop novel meat products.[30,54,79] However, PEF can also cause unfavorable changes which do not usually occur. Chilled raw meat doesn't undergo oxidation easily, i.e., it is oxidatively stable. But, slimming, cooking, and other processes intervene with the cell membrane of muscle cells that promote the reaction between heme/iron (lipid oxidation catalyst) and unsaturated fatty acids, thereby undergoing lipid oxidation and generating lipid hydroperoxides and free radicals. Lipid peroxides are vulnerable to breakdown into acids, alcohols, aldehydes, and ketones (secondary products), which contribute to rancid off-odor taste, also known as warmed-over taste.[16] Thus, PEF treatment may probably cause damage to cell membranes facilitating the initiation of lipid oxidation, eventually decreasing the meat's nutritional quality and health.

One of the earliest meat and fish preservation methods used salt, antimicrobial agent, and spices to develop a specific taste and aroma for the product. Cellular tissue disintegration caused by PEF may be used to improve the absorption process of these curing agents used in marinating fish or meat products. PEF treatment can accelerate the curing process and reduces the time required. Treatment of PEF on meat and fish tissues or protein gels is studied only to a lesser extent.[6,31,76] Due to improved absorption levels, PEF can reduce the additives concentration and the required process duration.

To this effect, McDonnell et al.[54] have identified the potential application of PEF to accelerate pork salting without affecting post-curing and water-binding capacity and underlined the need of optimizing the various treatment parameters.

Tenderness in meat is one factor that improves sensory quality and consumer acceptance. Aging is one of the critical choices for meat tenderization. Aging combined with treatment of PEF appears to increase the rate of proteolysis and enhance the tenderness of meat.[22] Studies on beef show that the application of PEF on muscles before aging increases proteolysis of myofibrillar proteins and tenderization. The effect of PEF (1.1–2.8 kV) on meat quality was studied by O'Dowd et al.[58] and reported no noteworthy improvement in the tenderness of meat during storage.

Bekhit et al.[10] reported that PEF (5–10 kV) has the potential to improve meat tenderness and cooking loss. Similarly, Suwandy et al.[77] applied PED (5–10Kv) at 20, 50, and 90 Hz frequencies on beef muscles. They revealed a 19% decrease in shear force and improved tenderness by amplifying desmin and troponin T degradation during 21 days of storage in refrigerator. Table 1.3 shows the PEF effect in processing meat and fish.

The repeated PEF application (1X,2X,3X) on beef *M. longissimus lumborum* muscles (LL) positively impacted the tenderization of meat. At the same time, PEF treatments (10 kV, 90 Hz, and 20 μs) were not affected on *M. semimembranosus* muscle (SM), using batch mode PEF systems with the meat fiber aligned parallel to the electrode.[77] Alahakoon et al.[1] reported an increase in tenderness, corresponding to the reduced cooking time for meat cuts rich in collagen using a 4 cm (6 cm long and 6 cm deep) triangular PEF chamber. This demonstrated the prospect of utilizing PEF for meat cuts rich in connective tissue.

The ability of PEF treatments for muscle cell electroporation, leading to faster meat drying at 4 °C, was assessed by Astráin-Redín et al.[4] It was found that the PEF impact on meat drying was determined by PEF parameters like the intensity of the electric field, pulse width, and specific energy. On a laboratory scale, the optimum condition that enhances the pork loin samples drying were: 1 kV/cm electric field strength, 200 μs pulse width, 28 kJ/kg, and total specific energy, which enhanced the drying rate (up to 60.4%). Recently a study was performed on the recovery of functional proteins from the waste meat of chicken by applying PEF. From that, protein fraction with antioxidant properties was obtained.[26]

The application of PEF technology to enhance antioxidant extracts from bones, gills, and heads of sea bream and sea bass was carried out by Franco et al.[23] Study results pointed that PEF-assisted water extraction improved the extract's antioxidant ability to water or methanol extracts. They suggested that PEF would be an eco-friendly and economical choice for antioxidant-extract production from fish offals. Table 1.3 presents effects of PEF in meat and fish processing.

TABLE 1.3 Effects of PEF in Meat and Fish Processing.

Source	Experimental conditions	Observations	Reference
Beef loins (LL) and topsides (SM)	Batch mode PEF treatment of meat fiber placed parallel to the electrodes. PEF strength of 0.27–0.56 kV/cm; voltages of 5 and 10 kV; three different frequencies of 20, 50, and 90 Hz; pulse width of 20 μs; a variety of pulse numbers	Meat tenderization and improving cooking loss	Bekhit et al. [10]
Beef muscle	PEF (5–10 kV) at 20, 50, and 90 Hz frequencies	19% decrease in shear force and improved tenderness	Suwandy et al. [77]
Cold boned beef *M. longissimus* et Lumborum	0.23–0.68 kV/cm PEF intensity; 2.5 and 10 kV voltage; 200 Hz frequency; 20 μs pulse width	Ultrastructural changes can occur in the low and high treatment of PEF on beef LL	Khan et al. [43]
Pork loin	1 kV/cm electric field strength, 200 μs pulse width, and total specific energy of 28 kJ/kg	Enhanced the drying rate up to 60.4%	Astráin-Redín et al. [4]
Turkey breast meat	Electric field intensity of 3 kV/cm; frequency 5 Hz; pulse width 20 μs; Total of 300 pulses	No major side effects on the lipid oxidation, cook loss, weight loss, texture, and color either on fresh or frozen turkey meat samples	Arroyo et al. [3]
Waste chicken meat	Two-step method; first step involving short pulses of high voltage, and second step involving long pulses of low voltage. Overall total specific energy of 38.4 J/g	Proteins with enhanced antioxidant activities were obtained	Gosh et al. [26]

1.4.2.3 IMPROVING MASS TRANSFER FOR FOOD PROCESSING OPERATIONS

Mass transfer mainly occurs in numerous operations in the food industry wherein the prime goal is to extract the compound of interest or introduce the compound into the food matrix. Most plant or animal-based food comprises tissues made up of cells with the basic eukaryotic organization. In eukaryotic cells, cytoplasm, nucleus, and various other cell organelles are enclosed by a plasma membrane. The cell wall forms the outermost layer surrounding the plasma membrane in plant cells. The rate of mass transfer relies on factors such as the concentration difference between the substances between two phases of the resistance experienced by the substance while moving from one phase to another. The movement of the substance from or into the food tissues is influenced by the plasma membrane, which acts as a semi-permeable barrier for the movement of substances. The migration of the substance in and out of the cell occurs by diffusion through the cell membrane. Conventional techniques used to enhance the mass transfer were found to deteriorate the cell's cytoplasmic membrane. The application of PEF improves mass transfer in both plant and animal cells without elevating the temperature and at a low cost of operation. Table 1.4 indicates the effects of PEF on enhancing mass transfer.

1.4.2.3.1 Effect of Mass Transfer: Juice Expression

The cells of fruits and vegetables contain a natural liquid called juice. The extraction of juice is mainly carried out by applying mechanical pressure. This pressure results in the breakdown of the plasma membrane, thus enabling the intracellular liquid discharge. Juices extracted from vegetables and fruits comprise physiologically and nutritionally valuable compounds. The juice extraction techniques must be gentle such that it doesn't affect these valuable compounds. Hence, PEF could be a potential innovative technique that could be used for juice expression.[66]

The PEF application for juice expression from fruits and vegetables such as grapes, apples, potatoes, and carrots has been widely studied.[14,66] Bazhal et al.[8] investigated the yield and quality of the juice extracted from fine-cut apples by treating it with PEF during compression and observed significant improvement in yield and quality.

Praporscic et al.[61] extracted juice from three white grapes (Muscadelle, Sauvignon, and Semillon) by applying PEF for 45 min. The mechanical

TABLE 1.4 Effect of PEF on Enhancing Mass Transfer.

Source	Experimental conditions	Observations	Reference
Juice expression			
Apple (both whole and sliced)	PEF intensity of 400 V/cm, 3 bars constant pressure, pulse duration of 100 μs, the gas between pulses = 10 ms	Increase in the juice yield from 44 g/100 g apple to 58 g/100 g apple in treated whole apple and 64 g/100 g apple in sliced apple	Grimi et al. [28]
Apple (fine cut)	Electric field strength (100–520 V/cm), pulse width of 100 μs. pulse repetition time of 10 ms, number pulses = 50 and 3 bars constant pressure	Enhanced juice yield, improved the quality of juice	Bazhal et al. [8]
White grapes (*Muscadelle, Sauvignon* and Semillon)	250–1000 V/cm electric field intensity, 100 μs pulse duration, pulse repetition time of 100 ms constant pressure of 5 bars	Increase in the juice yield at 45 min of pressing from 49–54% to 76–78%	Praporscic et al. [61]
Pigments extraction			
Cyanobacteria (*A. platensis*)—Anthocyanin extraction	The electric field strength in the range of 15–25 kV/cm, pulse width of 60–150 μs range and temperature range of 10–40 °C for PEF treatment	Maximum recovery of ~150 mg/g was obtained at an electric field strength of 15 kV/cm and treatment time of 150 μs. The increase in temperature from 10 to 40 °C reduced the electric field strength from 25 kV/cm to below 20 kV/cm	Martinez et al. [53]
Red cabbage—Anthocyanin extraction	2.5 kV/cm PEF instensity, pulse width of 15 μs, number of pulses = 50	2.15 fold surge in the total anthocyanin extraction	Bazhal et al. [8]
Red beet—colorants (mainly betalains)	375–1000 V/cm electric field strength, 0–0.2 s treatment duration	1000 V/cm electric field strength enhanced the betalains extraction rate and also decreased the time of extraction	Loginova et al. [48]
Yeast (*R. Glutinis*) — Carotenoids extraction	Three PEF treatment of 10, 15, and 20 kV/cm intensities, treatment duration between 15–300 μs	375 μg/g dry weight carotenoids extraction after incubating PEF treated cells for 24 h at 25 °C in pH 8.0 medium	Martinez et al. [52]

TABLE 1.4 *(Continued)*

Source	Experimental conditions	Observations	Reference
Winemaking			
Grape pomace (*Cabernet Sauvignon*)	5 kV/cm PEF intensity, 1 Hz frequency	Reduction in the maceration time from 268 to 72 h, increase in the anthocyanin concentration	Lopez et al. [49]
Merlot grapes	PEF of 500–700 V/cm intensity, pulse repetition time of 40–100 ms	Increased phenolic compounds extraction and enhancement of sensory attributes of wine	Delsart et al. [19]
Vegetable oil extraction			
Olive	2 kV/cm electric field strength and 25 Hz frequency	Increase in oil yield by 13.3%. Higher phenolic content, total tocopherols, and total phytosterols in the oil	Puertolas and de Maranon [63]
Rapeseed (hulled and non-hulled seeds)	Electric field strength of 5kV/cm and 60 pulses, and 7 kV/cm and 120 pulses, pulse duration of 30 μs	Maximum permeabilization (55%) at 7 kV/cm and 120 for hulled rapeseeds and 17% permeabilization at 5 kV/cm and 60 pulses for non-hulled rapeseeds. Higher concentration of antioxidants, tocopherols, polyphenols and phytosterols	Gunderjan et al. [29]
Dehydration			
Red beetroot tissue	PEF intensity of 200–600 V/cm, pulse treatment time of 10 μs, pulse width repetition of 200 μs, total 5–500 pulses	PEF pretreatment decreased temperature by ~20–25 °C	Shynkaryk et al. [75]
Potato tissue	Electric field strength less than 1500 V/cm, 1–30,000 pulses, pulse repetition time of 10^{-2}–100s, pulse duration of 10^{-5}–10^{-3}s	PEF strength of 300–400 V/cm is essential for drying the potato discs	Lebova et al. [46]
Carrot	Electric field intensity of 1 kV/cm and number of pulses = 20	Increase in the drying rate. Reduction in the peroxidase activity by 30–50%	Gachovska et al. [25]
Apple tissue	800 V/cm electric field strength, 100 μs pulse duration, 100 ms pulse prepetition time	Accelerated drying of the sample	Parniakov et al. [59]

TABLE 1.4 *(Continued)*

Source	Experimental conditions	Observations	Reference
Freezing			
Carrot discs	Electric field intensity of 1 kV/cm, duration between pulses = 25 μs, frequency of 4 Hz, number of pulsed = 100	Maintenance of texture firmness of carrot disc immersed in $CaCl_2$ after thawing	Shayanfar et al. [74]
Apple	Electric field strength (1.85–5 kV/cm), pulse number (10–100), treatment time (15–24 μs)	Reduction in the freezing time up to 3.5–17.2%	Wiktor et al. [84]

expression and juice characteristics such as absorbance and turbidity of all three grape juice were also determined. It was concluded that W ~ 20 kJ/kg energy at 750 V/cm electric field strength (E) was found to be optimum for maximum juice yield alongside improved juice quality. The PEF application increased the juice yield of all three grapes from 49–54% to 76–78%. The extraction of apple juice at a pilot scale from apple mash using PEF was carried out by Schilling et al.[71] The PEF treatment with 10 kJ/kg energy input at 3 kV/cm electric field intensity was employed for juice extraction. However, the result showed a decrease in the yield rather than an increase, but there was an enhanced release of polyphenols into the juice.

1.4.2.3.2 Effect of Mass Transfer: Extraction of Valuable Compounds

The extraction of valuable intracellular compounds such as pigments, oils, and polyphenols has been done using PEF. The effect of PEF in enhancing mass transfer is shown in Table 1.4.

- **Pigments:** Currently, the demand for pigments in the food industry is increasing to replace synthetic colorants. The health benefits of natural pigments and increased consumer demand for natural products have set the trend for its demand in the food industry. The pigments such as anthocyanins, carotenoids, betalains, and chlorophyll have been extracted using PEF. The pigments are extracted using organic or inorganic solvents based on their solubility. Natural colorants are extracted using gentle techniques due to their sensibility to light, oxygen, and temperature.[66] To date, applications of PEF in the recovery of pigments such as anthocyanin, betalains, carotenoids, and chlorophyll have been demonstrated.[66] Martinez et al.[53] reported the enhanced recovery of C phycocyanin (~150 mg/g) using PEF with an intensity of 25 kV/cm and 150 μs treatment time from *Arthrospira platensis* cells. Enhanced phycocyanin extraction was also obtained using PEF with 2.5 kV/cm electric field strength and pulse width of 50 μs from red cabbage.[24] Loginova et al.[48] investigated the application of PEF on the retrieval of natural colorants, mainly betalains. They stressed that implementing PEF with an electric field strength of 1000 V/cm accelerated the extraction of betalains and reduced time required for extraction. Martinez et al.[52] recovered up to 375 μg/g of carotenoids by PEF from *Rhodotorula glutinis* cells. The research studies ensure the potential of PEF as a beneficial innovative green technique for the enhanced extraction of colorants and their accelerated recovery.

- **Polyphenols during winemaking:** Wine is an alcoholic beverage derived from the fresh grape juice fermentation. Several techniques winemaking techniques have been developed for enhancing phenolic compound extraction. These techniques include flash release under vacuum, varied pre-fermentative maceration temperature, must or grape freezing, and enzymatic extraction. The ability of the phenolic compounds to modify flavor, color, and aging behavior and improve the sensory properties and quality of the wine has resulted in its enhanced demand in the wine industry during the preparation of wine.[19] Improvement in the phenolic compounds extraction by treating grapes with PEF during red winemaking has been the most-studied applications of PEF technology.[65] The initial work was carried out in extracting phenolics from different grapes varieties such as Graciano, Tempranillo, Granacha, Mazuello, and Caberet Sauvignon and suggested applying PEF before maceration-fermentation step accelerated and enhanced phenolic extraction.[49,66] Delsart et al.[19] also researched the recovery of the phenolics from Merlot grapes. They reported that applying PEF at 500–700 V/cm electric field intensity enhanced the recovery of phenolics from the grapes. The application of PEF doesn't cause any off-flavors or strange taste. Moreover, the treatment of grapes with PEF reduces maceration time and temperature during the elaboration of rose wine.[66,67]
- **Vegetable oils:** The oil naturally occurs in seeds such as rapeseeds, soyabean, sunflower seeds, or fruits. The average oil yield is estimated to be 25% from the oilseeds. Traditionally oil from oilseeds is extracted by mechanical pressing, centrifugation, and extrusion, and by solvent extraction using hexane or ethanol. Generally, the maximum oil yield is obtained by solvent extraction using hexane. Following oil extraction, oil is refined to remove unwanted materials. However, a conventional oil extraction technique possesses several disadvantages. Oil extraction by mechanical pressing may not result in the complete recovery of oil. Further, the solvent used for oil extraction is toxic to both humans and the environment.[64] Hence, these traditional oil extraction methods need to be replaced with new techniques that would help in the maximum oil recovery, causing no environmental damage. The PEF application in oil extraction from various oil seeds has been studied.[64] Puertolas and de Maranon[63] have reported that the treatment of the Arroniz olive paste treated with PEF of 2 kV/cm electric field strength enhanced the extraction yield by 13.3% without affecting the general and sensory characteristics of the oil. According to work by Guderjan

et al.,[29] applying PEF to the hulled rapeseed at 7 kV/cm and 120 pulses resulted in the formation of sufficient pores for the extraction of oil. The PEF treatment before the oil extraction increased the oil yield in hulled rapeseed.[29] Similarly, PEF treatment in the extraction of oil from other oilseeds such as canola seeds, sesame seeds, corn germ, and soyabean seeds resulted in a higher oil yield.[64] Therefore, PEF could emerge as one of the alternate novel techniques for extracting oil from oilseeds. Besides winemaking, juice expression, and vegetable oil recovery, PEF has also been used in the accelerated and increased bioactive compounds and biomolecules extraction from vegetables and fruits by improving mass transfer.[64]

1.4.2.3.3 Effect of Mass Transfer: Drying

The drying refers to the removal of the moisture from the food materials. Drying food material helps overcome microbial food spoilage and undesirable chemical reactions. Conventionally, freeze-drying and hot-air drying techniques are followed to remove moisture from food material. However, these techniques possess certain drawbacks. The conventional drying methods cause undesirable changes in the flavoring agents, pigments, and nutritionally valuable compounds. New hybrid technologies are being developed to dehydrate the food sample, but the PEF technique could also be an alternative for drying the food materials.[18]

High-intensity PEF has been reported to enhance the mass transfer during drying fruits and vegetable tissues. The PEF application results in the electroporation of tissue cells, facilitating moisture diffusion and enhancing drying.[46] Several scientific studies have been conducted on applying PEF for dehydrating food materials. Research on the PEF effect and temperature on the drying of beetroot tissue has been carried out by Shynkaryk et al.,[75] who concluded that the PEF application decreased the drying temperature of beetroot tissue by 20–25 °C.

Lebovka et al.[46] studied the influence of PEF on the convective drying of potato disks. The effects of thermal pretreatment before drying and even during PEF treatment of 300–400 V/cm intensity were investigated. The study showed that the thermal pretreatment mild temperature of 50°C increased the moisture effective diffusion coefficient, comparable with the samples pre-treated with PEF before drying.

Meanwhile, the thermal pretreatment at a high temperature of 70 °C did not affect the moisture removal from potato discs. This indicates that applying

PEF reduced the drying temperature by 20 °C. Further, the treatment of carrot with PEF of the electric field strength of 1 kV/cm before oven drying at 70 °C resulted in its increased drying rate.[25] The PEF pretreatment also resulted in a 30–50% reduction in the activity of the peroxidize enzyme in carrots.[25] Besides, PEF-assisted vacuum freeze drying of apple discs has also been attempted using PEF at 800 V/cm electric field strength. The PEF application accelerated the material's cooling and drying.[59] Hence, the PEF treatment would aid in improving the convective drying rate. Applying PEF would be beneficial in drying thermally sensitive food materials at moderate temperatures.

1.4.2.3.4 *Effect of Mass Transfer: Freezing*

In the food industry, while processing food materials, freezing is carried out so that the food retains its quality attributes such as organoleptic properties, color, firmness, etc. During freezing, the water present in the food material influences ice formation. Moreover, the thawing of the vegetables after the freezing softens their texture. Therefore, freezing process needs to be enhanced to maintain the textural quality of the product.[18] Few studies have been done on applying PEF in freezing vegetables, followed by an evaluation of texture after thawing. The PEF treatment during freezing modifies the diffusion of both intracellular and extracellular media due to the pore formation in the cell membrane.

Shayanfar et al.[74] investigated the effects of PEF on carrot discs before freezing on their initial quality parameters. The PEF with the electric field strength of 1 kV/cm and 100 pulses were applied to the carrot discs immersed in different solutions of trehalose, glycerol, tap water, and $CaCl_2$. The results suggested that using PEF to carrot disc immersed in $CaCl_2$ solution aided in the maintaining of the texture firmness even after thawing and resulted in higher mass loss values.

Wiktor et al.[84] studied the effects of PEF on the kinetics of immersion freezing and thawing and the quality of apples. They showed that applying PEF led to a 3.5–17.2% reduction in the total freezing time, and maximum mass loss was observed when the apple sample was treated with 50 pulses at 5 kV/cm. Thus, applying PEF would be crucial in aiding the freezing of the food material.

1.4.2.4 RECOVERY OF BIOMOLECULES FROM FOOD WASTE

The PEF treatment has also been beneficial for recovering compounds from food waste. The peels of many fruits and vegetables and generated by-products from the meat and fish processing are rich in colorants, bioactive compounds, proteins, minerals, natural polysaccharides, etc. However, the extraction of valuable compounds from this waste by traditional methods is very hectic, and those methods are not eco-friendly. PEF could be the best novel, innovative method for valuable compounds recovery from food waste. Applying PEF in the extraction of compounds from food waste would enhance and accelerate the recovery of target compounds and reduce the extraction time.

Table 1.5 shows the effect of PEF in the extraction of compounds from food waste. According to work carried out by Kouba et al.,[45] the maximum pigments yield was obtained from red prickly pear peel compared to whole fruit and pulp when treated with PEF before supplementary aqueous extraction (SAE). About 81.3 mg of colorants/100 g of fresh material was obtained from the peel with PEF pretreatment.

Luengo et al.[50] reported an increased total polyphenol extraction yield for orange peel from 20% to 159%, when the PEF electric field strength was increased from 1 to 7 kV/cm at 60 μs treatment time (20 pulses of 3 μs). Recently, some works have been carried out on extracting compounds from fish waste by applying the PEF.

He et al.[34] extracted chondroitin sulfate from fish bones by applying a high-intensity pulsed electric field (HIPEF) and obtained a maximum yield of 0.0069 g/mL. 3.24% NaOH, material-to-liquid ratio of 1:1.5 g/mL, 16.88 kV/cm electric field intensity, and pulse number of 9 was the extraction conditions employed for extracting chondroitin sulfate. It was also observed that the application of HIPEF reduced the extraction time and yielded more chondroitin sulfate than conventional extraction techniques. Apart from this, PEF was combined with a semi-ionic extraction method to recover collagen, chondroitin sulfate, and calcium from a fishbone. About 0.0038, 0.0198, and 0.0392 g/mL collagen, calcium, and chondroitin sulfate was extracted respectively.[33] Moreover, the PEF technique has also been used to extract protein, but the extraction was combined with the enzymatic extraction technique from the viscera of Abalone (*Haliothis discus Hannai ino*).[47]

1.4.3 OTHER APPLICATIONS OF PEF IN THE FOOD INDUSTRY

Over a decade, PEF found its predominant application in food industries (such as food preservation, tenderization of meat, curing fish and meat, cooking, juice expression, winemaking, and extracting valuable compounds). Besides, PEF has found other applications too. PEF technique has also been exploited for the treatment of sewage water and also to carry out electrophoresis.

TABLE 1.5 Effect of PEF on the Extraction of Compounds from Food Waste.

Source	Experimental conditions	Observations	Reference
Pear peel	8 kV/cm electric field intensity, 300 pulses	About 81.3 mg of colorants/100 g fresh material was obtained with PEF pretreatment followed by SAE	Koubaa et al. [45]
Orange peel	Electric field intensity ranging from 1 to 7 kV/cm, 5–50 pulses of 3 μs, 1 Hz frequency	Increase in the polyphenol extraction yield up to 159% with the increase in the electric field intensity from 1 kV/cm to 7 kV/cm	Luengo et al. [50]
Fishbone	Electric field strength in the range of 5–25 kV/cm, pulse number in the range of 6 to 10	Calcium, chondroitin sulfate, and collagen yield of 19.8, 39.268, and 3.895 mg/mL, respectively	He et al. [33]
Fishbone	PEF intensity of 10–20 kV/cm), 6–10 number of pulses	Chondroitin sulfate yield of 6.92 g/L at electric field strength of 16.8 kV/cm and pulse number of 9	He et al. [34]
Abalone viscera	Electric field intensity in the range of 5–20 kV/cm and Treatment time in 100–800 μs range	Maximum protein recovery at electric field intensity of 20 kV/cm and treatment duration of 600 μs	Li et al. [47]

Research has been carried out to demonstrate the feasibility of PEF technology for sewage water treatment. Various prototypes have also been designed for small-scale and pilot scale (400–2000 L/h) to study the effect of PEF on treating wastewater.[86] From these studies, it is evident that PEF technology was found to be successful in the treatment of wastewater. In molecular biology, the PEF has been used to separate DNA fragments by electrophoresis easily.

Birren et al.[11] reported that pulsed-field gel electrophoresis (PFGE) helped overcome conventional electrophoresis's drawbacks. The traditional technique of electrophoresis uses the static electric field to separate DNA fragments up to a maximum size of 20 kilobase pairs (kb). The application of PFGE resulted in the separation of DNA fragments from 1–50 kb size. The PFGE technique is currently used to produce DNA fingerprints of bacterial isolates.

1.5 MERITS AND DEMERITS OF PEF

PEF is emerging as a potential technology in food industries for food processing as well as food preservation. The PEF treatment has been successful in most food preservation, winemaking, juice expression, and recovery of compounds from food material. Apart from this, PEF has also been successfully employed to treat sewage water and DNA profiling. The PEF technique has the potential to substitute conventional methods, which harm the environment due to the heat generated during the extraction process and the use of a solvent which may hold the possible risk of getting leaked into the environment. The main advantages of PEF include high extraction yield, low toxicity, less duration of treatment, low operating costs, enhanced diffusion and mass transfer, reduced cost, and environmental pollution. Extraction of compounds using PEF treatment can be carried out at low temperatures without affecting the heat-sensitive compounds. Despite having several advantages that would help overcome conventional methods' drawbacks, the PEF technique has certain disadvantages too. The implementation of PEF in food industries is limited owing to its high initial cost and elevated processing cost.

Apart from the price, it is ineffective for decontaminating microorganisms from fresh meat. The microbial decontamination requires high-intensity electric field treatment, affecting meat's nutritional and sensory quality. The PEF treatment also fails to inactivate the bacterial spores and cannot be used to sterilize food and water. Recently novel HIPEF continuous extraction method has been developed for the enhanced recovery of bioactive compounds from food materials using a high-intensity electric field. However, no theories explain its underlying mechanism of enhancing mass transfer and cell inactivation.[87] To successfully implement PEF in food industries, user-friendly low-cost PEF systems need to be developed. Understanding the exact mechanism of PEF would contribute to further research.

1.6 FUTURE PROSPECTS OF PEF

PEF is a novel, innovative green technique with wide potential in the food industry, but still, considerable research needs to be carried out to determine its full potential. Currently, research on PEF technology is going on worldwide. Most of the research carried out at the laboratory level, and pilot scale has shown promising results. Even though, PEF technology has some drawbacks which need to be overcome for its complete utilization in the food industry. Future research must focus on determining the exact PEF process mechanism and developing a low-cost PEF system. This would help overcome the major hurdles in implementing the PEF system in industries. Moreover, research shows that combining the PEF technique and other green extraction techniques also helps overcome certain drawbacks associated with PEF.

1.7 SUMMARY

PEF has emerged as an effective technology for the processing and preservation of food materials without affecting its quality attributes. However, to determine the complete potential of PEF, substantial research needs to be carried out. The main hurdles hindering the PEF technique implementation in the food industry are investment cost and equipment development. On an industrial scale, the feasibility of the PEF system remains to be proven. The full utilization of PEF in food industries relies on developing low-cost, user-friendly PEF systems. Although much research has been carried out regarding PEF efficiency on microbial and enzyme inactivation, enhancement of mass transfer, and the effect on food quality attributes, detailed analysis needs to be explored to understand the mechanisms behind these functionalities. The successful industrial application of the PEF technique for the production of appropriate products and sound scale of production will be imperative to demonstrate the feasibility of the PEF technique and convince the potential end-users.

KEYWORDS

- **biomolecules recovery**
- **food preservation**
- **mass transfer**
- **non-thermal processing technique**
- **pulse electric field**

REFERENCES

1. Alahakoon, A. U.; Oey, I.; Silcock, P.; Bremer, P. Understanding the Effect of Pulsed Electric Fields on Thermostability of Connective Tissue Isolated from Beef Pectoralis Muscle Using a Model System. *Food Res. Int.* **2017,** *100,* 261–267.
2. Álvarez, I.; Raso, J.; Palop, A. Sala, F. J. Influence of Different Factors on the Inactivation of *Salmonella senftenberg* by Pulsed Electric Fields. *Int. J. Food Microbiol.* **2000,** *55* (1–3), 143–146.
3. Arroyo, C.; Eslami, S.; Brunton, N. P.; Arimi, J. M.; Noci, F.; Lyng, J. G. An Assessment of the Impact of Pulsed Electric Fields Processing Factors on Oxidation, Color, Texture, and Sensory Attributes of Turkey Breast Meat. *Poult. Sci.* **2015,** *94* (5), 1088–1095.
4. Astráin-Redín, L.; Raso, J.; Cebrián, G.; Álvarez, I. Potential of Pulsed Electric Fields for the Preparation of Spanish dry-cured Sausages. *Sci. Rep.* **2019,** *9* (1), 1–11.
5. Barba, F. J.; Parniakov, O.; Pereira, S. A.; Wiktor, A.; Grimi, N.; Boussetta, N.; Saraiva, J. A.; Raso, J.; Martin-Belloso, O.; Witrowa-Rajchert, D.; Lebovka, N. Current Applications and New Opportunities for the use of Pulsed Electric Fields in Food Science and Industry. *Food Res. Int.* **2015,** *77,* 773–798.
6. Barsotti, L.; Dumay, E.; Mu, T. H.; Fernandez Diaz, M. D.; Cheftel, J. C. Effects of High Voltage Electric Pulses on Protein-based Food Constituents and Structures. *J. Food Sci. Technol.* **2001,** *12,* 136–144.
7. Barsotti, L.; Merle, P.; Cheftel, J. C. Food Processing by Pulsed Electric Fields. I. Physical Aspects. *Food Rev. Int.* **1999,** *15* (2), 163–180.
8. Bazhal, M. I.; Lebovka, N. I.; Vorobiev, E. Pulsed Electric Field Treatment of Apple Tissue During Compression for Juice Extraction. *J. Food Eng.* **2001,** *50* (3), 129–139.
9. Bazhal, M. I.; Ngadi, M. O.; Raghavan, G. S. V.; Smith, J. P. Inactivation of *Escherichia coli* O157: H7 in Liquid Whole Egg Using Combined Pulsed Electric Field and Thermal Treatments. *LWT-Food Sci. Technol.* **2006,** *39* (4), 420–426.

10. Bekhit, A. E. D. A.; van de Ven, R.; Fahri, F.; Hopkins, D. L. Effect of Pulsed Electric Field Treatment on Cold-boned Muscles of Different Potential Tenderness. *Food Bioprocess Tech.* **2014,** *7* (11), 3136–3146.
11. Birren, B. W.; Lai, E.; Hood, L.; Simon, M. I. Pulsed Field Gel Electrophoresis Techniques for Separating 1-to 50-kilobase DNA fragments. *Anal. Biochem.* **1989,** *177* (2), 282–286.
12. Blahovec, J.; Kouřím, P. Combined Mechanical (DMA) and Dielectric (DETA) Thermal Analysis of Carrot at Temperatures 30–90°C. *J. Food Eng.* **2016,** *168,* 245–250.
13. Blahovec, J.; Kouřím, P.; Kindl, M. Low-temperature Carrot Cooking Supported by Pulsed Electric Field–DMA and DETA Thermal Analysis. *Food Bioprocess Tech.* **2015,** *8,* 2027–2035.
14. Bobinaitė, R.; Pataro, G.; Lamanauskas, N.; Šatkauskas, S.; Viškelis, P.; Ferrari, G. Application of Pulsed Electric Field in the Production of Juice and Extraction of Bioactive Compounds from Blueberry Fruits and Their by-products. *J. Food Sci. Technol.* **2015,** *52* (9), 5898–5905.
15. Bruno, S. F.; Ekorong, F. J. A. A.; Karkal, S. S.; Cathrine, M. S. B.; Kudre, T. G. Green and Innovative Techniques for Recovery of Valuable Compounds from Seafood by-Products and Discards: A Review. *Trends Food Sci. Technol.* **2019,** *85,* 10–22.
16. Byrne, D. V. Sensory Characterisation Studies on Warmed-over Flavor in Meat. PhD Thesis. Royal Veterinary and Agricultural University, Copenhagen, Denmark, 2000.
17. Castro, A. J.; Barbosa-Ca′nova, G. V.; Swanson, B. G. Microbial Inactivation of Foods by Pulsed Electric Fields. *J. Food Process. Preserv.* **1993,** *17* (1), 47 – 73.
18. Chauhan, O. P.; Unni, L. E. Pulsed Electric Field (PEF) Processing of Foods and its Combination with Electron Beam Processing. In *Electron Beam Pasteurization and Complementary Food Processing Technologies*; Pillai, S. D., Shayanfar, S., Eds.; Woodland Publishing Series in Food Science, Technology and Nutrition, 2015; pp 157–184.
19. Delsart, C.; Ghidossi, R.; Poupot, C.; Cholet, C.; Grimi, N.; Vorobiev, E.; Milisic, V.; Peuchot, M. M. Enhanced Extraction of Phenolic Compounds from Merlot Grapes by Pulsed Electric Field Treatment. *Am. J. Enol. Vitic.* **2012,** *63* (2), 205–211.
20. Doevenspeck, H. Influencing Cells and Cell Walls by Electrostatic Impulses. *Fleisch wirtschaft* **1961,** *13* (12), 968–987.
21. Ercolini, D.; Russo, F.; Ferrocino, I.; Villani, F. Molecular Identification of Mesophilic and Psychrotrophic Bacteria from Raw Cow's Milk. *Food Microbiol.* **2009,** *26* (2), 228–231.
22. Faridnia, F.; Ma, Q. L.; Bremer, P. J.; Burritt, D. J.; Hamid, N.; Oey, I. Effect of Freezing as Pre-treatment Prior to Pulsed Electric Field Processing on Quality Traits of Beef Muscles. *Innov. Food Sci. Emerg. Technol.* **2015,** *29,* 31–40.
23. Franco, D.; Munekata, P. E.; Agregán, R.; Bermúdez, R.; López-Pedrouso, M.; Pateiro, M.; Lorenzo, J. M. Application of Pulsed Electric Fields for Obtaining Antioxidant Extracts From Fish Residues. *Antioxidants* **2020,** *9* (2), 90–95.
24. Gachovska, T.; Cassada, D.; Subbiah, J.; Hanna, M.; Thippareddi, H.; Snow, D. Enhanced Anthocyanin Extraction from Red Cabbage Using Pulsed Electric Field Processing. *J. Food Sci.* **2010,** *75* (6), E323-E329.
25. Gachovska, T. K.; Simpson, M. V.; Ngadi, M. O.; Raghavan, G. S. V. Pulsed Electric Field Treatment of Carrots Before Drying and Rehydration. *J. Sci. Food Agric.* **2009,** *89* (14), 2372–2376.
26. Ghosh, S.; Gillis, A.; Sheviryov, J.; Levkov, K.; Golberg, A. Towards Waste Meat Biorefinery: Extraction of Proteins from Waste Chicken Meat with Non-Thermal Pulsed Electric Fields and Mechanical Pressing. *J. Clean. Prod.* **2019,** *208,* 220–231.

27. Glaser, R. W.; Leikin, S. L.; Chernomordik, L. V.; Pastushenko, V. F.; Sokirko, A. I. Reversible Electrical Breakdown of Lipid Bilayers: Formation and Evolution of Pores. *Biochem. Biophys. Acta* **1988,** *940*, 275–287.
28. Grimi, N.; Mamouni, F.; Lebovka, N.; Vorobiev, E.; Vaxelaire, J. Impact of Apple Processing Modes on Extracted Juice Quality: Pressing Assisted by Pulsed Electric Fields. *J. Food Eng.* **2011,** *103* (1), 52–61.
29. Guderjan, M.; Elez-Martínez, P.; Knorr, D. Application of Pulsed Electric Fields at Oil Yield and Content of Functional Food Ingredients at the Production of Rapeseed Oil. *Innov. Food Sci. Emerg. Technol.* **2007,** *8* (1), 55–62.
30. Gudmundsson, M.; Hafsteinsson. H. Effect of Electric Field Pulses on Microstructure of Muscle Foods and Roes. *Trends Food Sci. Technol.* **2001,** *12,* 122–128
31. Hafsteinsson, H.; Gudmundsson, M.; Arnarson, G. O.; Johnsson, A.; Siguroardottir, M. S. *High Electric Field Pulses: Food Safety; Quality and Critical Parameters*, Technological Institute of Iceland (IceTec): Iceland, 2000.
32. Harrison, S. L.; Barbosa-Cánovas, G. V.; Swanson, B. G. *Saccharomyces cerevisiae* Structural Changes Induced by Pulsed Electric Field Treatment. *LWT-Food Sci. Technol.* **1997,** *30* (3), 236–240.
33. He, G.; Yin, Y.; Yan, X.; Wang, Y. Semi-bionic Extraction of Effective Ingredient From Fishbone by High Intensity Pulsed Electric Fields. *J. Food Process Eng.* **2017,** *40* (2), 12392.
34. He, G.; Yin, Y.; Yan, X.; Yu, Q. Optimisation Extraction of Chondroitin Sulfate From Fish Bone by High Intensity Pulsed Electric Fields. *Food Chem.* **2014,** *164,* 205–210.
35. Hendrix, C. M.; Red, J. B. Chemistry and Technology of Citrus Juices and Byproducts. In *Production and Packaging of Non–Carbonated Fruit Juices and Fruit Beverages*; Ashurst P. R., Ed.; Springer publishers: Boston- USA, 1995; pp. 53–87.
36. Ho, S. Y.; Mittal, G. S.; Cross, J. D.; Griffiths, M. W. Inactivation of P. Fluorescens by High Voltage Electric Pulses. *J. Food Sci.* **1995,** *60* (6), 1337–1340.
37. Hülsheger, H.; Potel, J.; Niemann, E. G. Electric Field Effects on Bacteria and Yeast Cells. *Radiat. Environ. Biophys.* **1983,** *22,* 149–162.
38. Jaeger, H.; Balasa, A.; Knorr, D. Food Industry Applications for Pulsed Electric Fields. In *Electrotechnologies for Extraction from Food Plants and Biomaterial*; Vorobiev, E., Lebovka, N., Eds.; Springer-Verlag Publishers: New York – USA; pp 181–216.
39. Jaeger, H.; Janositz, A.; Knorr, D. The Maillard Reaction and its Control During Food Processing. The Potential of Emerging Technologies. *Pathol. Biol.* **2010,** *58,* 207–213
40. Jeantet, R.; Baron, F.; Nau, F.; Roignant, M.; Brule, G. High intensity Pulsed Electric Fields Applied to Egg White: Effect on *Salmonella enteritidis* Inactivation and Protein Denaturation. *J. Food Prot.* **1999,** *62* (12), 1381–1386.
41. Jeyamkondan, S.; Jayas, D. S.; Holley, R. A. Pulsed Electric Field Processing of Foods: A Review. *J. Food Prot.* **1999,** *62* (9), 1088–1096.
42. Joardder, M. U.; Masud, M. H. Food Preservation Techniques in Developing Countries. In*Food Preservation in Developing Countries: Challenges and Solutions*; Joardder M. U., Masud, M. H., Eds.; Springer publishers: Cham, Switzerland, 2019; pp 67–125.
43. Khan, A. A.; Randhawa, M. A.; Carne, A.; Ahmed, I. A. M.; Barr, D.; Reid, M.; Bekhit, A. E. D. A. Effect of Low and High Pulsed Electric Field on the Quality and Nutritional Minerals in Cold Boned Beef *M. longissimus et lumborum*. *Innov. Food Sci. Emerg. Technol.* **2017,** *41,* 135–143.

44. Knorr, D.; Geulen, M.; Grahl, T.; Sitzmann, W. Food Application of High Electric Field Pulses. *Trends Food Sci. Technol.* **1994,** *5* (3), 71–75.
45. Koubaa, M.; Barba, F. J.; Grimi, N.; Mhemdi, H.; Koubaa, W.; Boussetta, N.; Vorobiev, E. Recovery of Colorants from Red Prickly Pear Peels and Pulps Enhanced by Pulsed Electric Field and Ultrasound. *Innov. Food Sci. Emerg. Technol.* **2016,** *37,* 336–344.
46. Lebovka, N. I.; Shynkaryk, N. V.; Vorobiev, E. Pulsed Electric Field Enhanced Drying of Potato Tissue. *J. Food Eng.* **2007,** 78 (2), 606–613.
47. Li, M.; Lin, J.; Chen, J.; Fang, T. Pulsed Electric Field-Assisted Enzymatic Extraction of Protein from Abalone (*Haliotis Discus* Hannai Ino) Viscera. *J. Food Process Eng.* **2016,** *39* (6), 702–710.
48. Loginova, K. V.; Lebovka, N. I.; Vorobiev, E. Pulsed Electric Field Assisted Aqueous Extraction of Colorants From Red Beet. *J. Food Eng.* **2001,** *106* (2), 127–133.
49. López, N.; Puértolas, E.; Hernández-Orte, P.; Álvarez, I.; Raso, J. Effect of a Pulsed Electric Field Treatment on the Anthocyanin Composition and Other Quality Parameters of Cabernet Sauvignon freshly Fermented Model Wines Obtained after Different Maceration Times. *LWT-Food Sci. Technol.* **2009,** *42* (7), 1225–1231.
50. Luengo, E.; Álvarez, I.; Raso, J. Improving the Pressing Extraction of Polyphenols of Orange Peel by Pulsed Electric Fields. *Innov. Food Sci. Emerg. Technol.* **2013,** *17,* 79–84.
51. MacGregor, S. J.; Farish, O.; Fouracre, R.; Rowan, N. J.; Anderson, J. G. Inactivation of Pathogenic and Spoilage Microorganisms in a Test Liquid Using Pulsed Electric Fields. *IEEE Trans. Plasma Sci.* **2000,** *28* (1), 144–149.
52. Martínez, J. M.; Delso, C.; Angulo, J.; Álvarez, I.; Raso, J. Pulsed electric field-assisted Extraction of Carotenoids from Fresh Biomass of *Rhodotorula glutinis*. *Innov. Food Sci. Emerg. Technol.* **2018,** *47,* 421–427.
53. Martínez, J. M.; Luengo, E.; Saldaña, G.; Álvarez, I.; Raso, J. C. Phycocyanin Extraction Assisted by Pulsed Electric Field from *Artrosphira platensis*. *Food Res. Int.* **2017,** *99,* 1042–1047.
54. McDonnell, C. K.; Allen, P.; Chardonnereau, F. S.; Arimi, J. M.; Lyng, J. G. The use of Pulsed Electric Fields for Accelerating the Salting of Pork. *LWT-Food Sci. Technol.* **2014,** *59* (2), 1054–1060.
55. Min, S.; Zhang, H. *Effect of Water Activity on the Inactivation of Enterobacter cloacae by Pulsed Electric Field Treatment*. IFT Annual Meeting: Dallas–USA, 2000.
56. Monfort, S.; Gayán, E.; Raso, J.; Condón, S.; Álvarez, I. Evaluation of Pulsed Electric Fields Technology for Liquid Whole Egg Pasteurization. *Food Microbiol.* **2010,** *27* (7), 845–852.
57. Monfort, S.; Gayán, E.; Saldaña, G.; Puértolas, E.; Condón, S.; Raso, J.; Álvarez, I. Inactivation of *Salmonella Typhimurium* and *Staphylococcus aureus* by Pulsed Electric Fields in Liquid Whole Egg. *Innov. Food Sci. Emerg. Technol.* **2010,** *11* (2), 306–313.
58. O'Dowd, L. P.; Arimi, J. M.; Noci, F.; Cronin, D. A.; Lyng, J. G. An Assessment of the Effect of Pulsed Electrical Fields on Tenderness and Selected Quality Attributes of Post Rigour Beef Muscle. *Meat Sci.* **2013,** *93* (2), 303–309.
59. Parniakov, O.; Bals, O.; Lebovka, N.; Vorobiev, E. Pulsed Electric Field Assisted Vacuum Freeze-drying of Apple Tissue. *Innov. Food Sci. Emerg. Technol.* **2016,** *35,* 52–57.
60. Pothakamury, U. R.; Monsalve-González, A.; Barbosa-Cánovas, G. V.; Swanson, B. G. Inactivation of Escherichia Coli and Staphylococcus Aureus in Model Foods by Pulsed Electric Field Technology. *Food Res. Int.* **1995,** *28* (2), 167–171.

61. Praporscic, I.; Lebovka, N.; Vorobiev, E.; Mietton-Peuchot, M. Pulsed Electric Field Enhanced Expression and Juice Quality of White Grapes. *Sep. Purif. Technol.* **2007,** *52* (3), 520–526.
62. Prochownick, L.; Spaeth, F. Uber die keimtotende Wirkung des galvanischen Stroms. *Deutsche Medizinische Wochenschrift* **1890,** *26,* 564–565.
63. Puértolas, E.; de Marañón, I. M. Olive oil Pilot-production Assisted by Pulsed Electric Field: Impact on Extraction Yield, Chemical Parameters and Sensory Properties. *Food Chem.* **2015,** *167,* 497–502.
64. Puértolas, E.; Koubaa, M.; Barba, F. J. An Overview of the Impact of Electro Technologies for the Recovery of oil and High-value Compounds from Vegetable Oil Industry: Energy and Economic Cost Implications. *Food Res. Int.* **2016,** *80,* 19–26.
65. Puértolas, E.; López, N.; Condón, S.; Álvarez, I.; Raso, J. Potential Applications of PEF to Improve Red Wine Quality. *Trends Food Sci. Technol.* **2010,** *21* (5), 247–255.
66. Puértolas, E.; Luengo, E.; Álvarez, I.; Raso, J. Improving Mass Transfer to Soften Tissues by Pulsed Electric Fields: Fundamentals and Applications. *Ann. Rev. Food Sci. Technol.* **2012,** *3,* 263–282.
67. Puértolas, E.; Saldaña, G.; Álvarez, I.; Raso, J. Experimental design Approach for the Evaluation of Anthocyanin Content of Rosé Wines Obtained by Pulsed Electric Fields. Influence of Temperature and Time of Maceration. *Food Chem.* **2011,** *126* (3), 1482–1487.
68. Qin, B. L.; Chang, F. J.; Barbosa-Canovas, G. V.; Swanson, B. G. Non Thermal Inactivation of *S. cerevisiae* in Apple Juice Using Pulsed Electric Fields. *Food Sci. Technol.* **1995,** *28,* 564–568.
69. Raso, J.; Barbosa-Cánovas, G. V. Non Thermal Preservation of Foods Using Combined Processing Techniques. *Crit. Rev. Food Sci. Nutrit.* **2003,** *43* (3), 265–285
70. Sale, A. J. H.; Hamilton, W. A. Effects of High Electric Fields on Microorganisms, I: Killing of Bacteria and Yeasts. *Biochimica et Biophysica Acta* **1967,** *148,* 781–788.
71. Schilling, S.; Toepfl, S.; Ludwig, M.; Dietrich, H.; Knorr, D.; Neidhart, S.; Schieber, A.; Carle, R. Comparative Study of Juice Production by Pulsed Electric Field Treatment and Enzymatic Maceration of Apple Mash. *Eur. Food Res. Technol.* **2008,** *226* (6), 1389–1398.
72. Schoenbach, K. H.; Peterkin, F. E.; Alden, R. W.; Beebe, S. J. The Effect of Pulsed Electric Fields on Biological Cells: Experiments and Applications. *IEEE Trans. Plasma Sci.* **1997,** *25* (2), 284–292.
73. Sharma, P.; Bremer, P.; Oeya, I.; Everett, D. W. Bacterial Inactivation in Whole Milk Susing Pulsed Electric Field Processing. *Int. Dairy J.* *35* (1), 49–56.
74. Shayanfar, S.; Chauhan, O. P.; Toepfl, S.; Heinz, V. Pulsed Electric Field Treatment Prior to Freezing Carrot Discs Significantly Maintains Their Initial Quality Parameters after Thawing. *Int. J. Food Sci. Tech.* **2014,** *49* (4), 1224–1230.
75. Shynkaryk, M. V.; Lebovka, N. I.; Vorobiev, E. Pulsed Electric Fields and Temperature Effects on Drying and Rehydration of Red Beetroots. *Dry. Technol.* **2008,** *26* (6), 695–704.
76. Sitzmann, W.; Munch, E. W. Das ELCRACK Verfahren: Ein Neues Verfahren zur Verarbeitung Tierischer Rohstoffe. *Die Fleisch.* **1988,** 40 (2), 22–28.
77. Suwandy, V.; Carne, A.; van de Ven, R.; Bekhit, A. E. D. A.; Hopkins, D. L. Effect of Repeated Pulsed Electric Field Treatment on the Quality of Cold-boned Beef Loins and Topsides. *Food Bioprocess Tech.* **2015,** *8,* 1218–1228.
78. Syed, Q. A.; Ishaq, A.; Rahman, U. U.; Aslam, S.; Shukat, R. Pulsed Electric Field Technology in Food Preservation: A Review. *J. Nutrit. Health Food Eng.* **2017,** *6* (6), 168–172.

79. Toepfl, S.; Heinz, V.; Knorr, D. Application of Pulsed Electric Field Technology for the Food Industry. Chapter X. In *Pulsed Electric Field Technology for the Food Industry: Fundamentals and Applications*; Raso, J., Heinz, V., Eds.; Springer publishers: New York- USA, 2006; pp 197–221.
80. Toepfl, S.; Heinz, V.; Knorr, D. History of Pulsed Electric Field Treatment. In *Food Preservation by Pulsed Electric Field: from Research to Application*; Lelieveld, H. L. M., Notermans, S., de Haan, S. W. H., Eds.; Woodhead Publishing series in Food Science, Technology and Nutrition, 2007; pp. 264–289.
81. Tsong, T. Y. Electroporation of Cell Membranes. *Biophys. J.* **1991,** *60*, 297–306.
82. Vega-Mercado, H.; Martin-Belloso, O.; Chang. F. J.; Barbosa-Cinovas, G. V.; Swanson, B. C. Inactivation of *Escherichia coli* and *Bacillus subtilis* Suspended in Pea Soup Using Pulsed Electric Fields. *J. Food Process. Preserv.* **1997,** *20* (6), 501-510.
83. Vorobiev, E.; Lebovka, N. Enhanced Extraction From Solid Foods and Biosuspensions by Pulsed Electrical Energy. *Food Eng. Rev.* **2010,** *2,* 95–108
84. Wiktor, A.; Schulz, M.; Voigt, E.; Witrowa-Rajchert, D.; Knorr, D. The Effect of Pulsed Electric Field Treatment on Immersion Freezing, Thawing and Selected Properties of Apple Tissue. *J. Food Eng.* **2015,** *146,* 8–16.
85. Wouters, P. C.; Dutreux, N.; Smelt, J. P.; Lelieveld, H. L. Effects of Pulsed Electric Fields on Inactivation Kinetics of *Listeria innocua. Appl. Environ. Microbiol.* **1999,** *65* (12), 5364–5371.
86. Yadollahpour, A.; Rashidi, S.; Ghotbeddin, Z.; Jalilifar, M.; Rezaee, Z. Electromagnetic Fields for the Treatments of Wastewater: A Review of Applications and Future Opportunities. *J. Pure Appl. Microbiol.* **2014,** *8* (5), 3711–3719.
87. Yan, L. G.; He, L.; Xi, J. High Intensity Pulsed Electric Field as an Innovative Technique for Extraction of Bioactive Compounds—A Review. *Crit. Rev. Food Sci. Nutrit.* **2017,** *57* (13), 2877–2888.
88. Yin, Y. G.; Han, Y. Z.; Yong, H. Pulsed Electric Field Extraction of Polysaccharide from *Rana temporaria chensinensis* David. *Int. J. Pharm.* **2016,** *312,* 33–36.
89. Yonemoto, Y.; Yamashita, T.; Muraji, M.; Tatebe, W.; Ooshima, H.; Kato, J.; Kimura, A.; Murata, K. Resistance of Yeast and Bacterial Spores to High Voltage Electric Pulses. *J. Ferment. Bioeng.* **1993,** *75* (2), 99–102.
90. Zhang, Q. H.; Chang, F. J.; Barbosa-Canovas, G. V. Inactivation of Microorganisms in a Semisolid Model Food Using High Voltage Pulsed Electric Fields. *Lebensm.-Wiss. und-Technol.* **1994,** *27* (6), 538 – 543.
91. Zimmermann, U.; Pilwat, G.; Riemann, F. Dielectric Breakdown in Cell Membranes. *Biophys. J.* **1974,** *14,* 881–899.

CHAPTER 2

PULSED ELECTRIC FIELD FOR FOOD PROCESSING: HISTORY AND DESIGN ASPECTS

PREETI BIRWAL, MAHENDRA GUNJAL, MANDEEP SINGH AZAD, BHAVESH CHAVHAN, and KAWARDEEP KOUR

ABSTRACT

The pulsed electric field is an emerging technology for the processing of food. Although the technique is old, its applications have been highlighted in the last two decades. The equipment design involves understanding the electrical circuit and the workings of the transformer. The laboratory and pilot-scale works have shown promising results for successfully utilizing pulse electric fields for food preservation and processing. PEF equipment is designed to handle and process very high voltages of around 20–60 kV. Equipment for the PEF treatment usually consists of a pulse generator, a treatment chamber, and a system for control and data acquisition. Pulsed power is the main source of electroporation, which is achieved by the proper assembly of different electric components. The generation of high voltage is achieved using a step-up transformer and the output of high voltage is controlled by the output pulse through the switching circuit. Proper design of the treatment chamber is important in order to ensure effective cellular permeabilization during PEF treatment. The design aspects of the treatment chamber include chamber geometry, size, and construction material. Four different wave shapes, namely, exponential, triangle, square, and oscillating decay have been extensively reported in PEF systems applied to inactivate the microorganisms. This chapter compiles the information reviewed to

Nonthermal Pulsed Electric Field Technology for the Food Industry.
Preeti Birwal, Megh R. Goyal, Baldev Singh Kalsi, & Ankita Kataria (Eds.)

DOI: 10.1201/9781003605904-3

understand the PEF circuit and components, their applications, and the effect of design and operating parameters on treated products.

2.1 HISTORICAL BACKGROUND

Pulsed electric field (PEF) is a non-thermal food processing technology, which is gaining attention among researchers, academicians, and food processing unit's small-scale and large-scale sectors.[47] The PEF and its implementation in food processing is considered an emerging technology; its idea can be traced back to the 20th century. It was primarily used for the electropure process, which inactivates microbes and enzymes in food materials.[24] However, PEF took three decades to be used in food processing lines at a commercial level. The pioneering research in PEF was mainly credited to the German engineer Heinz Doevenspeck.[50]

In the German Patent Office in the 1980s, Krupp was the first to file a patent on the use of PEF treatment on food. This patent application claimed that the process of inactivating microorganisms in milk and fruit juices was based on the development of the "Elcrack" and "Elsteril" techniques. Several food industries have constructed scaled-up prototypes and industrial facilities depending on the ELCRACK® technology for handling food commodities.[41] Industrial adoption of PEF technology can be seen in the later half of the 1990s which led to commercialization of PEF treatment for pasteurization of the food products. Hulsheger and Niemann published several papers on the effect of PEF on bacterial inactivation in the early 1980s.[35] In 1992, PEF preservation technology was recognized as a processing technique that could provide fresh-like food with better organoleptic properties to the consumer over traditionally processed food.[67]

The FDA approved the PEF technique created by PurePulse Technologies, USA (CoolPure R) for the treatment of pumpable foods in 1995, and it was the first attempt to make the PEF treatment legal in food processing industries. In 2005, the first commercial adoption of PEF technology was introduced by Genesis Juice Cooperative, USA, where cold pasteurization of fruit juices was attempted at a commercial scale.[15] Currently, various research groups across the world are working on the different applications of the PEF technology, namely, Canada, Netherlands, Belgium, China, France, Germany, New Zealand, Iceland, Japan, Australia, Sweden, Switzerland, United Kingdom, USA, and Taiwan.

This chapter compiles information from the literature to understand the PEF circuit and components, its applications, and the effect of design and operating parameters on treated products.

2.2 PRINCIPLE OF OPERATION

In the PEF processing method, food materials are placed or passed between two electrodes and it uses high electric voltage in the range of 20–80 kV/cm for some duration of time.[1,40] The ranges of operating parameters[10] are given in Table 2.1.

TABLE 2.1 Operating Parameters for Pulsed Electric Field Food Processing.

Parameter	Operating range
Current	100 A–10 mA
Energy	10 j–100 mj
Power densities	1w–1 mw/cm^2
Pulse width	100 ps–100 μs
Voltages	1 kv–10 mv

Several changes occurred inside the cell membrane when bacterial cells were exposed to the high electric voltage. When a high electric voltage is applied, it reduces the cell resistance and increases its conductance—this process is known as electroporation.[47] When a cytoplasmic microbial membrane is subjected to a higher voltage for a longer period, pores begin to form in a membrane.[76]

Electroporation and disruption (swelling of cells, shrinking, and finally lysis) are major working principles behind the degradation of bacterial cells when exposed to the treatments of PEF. Their mechanism of cell inactivation is shown in Figure 2.1.

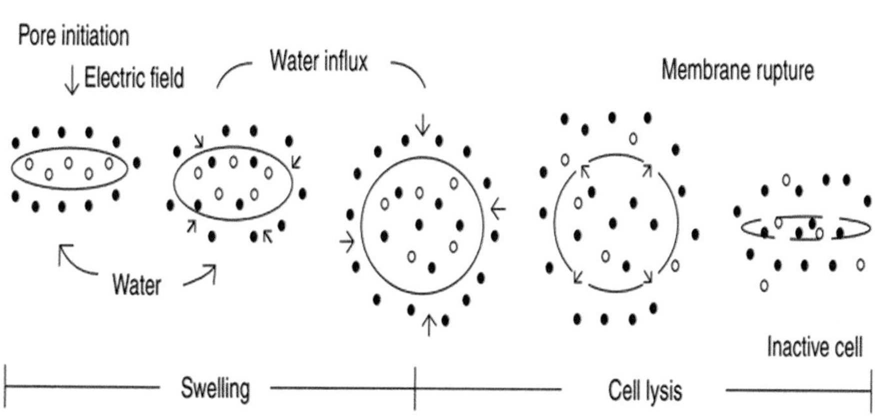

FIGURE 2.1 Mechanism of cell inactivation.

The electric field has a direct relationship between cell electroporation and electric field and can be easily regulated by varying the intensity of electric field strength under controlled conditions. Exposure of bacterial cells to electric fields brings about detrimental changes in cell membranes. Gradual loss in semi-permeability and becoming porous cells resulted in permanent membrane damage whereby microorganism loses their activity.

A few of the theories put forth by researchers to explain the mechanism of inactivating microbial cells include the increase in transmembrane potential, osmotic imbalance, compression-induced electromechanical alteration of the cell membrane, and macromolecular changes in the lipids or protein constituents.

The dielectric breakdown theory has been the most accepted model.[81] The ions within a cell move in the direction of the applied field, when a cell suspension is exposed to an electric field, according to the theory that focuses on the effects of electric fields (Figure 2.1). Deposition of ions on the surface of membrane and accumulation of charges, increase the electromechanical stress and ultimately leads to compression pressure that causes the decrease of the membrane thickness.[81] The electro-compressive force (P_e), per unit area of the membrane is calculated as below:

$$\text{Electrocompressive force}(P_e) = \frac{\varepsilon_m \varepsilon_0 \Delta\varphi^2}{2\delta^2} \tag{2.1}$$

where δ = thickness of the membrane; ε_0 = electrical conductivity of the medium; ε_m = dielectric constant of the empty space; and $\Delta\varphi_c$ = critical potential difference.

Further increase in the intensity of the applied electric field disturbs the critical membrane potential resulting in a breakdown of the cell membrane. For the electrical breakdown of cell membrane, the viscoelastic properties (surface tension and viscosity of membrane) are used to explain.[7]

Grenier[33] highlights another model of how an electric field affected biological cell membranes; charges between cell membranes led to membrane thinning. The balance between electrical compressive and restoring forces determines the membrane thickness at a particular membrane potential. Pore formation is thought to occur when there is no change in thickness with an increase in membrane potential and compression prevails over restoring forces. This results in a loss of membrane stability and the microorganisms die.

The electric field strength of any transmembrane ranges from 1 to 20 kV/cm and it depends on the radius of the cell.[17] The osmotic imbalance created due to leakage of ions molecules and other molecules induced by PEF treatment was believed to be the cause of membrane rupture.[39] It has been postulated that the osmotic pressure on cytoplasmic content plays an important role in the swelling of the cell and rupture of the cell membrane which ultimately leads to cell lysis and death.[72] Vega-Mercado et al.[75] investigate another theory of the osmotic imbalance; an investigation on the effects of pH and ionic strength on the inactivation of *E. coli* concluded that membrane permeability increases due to mechanical compression and poration as the main cause.

2.3 DESIGN ASPECTS OF PEF SYSTEMS

Pulsed power is the main source of electroporation, which is achieved by the proper assembly of different electric components. The pulsed power supply system primarily converts a low peak power to a high peak power and a long duration pulse to a short duration. To achieve this scaling up and down of power and pulse duration as desired, different components, such as high voltage power suppliers, energy storage devices, pulse generation units, switches, and controllers, have been reported in PEF systems.[43]

2.3.1 VOLTAGE GENERATORS

PEF equipment is designed to handle and process very high voltages around 20–60 kV. Equipment for the PEF treatment is usually made up of different systems such as a pulse generator, treatment chamber, and a system for control and data acquisition.[42] The capacitor immediately released the stored energy at a very high power level. The high-voltage handling switches should be particularly dependable for high power and steady for high repetition rate and control of the discharge. The generation of high voltage is achieved using a step-up transformer and the output of high voltage is controlled by the output pulse through a switching circuit. The required primary and secondary voltages and number of turns calculations are given in Table 2.2. The electrodes are attached to the treatment chamber and they produce high voltage pulse.[54]

A frequency-based transformer triggered using a switching system was used to convert low voltage to high voltage and very low current to low current.[43] In the system, reported high voltage AC power was rectified and converted to DC power. Another system to produce high voltage using capacitors was published.[80] This system used high-frequency AC inputs and gave a commanding charge at a greater repetition rate than a DC power supply.

TABLE 2.2 Formulas to Design the Transformer.

Parameter	Formula
Primary volt – Amps (PVA)	$\frac{\text{Secondary volt} - \text{Amps(SVA)}}{0.9(\text{Assuming efficiency } 90\%)}$
Secondary voltage – Amps (SVA)	Secondary voltage (V_s) × Secondary current (I_s)
Primary voltage (V_p) – Amps (SVA)	$\frac{\text{Secondary voltage}(V_S)}{\text{Turns ratio}(\frac{n_2}{n_1})}$
Primary current (I_p)	$\frac{\text{Primary volt} - \text{Amps(PVA)}}{\text{Primary voltage}(V_p)}$
The cross – sectional area of the core	1.15 × (PVA)
Turns per volt (T_{pv})	$\frac{1}{(4.44 \times \text{Core area} \times \text{Frequency} \times \text{Flux density})}$

According to research studies, the frequency of the capacitor's charging and discharging cycles determines the voltage generators' complete power rating. The electrical resistance of the charging resistor and number, and the size of a charged capacitor were the additional two elements that affected the power needed to charge the capacitor bank. It was discovered that charging a larger capacitor than a smaller one required more time and energy. Similar to this, a smaller charging resistor speeds up charging but requires more power.[47]

2.3.2 SWITCHES

Thyratron switches are typically recommended as the best option to manage high-voltage and quick turn-on. While the turning off of the thyratron switches is frequently reliant on a resistor-capacitor circuit with a continuous load of resistance, the turn-on is controllable. Resistance-capacitor (RC) circuits are capable of generating exponential-decay pulse and thyratron-based pulsed power supplies are integrated with a pulse forming network (PFN) in order to deliver a square pulse.[30] The semiconductor switches have the controllable option (turn-on and turn-off) and they can generate a square wave of different controllable widths. Other advantages of semiconductor switches include high repetition rates and extended life.[2,28]

Among the semiconductor switches reported in pulsed power supply systems, the common options listed include insulated gate bipolar transistors (IGBTs) and metal oxide semiconductor field effect transistors (MOSFETs) while solid-state semiconductor switches have been projected as the future of high-power switching.[8] MOSFETs are said to be capable of faster-switching systems, and low voltage and current rating compared to IGBT and have wide applications in industry.[32] MOSFET-based circuits are less complex compared to others.[14] However, to increase the switch capacity, several units are often connected in series and in parallel; this results in a significant increase in costs.

Hess et al.[31] designed a system where MOSFETs were connected in series to increase the voltage. Sundararajan et al.[68] described a power supply connecting four MOSFETs in series to generate 1500 V. Aly[3] used a single IGBT circuit to generate power and studied the same on the viability of *E. coli* under 1.3 and 1.5 MV/m for 4 ms and 32 ms, respectively.

2.3.3 TREATMENT CHAMBER

The PEF treatment process for more effeteness to ensure successful cellular permeabilization and the treatment chamber must be designed properly. Design components of the treatment chamber include chamber geometry, size, and construction material. Over the years, many treatment chamber designs have been put forth; the most frequently used designs include the parallel, coaxial, and co-linear configurations, as well as the treatment chamber designs for concentric cylinders, cones, converging electric field, and co-field (Figure 2.2).[12,44,69,78] The design of the process chamber is highly dependent on the uniformity of the process.[25] Uniform electric field and temperature ensure a uniform treatment.[69] Batch, semicontinuous, and continuous PEF treatment chambers of different designs have been reported to be used in industrial applications.[34] The entire unit also requires auxiliary equipment including pumps in semicontinuous and continuous operation and a temperature controller in addition to the primary treatment chambers part.[69]

The primary standard to adhere to in the treatment chamber's design is the prevention of food dielectric breakdown, which causes the formation of a spark inside the chamber.[7,10] The food dielectric breakdown is attributed to the local enhancement of the electric field or the presence of air bubbles.

It must be washable (cleaning-in-place) or autoclavable and chemically inert concerning foods when choosing construction materials for the treatment chamber. The good quality material for electrodes is stainless steel, which has been used widely.[11,21] To withstand the strength of the applied electric field, it's also recommended that insulation utilized in the treatment chamber have higher dielectric strength.[47]

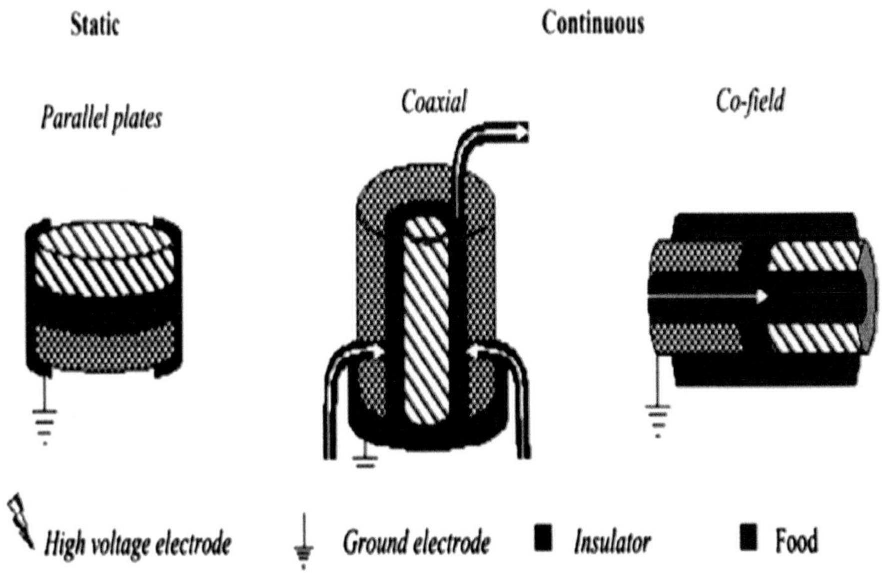

FIGURE 2.2 Static parallel type, continuous coaxial, and co-field treatment chambers.

Parallel plate electrodes with a small gap between them provide a uniform electric field and represent the most practical choice of batch chambers.[5,19,26,46,58] In fact, one of the earlier reports is a straightforward design recommending the electrodes as the tube used for the transportation of the product.[27]

The electrodes with a disc shape and round edges have also been shown to decrease electric field enhancement, hence reducing the risk of dielectric breakdown.[79] During the processing of foods, the polished electrode surfaces reduce the risk of dielectric breakdown.[6] The co-linear configuration was reported to be less uniform in its distribution of the electric field; however, with minor modifications in electrode configuration and isolator geometry, the electric field uniformity could be significantly improved to enhance the treatment performance.[55]

2.3.4 WAVE SHAPE

Four different wave shapes, namely, exponential, triangle, square, and oscillating decay have been extensively reported in PEF systems applied to inactivate the microorganisms. Square/rectangle pulses are generated with MOSFET-driven circuits and IGBT semiconductors. Square pulses are the most desirable amongst the pulses of the same energy as the lethality of the square wave is maximum.[79] The bipolar pulses in square waves are more lethal than monopolar pulses because they induce more stress due to the alternating changes in movement and their enhanced electric breakdown.[7,23,56]

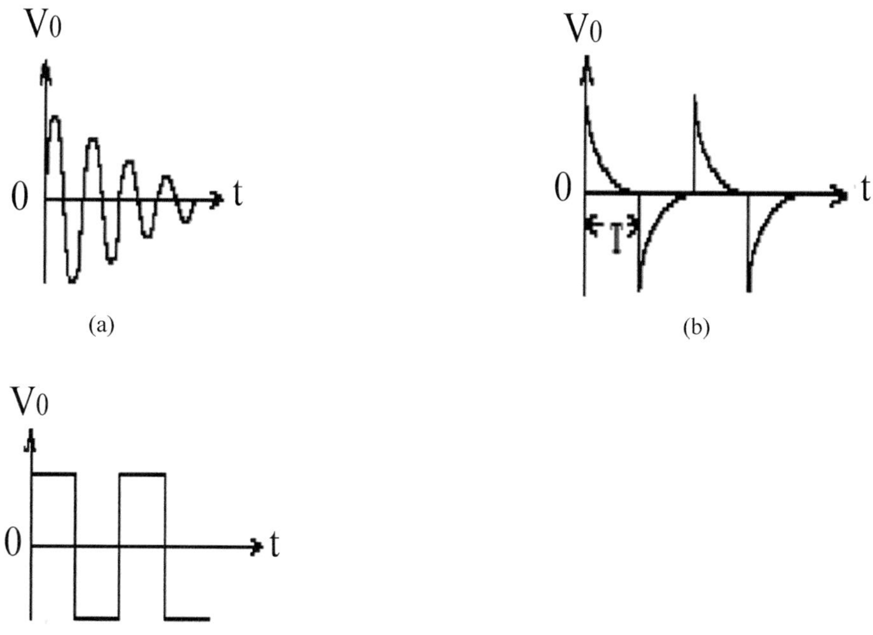

FIGURE 2.3 The waveforms generated in PEF system, where (a) is oscillatory decay pulse; (b) is an exponential decay pulse; and (c) is square wave.

Minimum energy utilization, reduction in deposition of solids on the surface of the electrode, and decreased food electrolysis are the characteristics and advantages of bipolar pulses. Amiali et al.[4] in their experimental study applied instant 16-reversal square wave pulses and showed that these waveforms were more effective at pasteurizing egg products. The poorer efficiency of the exponential pulse is attributed to the hindrance created in liquid food that restricts the exposure of bacterial cells to an electric field continuously for a long time. It is reported that voltage surging exponential pulses show high surging rates but slow decaying leading to undesirable heating of the liquid.[36]

The waveform of different shapes is represented in Figure 2.3. The pulse forming network (PFN) is a technique for producing electrical pulses with a specific duration, voltage, and shape.[61] Resistor-inductor-capacitor (RLC) circuits are one of the alternate options; their efficiency is reported to be between 40 and 50% higher than that of RC circuits, which is around 38%.[54]

2.4 BASIC ELECTRICAL PARAMETERS FOR PEF TREATMENT

2.4.1 *TREATMENT TIME*

It is the total time the food is exposed to the electrical pulse and is expressed as the product of the number of pulses and individual pulse width.

$$t = n\tau_p \tag{2.2}$$

where τ_p: pulse duration; and n: number of pulses applied to the food.

In eq 2.2, the treatment time is directly proportional to the number of pulses and pulse duration. While increasing pulse duration may increase the temperature of the food, increasing the number of pulses may increase the overall energy consumption.[7]

2.4.2 *ELECTRIC FIELD STRENGTH*

For homogeneous samples (i.e., permittivity is constant across the sample), the average electric field strength E is computed as follows.

$$E = \frac{V}{d} \tag{2.3}$$

where V: voltage across the food sample; and d: distance between the electrodes.

From eq 2.3, it can be seen that with increasing gap between the electrodes, a higher voltage is required to obtain the desired electric field for membrane breakdown.[7]

2.5 APPLICATION OF PEF TREATMENT IN FOOD PROCESSING

PEF technology has been found as a potential alternative method for the alteration of enzyme activity, enhancement of metabolite extraction and drying rates, and various applications in biotechnology and genetic engineering.[47] Multiple studies have demonstrated the effectiveness of PEF technology in producing superior food, including milk and milk products, soups, eggs, yogurt, juice, water, beer, and others with a safe, fresher, and maintaining the nutrient profile of these food products.[37,63]

PEF is an alternative non-thermal technology for liquids and semi-solid foods in place of traditional heat pasteurization with the main purpose of inactivating pathogenic bacteria has also been reported.[13,52] PEF preservation technology is associated with advantages such as reduced energy requirement and better energy efficiency compared to thermal processing.

This technology will find its place in the industry at a commercial scale due to the advancements in the design and development of cost-effective equipment.

2.5.1 PEF TREATMENT OF MILK

The milk processing industry is a vital part of processing foods and a significant part of the PEF study was dedicated to studying the effect of PEF on milk and dairy products. The research organization has developed this technology as a substitute for the conventional methods of pasteurizing liquid food commodities, providing it with increasing significance on a global level.

Perishable commodities such as milk are popular and commercialized liquid food products that act as an excellent growth medium for microbes. Milk was studied extensively to understand the PEF treatment effects on quality attributes such as microorganism's inactivation.[9] Researchers examined the physicochemical properties and shelf life of PEF-pasteurized raw milk. The PEF treatment conditions were used for both skim milk and whole milk, with electric field strengths ranging from 30.76 to 53.84 kV/cm at the different pulses (12, 21, and 30) and temperatures ranging from 20, 30, and 40°C with a 2 μs monopolar pulse width. After processing, there were slight differences in the physicochemical properties based on the lipid and protein composition. After processing at 46.15 kV/cm, temperature (20–60°C), and 30 pulses, the shelf stability of skim milk and whole milk were studied. The pH and mesophilic and psychrophilic loading during storage at 4 and 21°C were assessed. Milk samples treated with PEF and kept at 4°C exhibited greater stability. However, samples stored at 21ºC showed signs of early spoilage. In both samples, the growth of mesophilic bacteria was delayed but psychrophilic bacteria grew rapidly.

Odriozola-Serrano et al.[51] evaluated the shelf life of PEF-treated milk against thermal pasteurized milk stored under refrigerated conditions and concluded that PEF treatment extended (by a few more days) the shelf life of fluid milk under refrigeration. With process parameters (36.7 kV/cm and 40 pulses delivered over 25 min, the effect of PEF treatment on homogenized milk infected with *Salmonella Dublin* was studied. The findings verified that the pathogen was rendered inactive even after a storage period of 8 days at 7–9°C.[19]

Dutreux et al.[20] studied the inactivation *Listeria monocytogenes* through PEF and compared it with pasteurized whole-fat milk, 2% fat milk, and skim milk. A decrease in the *L. monocytogenes* population at storage temperature 25°C 1–3 log cycles were reported and 50°C 4 log cycles were reported, without any significant differences among the three milks. Similarly, the effect of PEF-treatment of 158 μs at 41 kV/cm and at 37°C on the viability of *E. Coli* O1:K1:H7 in pasteurized skim milk was found to result in a 4 log reduction.[59]

Miranda et al.[45] investigated the bactericide nisin-induced sensitization by inactivating *Listeria innocua* suspended in skim milk. The population of *L. innocua* was reported to have by applying PEF treatment (30, 40, and 50 kV/cm) by 2.5 log. The decrease of 2, 2.7, and 3.4 log reduction cycles of *L. innocua* species were accomplished using the same PEF intensities and after exposure to 10 IU nisin/mL; the additional inactivation was attributable to the action of post-PEF exposure to nisin. *Pseudomonas* spp. in milk was said to be very effectively inactivated by PEF, extending the milk's shelf life at a temperature of 4°C for 8 days.[20]

The milk contains 2% fat and it is treated with two steps of seven pulses each and another one step of six pulses with an electric field (40 kV/cm) the shelf life of milk was increased up to 2 weeks under refrigeration conditions;[55] PEF treatment did not alter the milk's physicochemical properties, and there were no significant changes among heat pasteurized and PEF treated milk in terms of sensory parameters.

The size of casein micelles and fat globules in raw milk was reported to be unaffected by PEF processing, while a slight reduction in these sizes was seen when milk was treated at high temperatures.[67] According to some reports, PEF treatments can inactivate milk's spoilage microorganisms at least to the same degree as HTST pasteurization.

The PEF treatments are limited to the inactivation of microorganisms exclusively, according to the current review of the literature. However, it's feasible that more uses in dairy processing and products may be investigated after the impact of PEF on the chemical and physical properties of milk is fully understood.

2.5.2 DAIRY-BASED BEVERAGES AND PRODUCTS

The effects of PEF treatment on the processing of milk-based beverages were reported by several researchers. PEF treatments were tested with several dairy drinks in combination with fruit juices. Fruits are known to be good antioxidants and deliver many health-promoting benefits. Fruit-based dairy drinks are often made using different types of preservatives such as sugar, citric acid as an acidifier, and pectin as a stabilizer with a small amount of water.[29]

The milk (20%) and orange juice (50%) were exposed to at 15 kV/cm for 700 μs, there was a 3.83 log reduction in *E. coli* observed.[60] Schoenbach et al.,[64] studied the inactivation of *Salmonella* spp. using PEF treatment in an orange, juice (30%), and milk (20%) beverage.

Salvia-Trujillo et al.[62] researched a beverage mix of an orange (30%), apple (10%), and mango (10%) with whole milk. When *Listeria innocua* was exposed to bipolar 4 μs square wave pulses at 200 Hz and 35 kV/cm, there was a reduction in the 5-log cycle of the bacteria. The shelf life of the dairy-based fruit beverage after 56 days at 4°C was ensured by heat treatments in combination with high-intensity PEF without affecting its pH, acidity, and soluble solid content.

In another study, *S. cerevisiae* inoculated yogurt was quickly inactivated by an electric field compared to *Lactobacilli.* When it was exposed to an electric field at 45°C when inoculated with yeast, yogurt's shelf life was extended up to 10 days when kept at 4°C temperature. Similar to this, treatment at 55°C produced a one-month shelf life under storage at 4°C temperature.[19]

The use of heat and PEF treatments on the yogurt, drinks made with yogurt, and flavored milk the quality parameters such as organoleptic qualities did not significantly change. The yogurt-based beverage is made with different flavors that have undergone a combination treatment of heat (60°C, 30 s) and PEF (30 kV/cm, 32 s). This processed beverage kept at 4°C, it demonstrated a 90-day storage life with 2–4 log reductions in both total viable counts and moulds observed.[19] When *C. sakazakii* inoculated infant formula milk used treatment PEF from 10 to 40 kV/cm and from 60 to 3895 μs, a reduction of 1.2 log was achieved at 40 kV/cm for 360 μs.[53]

Numerous workers have assessed the pre-treatment with PEF of milk intended for manufacturing cheese. Dunn[18] indicated no significant change observed in both physical and chemical changes in milk quality attributes pre-treated with PEF before cheese making. Further, a 3-log reduction was achieved immediately after the treatment when *E. coli* was used as the challenge bacteria in the milk, facilitating the manufacture of cheese with a good sensory profile by minimizing the heat treatment.

While there was no discernible change in the adhesiveness and cohesiveness of Cheddar cheese from milk pasteurized using PEF, it did produce cheese that was hard and springier than cheese made from milk that had undergone normal pasteurized[66] Cheese milk was treated with PEF, and its effect was compared with pasteurized milk. It was concluded that a similar flavor in cheddar cheese in PEF-treated milk as that of raw milk, and superior to that of pasteurized milk was achieved. This indicates that PEF could replace the conventional pasteurization process in the case of cheese-making.[37]

2.5.3 JUICES

An extensive study has been carried out on fruit juices such as apple and orange juices to know the effect of PEF. PEF treatment improved the shelf life of fruit juices and its sensory attributes were found superior.

The PEF technology is most effective against the killing of microorganisms, improving juice extraction, raising pressing efficiency, and accelerating the dehydration and drying processes of food materials.[77] During treatment of the apple juice, energy utilized in PEF was 90% less than the amount of energy used in high temperature and short time-processing methods.

In a study, two parallel plate electrodes and a dielectric space insulator the orange juice were employed as the treatment mediums. The electrolyte solution comprising sodium hydroxide, sodium carbonate, potassium carbonate, and potassium hydroxide used to facilitate the electrical conduction between electrodes and ion-permeable membranes was separated from the food by acrylic acid and sulfonated polystyrene. Circulation was done continuously to lead to less deposition.[19] Freshly squeezed high-pulp orange juice was subjected to 35 pulses of a 35 kV/cm electric field, which led to a 5-log reduction in naturally occurring microbial load. In contrast to untreated juice, which was found to be undesirable after 4 days changes in juice color and flavor were acceptable for at least 10 days.

Torregrosa et al.[70] studied the effect of HIPEF juice on an orange–carrot mixture (80:20 v/v). A juice mixture was also pasteurized traditionally at 98°C for 21 s. The samples treated with an electric strength of 25 and 30 kV/cm were found to have high concentrations of carotenoids and vitamin A as compared to pasteurized juice. The degradation kinetics of ascorbic acid was determined in orange–carrot juice treated by different electric field intensities (25, 30, 35, and 40 kV/cm) and treatment times (30–340 μs). The 25 kV/cm showed a minimum degradation rate of ascorbic acid in the treated orange–carrot juice.[71]

Comparison of the effect of pasteurization and PEF treatment on color, browning, and hydroxyl methyl furfural (HMF) content of refrigerated orange juice revealed no significant difference in the HMF content of the juice between the treatments, PEF-treated orange juice showed lower non-enzymatic browning than the pasteurized samples.[16] Investigation on the effect of pulsed electric field (PEF) treatment on the inactivation of *E. coli* O157:H7 in apple juice and cider at electric field strengths of 34, 31, 28, 25, and 22 kV/cm and a mean total treatment time of 166 μs revealed a 4.5 log reduction with no change in color and vitamin C content.[22]

2.5.4 EGGS, SOUP, AND OTHER PRODUCTS

PEF treatment of eggs was studied extensively by various researchers. In an experiment, egg products were subjected to a static parallel electrode treatment chamber with 25 exponentially decaying pulses with peak voltages of around 36 kV.[19] The shelf life of egg products significantly improved when PEF treatment was applied to liquid eggs and heat-pasteurized liquid egg products.

Inactivation of *Salmonella* using PEF treatment in egg products including liquid whole egg, yolk, and egg white products have been investigated and a 3–4 log reduction in the population of the organism, depending on the treatment conditions (25–45 kV/cm, 30–105 μs, 369–441 kJ/kg) was achieved.[48,49] When pea soup was treated with PEF in two steps using 16 pulses at 35 kV/cm, it did not appear to affect the quality parameters such as sensory, physical, and chemical characteristics of the soup when it was stored at room temperature for up to four weeks.[74]

Ulmer et al.[73] examined the inactivation of beer spoilage organism *Lactobacillus plantarum* and its sub-lethal injury in a model system using different PEF exposure to 10–19 kV/cm and total energy inputs of 13–42 kJ/kg. A ready to serve (RTS) tender coconut water drink which was blended with nanari extract was examined for its physiochemical changes. It improved responses of CIE color (*L*, a*,* and *b*)* values, pH, ^{0}Brix as well as its sensory overall acceptability. In the prepared beverage store under ambient storage conditions (27–30°C), PEF-treated tender coconut water–nannari blended beverage was stable for up to 120 days.[38]

2.6 SUMMARY

The PEF is an emerging novel technology with a numerous application. The effects of PEF on food properties are not significant while the extension in shelf life is highly reported. Tremendous work in the research field to make this application commercial worldwide can be forecasted after concluding the novelty of the technology.

KEYWORDS

- **design**
- **wave length**
- **switches**
- **quality**
- **intensity**
- **electrical potential**

REFERENCES

1. Ade-Omowaye, B. I. O.; Angersbach, A.; Taiwo, K. A.; Knorr, D. Use of Pulsed Electric Field Pre-Treatment to Improve Dehydration Characteristics of Plant-Based Foods. *Trends Food Sci. Technol.* **2001,** *12* (8), 285–295.
2. Alkhafaji, S. R.; Farid, M. An Investigation on Pulsed Electric Fields Technology Using New Treatment Chamber Design. *Innovat. Food Sci. Emerg. Technol.* **2007,** *8* (2), 205–212.
3. Aly, R. E.; Joshi, R. P.; Stark, R. H.; Schoenbach, K. H.; Beebe, S. J. In *The Effect of Multiple, Microsecond Electrical Pulses on Bacteria*, PPPS-2001 Pulsed Power Plasma Science 2001. 28th IEEE International Conference on Plasma Science and 13th IEEE International Pulsed Power Conference. Digest of Papers (Cat. No. 01CH37251), 2001; Vol. 2, pp 1114–1117. IEEE.

4. Amiali, M.; Ngadi, M. O.; Raghavan, V. G.; Nguyen, D. H. Electrical Conductivities of Liquid Egg Products and Fruit Juices Exposed to High Pulsed Electric Fields. *Int. J. Food Propert* **2006,** *9* (3), 533–540.
5. Angersbach, A.; Knorr, D. High Intensity Electric Field Pulses as Pre-Treatment for Affecting Dehydration Characteristics and Rehydration Properties of Potato Cubes. *Nahrung (Germany)* **1997,** *55,* 143–146.
6. Angersbach, A.; Heinz, V.; Knorr, D. Effects of Pulsed Electric Fields on Cell Membranes in Real Food Systems. *Innovat. Food Sci. Emerg. Technol.* **2000,** *1* (2), 135–149.
7. Barbosa-Canovas, G. V.; Gongora-Nieto, M. M.; Pothakamury, U. R.; Swanson, B. G. Fundamentals of High-Intensity Pulsed Electric Fields (PEF). In *Pulsed Electric Fields for Food Processing Technology*; Academic Press: San Diego, CA, USA, 1999; pp 119–125.
8. Bartos, F. J. Medium Voltage AC Drives [Online]. 2000, Control Engineering Online, No. 20. http://www.Controleng.com/ Archives /2000/ctl0201.00/000203.htm (accessed 30 Jun 2023).
9. Bermudez-Aguirre, D.; Fernandez, S.; Esquivel, H.; Dunne, P. C.; Barbosa-Canovas, G. Milk Processed by Pulsed Electric Fields: Evaluation of Microbial Quality, Physicochemical Characteristics, and Selected Nutrients at Different Storage Conditions. *J. Food Sci.* **2011,** *76* (5), S289–S299.
10. Birwal, P.; Gajanan, P. D.; Menon, R. R. Non-Thermal Processing of Dairy Beverages. In *Milk Based Beverages*; Alexandru Grumezescu, Alina-Maria Holban. Ed.; The Science of Beverages, 2019; Vol. 9, pp 397–426.
11. Bouzrara, H.; Vorobiev, E. Beet Juice Extraction by Pressing and Pulsed Electric Fields. *Int. Sugar J.* **2000,** *102* (1216), 194–200.
12. Bushnell, A. H.; Dunn, J. E.; Clark, R. W.; Pearlman, J. S. U.S. Patent 5,235,905. Washington, DC: U.S. Patent and Trademark Office, 1993.
13. Butz, P.; Tauscher, B. Emerging Technologies: Chemical Aspects. *Food Res. Int.* **2002,** *35* (2–3), 279–284.
14. Chaney, A.; Sundararajan, R. Simple MOSFET-Based High-Voltage Nanosecond Pulse Circuit. *IEEE Trans. Plasma Sci.* **2004,** *32* (5), 1919–1924.
15. Clark, J. P. Pulsed Electric Field Processing. *Food Technol.* **2006,** *1,* 66–67.
16. Cortes, C.; Esteve, M. J.; Frigola, A. Color of Orange Juice Treated by High Intensity Pulsed Electric Fields During Refrigerated Storage and Comparison with Pasteurized Juice. *Food Control* **2008,** *19* (2), 151–158.
17. Coster, H. G.; Zimmermann, U. The Mechanism of Electrical Breakdown in the Membranes of *Valonia Utricularis*. *J. Membr. Biol.* **1975,** *22* (1), 73–90.
18. Dunn, J. Pulsed Light and Pulsed Electric Field for Foods and Eggs. *Poultry Sci.* **1996,** *75* (9), 1133–1136.
19. Dunn, J. E.; Pearlman, J. S. Methods and Apparatus for Extending the Shelf Life of Fluid Food Products. U. S. Patent 4, 695, 472, 1987.
20. Dutreux, N.; Notermans, S.; Wijtzes, T.; Gongora-Nieto, M. M.; Barbosa-Canovas, G. V. G. Pulsed Electric Fields Inactivation of Attached and Free-Living *Escherichia Coli* and *Listeria Innocua* Under Several Conditions. *Int. J. Food Microbiol.* **2000,** *54* (1–2), 91–98.
21. Eshtiaghi, M. N.; Knorr, D. High Electric Field Pulse Pre-Treatment: Potential for Sugar Beet Processing. *J. Food Eng.* **2002,** *52* (3), 265–272.
22. Evrendilek, G. A.; Jin, Z. T.; Ruhlman, K. T.; Qiu, X.; Zhang, Q. H.; Richter, E. R. Microbial Safety and Shelf-Life of Apple Juice and Cider Processed by Bench and Pilot Scale PEF Systems. *Innovat. Food Sci. Emerg. Technol.* **2000,** *1* (1), 77–86.

23. Evrendilek, G.A.; Zhang, Q. H. Effects of Pulse Polarity and Pulse Delaying Time on Pulsed Electric Fields-Induced Pasteurization of *E. Coli* 0157:H7. *J. Food Eng.* **2005,** *68,* 271–276.
24. Fetterman, J. C. The Electrical Conductivity Method of Processing Milk. *Agric. Eng.* **1928,** *9* (4), 107–108.
25. Fiala, A.; Wouters, P. C.; Van Den Bosch, E.; Creyghton, Y. L. Coupled Electrical-Fluid Model of Pulsed Electric Field Treatment in a Model Food System. *Innovat. Food Sci. Emerg. Technol.* **2001,** *2* (4), 229–238.
26. Fincan, M.; DeVito, F.; Dejmek, P. Pulsed Electric Field Treatment for Solid–Liquid Extraction of Red Beetroot Pigment. *J. Food Eng.* **2004,** *64* (3), 381–388.
27. Getchell, B. E. Electric Pasteurization of Milk. *Agric. Eng.* **1935,** *16* (10), 408–410.
28. Gongora-Nieto, M. M.; Sepulveda, D. R.; Pedrow, P.; Barbosa-Cánovas, G. V.; Swanson, B. G. Food Processing by Pulsed Electric Fields: Treatment Delivery, Inactivation Level, and Regulatory Aspects. *LWT-Food Sci. Technol.* **2002,** *35* (5), 375–388.
29. Granato, D.; Branco, G. F.; Cruz, A. G. Probiotic Dairy Products as Functional Foods. *Compr. Rev. Food Sci. Food Saf.* **2010,** *9,* 455–470.
30. Grenier, J. Design of a MOSFET-Based Pulsed Power Supply for Electroporation. Master's Thesis, University of Waterloo, 2006.
31. Hess, H. L.; Baker, R. J. Transformer Less Capacitive Coupling of Gate Signals for Series Operation of Power MOS devices. *IEEE Trans. Power Electron.* **2000,** *15* (5), 923–930.
32. Hickman, B.; Cook, E. *Evaluation of MOSFETs and IGBTs for Pulsed Power Applications*; Lawrence Livermore National Lab: CA, 2001.
33. Ho, S. Y.; Mittal, G. S. Electroporation of Cell Membranes: A Review. *Crit. Rev. Biotechnol.* **1996,** *16* (4), 349–362.
34. Huang, K.; Wang, J. Designs of Pulsed Electric Fields Treatment Chambers for Liquid Foods Pasteurization Process: A Review. *J. Food Eng.* **2009,** *95* (2), 227–239.
35. Hulsheger, H.; Niemann, E. G. Lethal Effects of High-Voltage Pulses on *E. Coli* K12. *Radiat. Environ. Biophys.* **1980,** *18* (4), 281–288.
36. Jambari, H.; Naziha, A.; Azli, M. A. M. P. Comparison of Pulsed Electric Field Generation Techniques for Microbial Inactivation Application. *J. Theor. Appl. Inform. Technol.* **2011,** *33* (1), 22–30.
37. Juan, L.Y.; Ngadi, M.; Raghavan, V. Proteolysis of Cheese Slurry Made from Pulsed Electric Field-Treated Milk. *Food Bioprocess Technol.* **2012,** *5* (1), 47–54.
38. Kathiravan, T.; Kumar, R.; Lakshmana, J. H.; Kumaraswamy, M. R.; Nadanasabapathi, S. Pulsed Electric Field Processing of Functional Drink Based on Tender Coconut Water (*Cococus nucifera* L.) and Nannari (*Hemidesmus Indicus*) Blended Beverage. *Croat. J. Food Sci. Technol.* **2014,** *6* (2), 84–96.
39. Kinosita J, K.; Tsong, T. Y. Voltage-Induced Pore Formation and Hemolysis of Human Erythrocytes. *Biochim. Biophys. Acta (BBA)-Biomembr.* **1977,** *471* (2), 227–242.
40. Knorr, D.; Angersbach, A.; Eshtiaghi, M. N.; Heinz, V.; Lee, D. U. Processing Concepts Based on High Intensity Electric Field Pulses. *Trends Food Sci. Technol.* **2001,** *12* (3–4), 129–135.
41. Kumar, Y.; Patel, K. K.; Kumar, V. Pulsed Electric Field Processing in Food Technology. *Int. J. Eng. Stud. Tech. Approach,* **2015,** *1* (1), 6–17.
42. Loeffler, M. J. Generation and Application of High Intensity Pulsed Electric Fields. In *Pulsed Electric Fields Technology for the Food Industry*; Springer: Boston, MA, 2006; pp 27–72.

43. Mankowski, J.; Kristiansen, M. A Review of Short Pulse Generator Technology. *IEEE Trans. Plasma Sci.* **2000,** *28* (1), 102–108.
44. Matsumoto, Y.; Satake, T.; Shioji, N.; Sakuma, A. In *Inactivation of Microorganisms by Pulsed High Voltage Application*, Conference Record of IEEE Industrial Applications Society Annual Meeting, 1991; pp 652–659.
45. Miranda, M. L. C. Inactivation of Listeria Innocua by Pulsed Electric Fields and Nisin; Doctoral Dissertation, Washington State University, 1998.
46. Mizuno, A.; Hori, Y. Destruction of Living Cells by Pulsed High-Voltage Application. *IEEE Trans. Ind. Appl.* **1988,** *24* (3), 387–394.
47. Mohamed, M. E.; Eissa, A. H. A. Pulsed Electric Fields for Food Processing Technology. In *Structure and Function of Food Engineering*; InTech Open, 2012; pp 276–306.
48. Monfort, S.; Gayán, E.; Condón, S.; Raso, J.; Álvarez, I. Design of a Combined Process for the Inactivation of *Salmonella Enteritidis* in Liquid Whole Egg at 55°C. *Int. J. Food Microbiol.* **2011,** *145* (2–3), 476–482.
49. Monfort, S.; Gayan, E.; Saldana, G.; Puertolas, E.; Condon, S.; Raso, J.; Alvarez, I. Inactivation of *Salmonella Typhimurium* and *Staphylococcus Aureus* by Pulsed Electric Fields in Liquid Whole Egg. *Innovat. Food Sci. Emerg. Technol.* **2010,** *11* (2), 306–313.
50. Notermans, S. *An Introduction-Food Preservation by Pulsed Electric Fields Preservation of Food by Pulsed Electric Fields*; Lelieveld H.L.M., Ed.; CRC Press: Boca Raton, FL, 2007; pp 1–7.
51. Odriozola-Serrano, I.; Bendicho-Porta, S.; Martín-Belloso, O. Comparative Study on Shelf Life of Whole Milk Processed by High-Intensity Pulsed Electric Field or Heat Treatment. *J. Dairy Sci.* **2006,** *89* (3), 905–911.
52. Oziembłowski, M.; Kopeć, W. Pulsed Electric Fields (PEF) as an Unconventional Method of Food Preservation. *Polish J. Food Nutr. Sci.* **2005,** *14* (55), 31–35.
53. Pina-Perez, M. C.; Silva-Angulo, A. B.; Rodrigo, D.; Martínez-López, A. Synergistic Effect of Pulsed Electric Fields and Cocoan OX 12% on the Inactivation kinetics of *Bacillus Cereus* in a Mixed Beverage of Liquid Whole Egg and Skim Milk. *Int. J. Food Microbiol.* **2009,** *130* (3), 196–204.
54. Putri, R. I.; Syamsiana, I. N. Design of High Voltage Pulse Generator for Pasteurization by Pulse Electric Field (PEF). *Int. J. Comput. Electric. Eng.* **2010,** *2* (5), 916–920.
55. Qin, B. L. Food Pasteurization Using High-Intensity Pulsed Electric Fields. *Food Technol.* **1995,** *49* (12), 55–60.
56. Qin, B. L.; Zhang, Q.; Barbosa-Canovas, G. V.; Swanson, B. G.; Pedrow, P. D. Inactivation of Microorganisms by Pulsed Electric Fields of Different Voltage Waveforms. *IEEE Trans. Dielectr. Electric. Insulat.* **1994,** *1* (6), 1047–1057.
57. Qin, B.L.; Zhang, Q.; Barbosa-Cánovas, G. V.; Swanson, B. G.; Pedrow, P. D. Pulsed Electric Field Treatment Chamber Design for Liquid Food Pasteurization Using a Finite Element Method. *Trans. ASAE* **1995,** *38* (2), 557–565.
58. Rastogi, N. K.; Eshtiaghi, M. N.; Knorr, D. Accelerated Mass Transfer During Osmotic Dehydration of High Intensity Electrical Field Pulse Pre-Treated Carrots. *J. Food Sci.* **1999,** *64* (6), 1020–1023.
59. Reina, L. D.; Jin, Z. T.; Zhang, Q. H.; Yousef, A. E. Inactivation of *Listeria Monocytogenes* in Milk by Pulsed Electric Field. *J. Food Protect.* **1998,** *61* (9), 1203–1206.
60. Rivas, A.; Sampedro, F.; Rodrigo, D.; Martínez, A.; Rodrigo, M. Nature of the Inactivation of Escherichia Coli Suspended in an Orange Juice and Milk Beverage. *Eur. Food Res. Technol.* **2006,** *223* (4), 541–545.

61. Sale, A. J. H.; Hamilton, W. A. Effects of High Electric Fields on Microorganisms. Killing of Bacteria and Yeasts. *Biochim. Biophys. Acta (BBA)-Gen. Subj.* **1967,** *148* (3), 781–788.
62. Salvia-Trujillo, L.; Morales-De La Pena, M.; Rojas-Graü, A.; Martín-Belloso, O. Changes in Water-Soluble Vitamins and Antioxidant Capacity of Fruit Juice–Milk Beverages as Affected by High-Intensity Pulsed Electric Fields (HIPEF) or Heat During Chilled Storage. *J. Agric. Food Chem.* **2011,** *59* (18), 10034–10043.
63. Sampedro, F.; McAloon, A.; Yee, W.; Fan, X.; Zhang, H. Q.; Geveke, D. J. Cost Analysis of Commercial Pasteurization of Orange Juice by Pulsed Electric Fields. *Innovat. Food Sci. Emerg. Technol.* **2013,** *17,* 72–78.
64. Sampedro, F.; Rodrigo, D.; Martinez, A. Modelling the Effect of pH and Pectin Concentration on the PEF Inactivation of *Salmonella Enterica Serovar Typhimurium* by Using the Monte Carlo Simulation. *Food Control* **2011,** *22,* 420–425.
65. Schoenbach, K. H.; Katsuki, S.; Stark, R. H.; Buescher, E. S.; Beebe, S. J. Bioelectrics-New Applications for Pulsed Power Technology. *IEEE Trans. Plasma Sci.* **2002,** *30* (1), 293–300.
66. Sepulveda, D. R.; Ortega-Rivas, E.; Barbosa-Canovas, G. V. Quality Aspects of Cheddar Cheese Obtained with Milk Pasteurized by Pulsed Electric Fields. *Food Bioproducts Process.* **2000,** *78,* 65–71
67. Shamsi, K.; Sherkat, F. Application of Pulsed Electric Field in Non-Thermal Processing of Milk. *Asian J. Food Agro-Ind.* **2009,** *2* (3), 216–244.
68. Sundararajan, R.; Shao, J.; Soundarajan, E.; Gonzales, J.; Chaney, A. Performance of Solid-State High-Voltage Pulses for Biological Applications- A Preliminary Study. *IEEE Trans. Plasma Sci.* **2004,** *32* (5), 2017–2025.
69. Toepfl, S.; Heinz, V.; Knorr, D. High Intensity Pulsed Electric Fields Applied for Food Preservation. *Chem. Eng. Process.: Process Intensificat.* **2007,** *46* (6), 537–546.
70. Torregrosa, F.; Cortés, C.; Esteve, M. J.; Frígola, A. Effect of High-Intensity Pulsed Electric Fields Processing and Conventional Heat Treatment on Orange-Carrot Juice Carotenoids. *J. Agric. Food Chem.* **2005,** *53* (24), 9519–9525.
71. Torregrosa, F.; Esteve, M. J.; Frígola, A.; Cortés, C. Effect of High-Intensity Pulsed Electric Fields Processing and Conventional Heat Treatment on Orange-Carrot Juice Carotenoids. *J. Food Eng.* **2004,** *73* (4), 339–345.
72. Tsong, T. Y. On Electroporation of Cell Membranes and Some Related Phenomena. *Bioelectrochem. Bioenerget.* **1990,** *24* (3), 271–295.
73. Ulmer, H. M.; Heinz, V.; Gänzle, M. G.; Knorr, D.; Vogel, R. F. Effects of Pulsed Electric Fields on Inactivation and Metabolic Activity of *Lactobacillus Plantarum* in Model Beer. *J. Appl. Microbiol.* **2002,** *93* (2), 326–335.
74. Vega-Mercado, M. Martín-Belloso, O.; Chang, F. J.; Barbosa-Canovas, G. V.; Swanson, B. G. Inactivation of Escherichia Coli and Bacillus Subtilis Suspended in Pea Soup Using Pulsed Electric Fields. *J. Food Process. Preserv.* **1996b,** *20* (6), 501–510.
75. Vega-Mercado, M.; Pothakamury, U. R.; Chang, F. J.; Barbosa-Canovas, G. V.; Swanson, B. G. Inactivation of *Escherichia Coli* by Combining pH, Ionic Strength and Pulsed Electric Fields Hurdles. *Food Res. Int.* **1996a,** *29* (2), 117–121.
76. Weaver, J. C. Electroporation of Cells and Tissues. *IEEE Trans. Plasma Sci.* **2000,** *28* (1), 24–33.
77. Wouters, P. C.; Dutreux, N.; Smelt, J. P.; Lelieveld, H. L. Effects of Pulsed Electric Fields on Inactivation kinetics of *Listeria Innocua. Appl. Environ. Microbiol.* **1999,** *65* (12), 5364–5371.

78. Yin, Y.; Zhang, Q. H.; Sastry, S. K. U.S. Patent No. 5,690,978. Washington, DC: U.S. Patent and Trademark Office, 1997.
79. Zhang, Q.; Barbosa-Canovas, G. V.; Swanson, B. G. Engineering Aspects of Pulsed Electric Field Pasteurization. *J. Food Eng.* **1995,** *25* (2), 261–281.
80. Zhang, Q.; Qin, B. L.; Barbosa-Canovas, G. V.; Swanson, B. G.; Pedrow, P. D. Batch Mode Food Treatment Using Pulsed Electric Fields. U. S. Patent 5,549,041, 1996.
81. Zimmermann, U. Electric Breakdown, Electro Permeabilization and Electrofusion. *Rev. Physiol. Biochem. Pharmacol.* **1986,** *105,* 196–256.

PART II

ADVANCES AND APPLICATIONS OF PULSED ELECTRIC FIELD

CHAPTER 3

INACTIVATION KINETICS OF MICROORGANISMS BY PULSED ELECTRIC FIELD

GOPINATH MUMMALETI, SRINIVAS YARRAKULA, CHAYANIKA SARMA, HIMABINDU GAZULA, SAGARIKA NUKASANI, KRISHNAVENI RAJAMAHENDRAVARAPU, and VIMALA BEERA

ABSTRACT

Pulsed electric field (PEF) is an efficient non-thermal food processing technique that can effectively inactivate microorganisms and retain the natural flavors and nutritional quality of foods. The PEF is effective in reducing the pathogenic organisms by up to 2–3 log when used alone, and when combined with other processing technologies, more than 5-log reduction was obtained. This chapter summarizes the mechanism of the inactivation of microbes using PEF, factors affecting the inactivation of microorganisms, the effect of PEF on different microorganisms and spores, inactivation kinetics and models, PEF application in food processing, microbial stress response toward PEF, combination techniques for microbial inactivation, and challenges of PEF in inactivation of microorganisms. The application of PEF on various foods for the inactivation of various microorganisms was discussed. The sublethal injury and its effect on food safety were discussed. The inactivation kinetics of various microorganisms and kinetic models developed for PEF processing were also discussed. The combination of PEF with other processing techniques enhances the rate of inactivation of microorganisms. Overall, the PEF emerges as a promising non-thermal food processing technique for pasteurizing and producing safe foods.

Nonthermal Pulsed Electric Field Technology for the Food Industry.
Preeti Birwal, Megh R. Goyal, Baldev Singh Kalsi, & Ankita Kataria (Eds.)

DOI: 10.1201/9781003605904-5

3.1 INTRODUCTION

Various microorganisms commonly contaminate food materials during multiple steps in the supply chain, such as harvesting, storage, handling, processing, and distribution. The foods are contaminated by different organisms depending on the composition of the food, such as water activity (a_w), pH, nutrient content, and environmental factors such as temperature, relative humidity, and gas composition.[71] The microorganisms contaminate foods from various sources such as soil, water, animals, sewage, and air. Improper handling and sanitation of food processing plants, storage structures, and personnel hygiene are some crucial factors responsible for food contamination.[131]

The microorganisms from the environment interact with food material, utilize nutrients in the food, increase in number, produce various enzymes, and cause various deteriorative changes leading to spoilage of foods. The spoilage of foods can be identified through visible changes such as color change, slime formation, yeast, and mold growth and from various off-odors produced due to spoilage.[25] The pathogenic microorganisms grown in foods cause various diseases in humans, and some pathogens may even lead to the death of individuals on consumption of contaminated food. Foodborne illnesses and outbreaks are increasing every year due to various pathogenic microorganisms, which need to be focused on protecting public health. Microbial contamination leads to enormous food wastage worldwide and severe economic loss.[39]

The production of safe, hygienic food is essential to protect public health and reduce food wastage, hunger, and economic loss. Food preservation can be done through various processing techniques, controlling environmental conditions, and adding various food preservatives that inhibit the growth and survival of microorganisms.[110] Thermal processing techniques are commonly employed for the destruction of microorganisms in foods. The food material is heated to a sufficient temperature for pasteurization to kill all the pathogenic organisms or sterilization to kill all the microorganisms in the food.[61] The treated foods are further stored under controlled environmental and hygienic conditions to avoid contamination and the growth of microorganisms. Although thermal treatment is a highly efficient processing method, it shows adverse effects on food, such as browning, denaturation of proteins, and loss of heat-sensitive nutrients and flavors.[13] The researchers are continuously exploring alternate processing techniques to avoid the adverse effects of thermal processing.[95]

Non-thermal food processing technologies are gaining importance due to their potential effect on microbial destruction. These innovative methods have shown minimal-to-negligible influence on the sensory attributes and nutritional composition of food items.[5] The pulsed electric field (PEF) technique is a highly promising method among a range of non-thermal technologies and has undergone extensive research on various food products. The technique involves generating short, high-voltage electric fields (5–50 kV/cm) in pulses between the electrodes, effectively inactivating microorganisms. Researchers and food companies have been drawn to investigate the feasibility of commercializing PEF and generating high-quality food items that meet the needs of customers.[110]

The process by which PEF inactivates microorganisms is still unclear, but it is assumed that the electric field produces pores in the cell membrane, causing cell swelling, rupturing, and leading to microbial death. The free radicals produced by electrolysis and heat produced by electrical energy were also believed to cause the inactivation of microorganisms.[115] The optimum design of the PEF chamber is a crucial step, and many designs were developed and studied the effects of various solid and liquid food materials. Factors such as velocity, electric field distribution, and temperature inside the PEF chamber profoundly affect the inactivation of microorganisms.[73,131] The combination of various parameters was studied, and numerical simulation models were developed to improve the efficiency of PEF design.[5,48,77] The safety of foods and extending the shelf-life are the crucial objectives in developing any processing technique. This can be accomplished by developing various numerical simulations and mathematical models with different process and environmental parameters.[71]

This chapter extensively reviews the mechanism of microbial inactivation by PEF and factors affecting microbial inactivation. The microbial kinetic models developed and used for defining the process parameters and their effect on microbial inactivation were included. The application of PEF on various foods, its effect in combination with other technologies, microbial tolerance, and stress response were incorporated. The chapter concludes with the challenges of PEF processing and the need for further study.

3.2 FUNDAMENTALS OF INACTIVATION OF MICROORGANISMS

The successful inactivation of microorganisms and the production of safe food is the primary objective behind developing any novel technology. Cell death is commonly owed to structural or physiological damage. Structural damage implies the destruction of the cell membrane, ribosomal changes and protein destruction, and nucleic acid damage or conformational changes. The physiological damage includes affecting the cell permeability and inactivation of enzymes responsible for various functions leading to cell death.[92] These damages can co-occur when exposed to any treatment, and it is tough to predict the type of damage caused and responsible for cell death. Heat treatment damages the cell membrane, nutrients, ribosome, protein aggregation, and DNA breakdown, and inactivates the enzymes simultaneously, leading to cell inactivation. The ionizing radiations directly show their effect on target molecules due to radiation absorption and indirectly produce free radicals that react and break the DNA molecule leading to cell death.[106]

Cell death is caused by the pressure generated through ultrasound and high hydrostatic pressure processing, as it damages the phospholipid bilayer, inducing cavitation in the cell membrane followed by impairment of the proteins.[2] The PEF shows a direct effect on cell membranes and organelles through electroporation that causes pore formation and cell death. The bacteria, after treatment, were studied under an electron microscope and observed to have several morphological changes, such as disruption of organelles, cell membrane, and surface roughness.[24]

The processing technique or preservatives may not be completely lethal to all the organisms in the foods when used singly due to various environmental factors. The use of one preservation technique is not sufficient in such conditions to produce safe food. A combination of treatments and preservatives is required to efficiently damage various physiological functions of the microbes and kill them. The lethality of each treatment should be studied to clearly understand the level of inactivation and should be combined with other treatments.

The sublethal effects might lead to the reactivation of microorganisms by repairing the cell damage, which is more dangerous for food safety. The sublethal injury due to individual treatment should be carefully studied to understand the type of injury so that combination treatment can be selected carefully to completely inactivate the organism.[65] Most of the sublethal injuries occur in the cell membrane that can be repaired by the microorganism itself and remains active. In such instances, the other treatments affect the homeostasis and physiological activity of organisms so that the combined effect can inactivate the organisms completely, and safer food can be produced.[83]

3.3 MECHANISM OF MICROBIAL INACTIVATION BY PEF

The application of high-intensity PEF on microorganisms causes permanent or temporary permeation of the cell membrane, and this phenomenon is usually called "electro-permeabilization" or "electroporation." The PEF also causes electrofusion of cells in which fusion of more than one cells resulting into a single big cell occurs.[82] The application of a high-intensity electric field induces a permanent, irreversible alteration in the pores of the cell membrane. This permanent electroporation damages the structural integrity of the cell membrane and cell organelles leading to cell lysis (Figure 3.1). When an intensive electric field is applied, the potential across the transmembrane rises which affects the cell membrane and forms electrophoresis.[57] The resulting electrophoresis can be either temporary or permanent, influenced by parameters such as the number of pulses, frequency, intensity of the electric field, cell size, shape, and membrane potential. The formation of pores on the membrane is contingent upon the treatment duration and the strength of the electric field, determining their size and quantity.

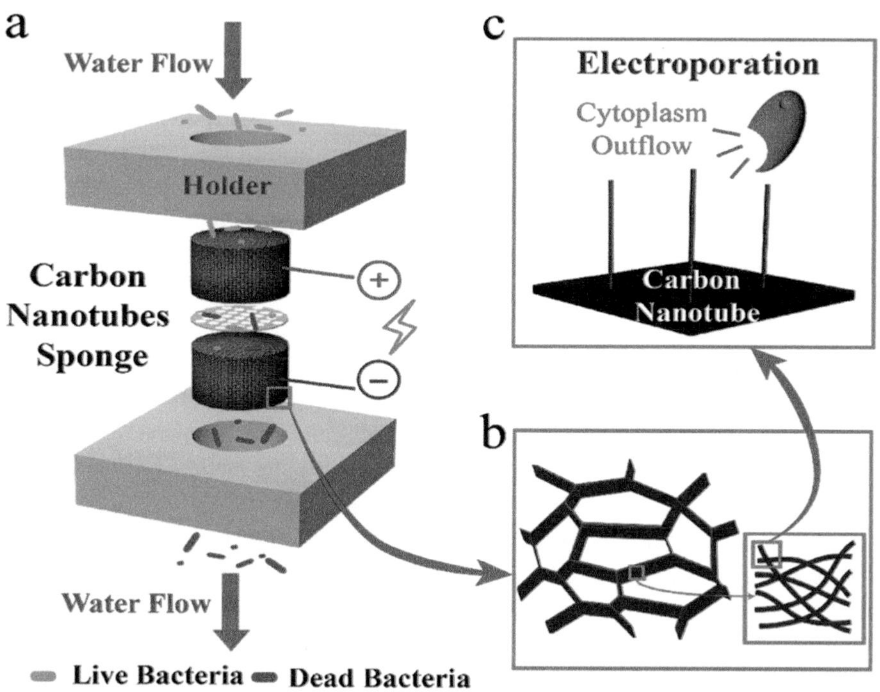

FIGURE 3.1 Electroporation of microorganism.

The critical or threshold electric field denotes the level of intensity at which membrane breakdown happens. The amount, as well as the size of pores formed, is extremely small in case the electric field is near the threshold level and the duration of treatment is small.[51] Under such conditions, electroporation is rendered irreversible, and the cell membrane's structure is repaired, leading to its functional recovery. But when the high-intensity electric field is applied for a longer treatment time, the quantity as well as size of pores is increased, causing mechanical destruction of the cell.[26] The pores formed on the membranes facilitate the transfer of cell organelles and constituents without any agent leading to the inactivation of microorganisms.[12]

The electric field also influences the phospholipid membrane of the cell. Due to dipolar rotation, the phospholipids located between the monolayers of the cell's membrane tend to reorient, causing conformational changes in the membrane structure. As a result, the semipermeable nature of the cell membrane is lost, which inactivates the cells.[21] When cells undergo exposure to an electric field, produce hydrophilic pores that conduct the electricity and produce Joule heating which causes a lipid bilayer thermal transition and modifies the semipermeable nature of the cell membrane. The trans-membrane potential also affects the opening and closing of protein channels present in the cell membranes.[17]

The Joule heating causes permanent denaturation of the protein channels and forms pores in the cell membrane.[77] The intracellular voltages affect the organelles of the cells; the PEF modifies the actin in the cytoskeleton, permeabilizes mitochondria, releases calcium stores, and affects the cell function causing apoptosis of the cell.[37] The application of the electric field mainly affects the semipermeable nature of the cell membrane owing to the destruction of a phospholipid bilayer in the cell membrane and protein channels.

The structure of major enzymes responsible for the functioning and metabolism of the cells gets affected and inactivates when exposed to an electric field.[82] The inactivation of enzymes leads to the dysfunction of metabolic reactions in the organisms and destroys the organisms. The electric field also affects the genetic material DNA and RNA in the microbial cell and affects the replication, transcription, and translation that affect the growth and multiplication of microorganisms.[24]

3.4 FACTORS AFFECTING MICROBIAL INACTIVATION BY PEF

3.4.1 PEF PROCESSING PARAMETERS

The PEF process parameters that are majorly responsible for the inactivation of microorganisms are electric field strength, frequency, width and shape of the pulse, specific energy, temperature, and time of treatment. Among these process parameters, treatment time, electric field strength, and pulse energy are process controls of PEF, and the intensity of these parameters is responsible for the lethality of microorganisms.[129] The period of processing is defined by the width and frequency of pulses, whereas time and temperature rarely affect the inactivation of microbes. The intensity of the electric field is the most important element in determining microbial lethality.

Critical field strength (E_c) characterizes the resistance of microorganisms, and it increases with field strength intensity. Another crucial component influencing microorganism inactivation is the distribution of electric field intensity throughout the treatment chamber.[80] The homogenous distribution of the electric field is required to enhance the feasibility of the field applied; heterogeneous distribution results in the presence of arcing zones owing to elevated electric field intensities. The distribution of field strength is affected by the electrode's shape, gap inside the chamber, and gas impurities.[9]

The shape and width of the pulses are critical parameters that majorly affect the inactivation of microorganisms. Exponential or square wave pulses are typically used in PEF. Square wave pulses with similar electric fields and energy were reported to be effective in microbial inactivation. Monopolar or bipolar pulses are produced, with bipolar pulses consisting of both a positive and a negative pulse. The microbial inactivation increased by 50–60% when square wave bipolar pulses were applied compared to monopolar pulses.[29]

The impact of pulse width on microbial inactivation was reported differently by different authors. Few authors reported that the inactivation of microbes is independent of pulse width, and some reported that inactivation is higher when longer pulses are applied. The longer pulses were observed to inactivate the microorganisms efficiently when final temperatures are higher, and efficiency is lower at low temperatures.[34] The shorter pulses with higher field strength and the same specific energy were highly effective compared to wider pulses with lower electric field strength. The wider pulses were stated to have the electrolytic effect that produces chemical species that are lethal to microorganisms.

The duration of exposure of microorganisms to electric field intensity is referred to as treatment time. The treatment time depends on the electric field strength, number, and width of the pulse. The increase in the frequency of the pulse decreases the time of processing.[86] However, the increase in frequency increases the temperature, which will affect the properties of the food material. The microbial inactivation increases with an increase in specific energy applied. Specific energy, as an integral control factor of the PEF, is calculated by taking into consideration the resistance of the treatment chamber, which is influenced by the conductivity and size of the chamber, as well as the electric field intensity and treatment duration.[7]

The electric field strength has a higher impact compared to specific energy; the specific energy should be expressed in conjunction with the strength of the electric field and cannot be specified alone. Increasing the electric field strength can effectively inactivate the microorganisms within a short time and lower energy. The temperature of the treatment medium also influences microbial inactivation effectiveness. The importance of applying PEF to food materials is to avoid the detrimental effects of thermal treatment on the properties of the foods.[4] However, the effect of PEF increases with the increase in temperature of the treatment medium. Temperature increases the rate of inactivation because the minimal transmembrane potential needed to cause dielectric rupture of cell membranes decreases with increasing temperature. The decrease is caused by the phase shift of phospholipids from gel to liquid at higher temperatures.

Moody et al.[77] studied the inactivation of *Escherichia coli* in apple juice when treated with PEF and the result showed that there is no significant reduction in *E. coli* when PEF treatment is given at room temperature (20 °C). However, when PEF treatment was given at 40 °C, the enhancement in the inactivation of *E. coli* was observed.

The effect of an increase in electrode treatment to inactivate *Listeria monocytogenes* STCC 5672 in solid samples was studied by Astráin-Redín et al.,[9] and the result showed that a 5-log reduction was observed at the cold spot of the cylinder. The treatment time of 68 s was required for inactivation at room temperature, and it decreased to 26 s when electrodes were tempered to 39 reaching 71 °C temperature. The inoculation of apple juice with *E. coli* O157:H7 was followed by a treatment at electric field strengths of 22, 25, 28, 31, and 34 kV/cm at 38 °C, the reduction in *E. coli* was observed with an increase in field strength, and 4.5-log reduction was observed at 34 kV/cm electric field strength.

The application of PEF processing for the inactivation of *Staphylococcus aureus* in orange juice has shown that *S. aureus* decreased by 5.891–5.924 log at 20 kV/cm, 5.891–5.924 log at 30 kV/cm, and 5.917–5.923 log at 40 kV/cm.

The influence of electric field strength, specific energy, the conductivity of the treatment medium, and treatment time on the inactivation of *L. monocytogenes* was explored by Álvarez et al.[4] The maximum of 4.77-log reduction in *L. monocytogenes* was seen at 28 kV/cm, 3490 kJ/kg, and 2000 μs. The electric field strength and specific energy significantly affected the survival of *L. monocytogenes*, and an increase in these parameters increased the log reduction. Table 3.1 shows the inactivation level of various organisms at different process parameters of PEF

3.4.2 MICROBIAL STRAIN AND MORPHOLOGY

PEF inactivation is influenced by the innate resistance of microbial species or strains. Gram-positive organisms are more resistant than the Gram-negative ones, and bacteria are more resistant than yeast. The resistance of microorganisms can be investigated using the fluorescence-staining technique. The relation between membrane permeabilization and intracellular organelles leakage due to PEF treatment on *Saccharomyces cerevisiae* was studied by Aronsson et al.[8] The rapid uptake of Propidium iodide by non-lethal PEF suggests that permeabilization is not correlated to PEF resistance. The morphological changes observed using transmission electron microscopy and the correlation between the level of inactivation and large amounts of intracellular components leaked indicate that along with membrane permeabilization, another mechanism of inactivation is also involved in yeast cells. The resistance of microorganisms is related to the size and shape of the cell.

The *L. monocytogenes* are the smallest vegetative bacteria that showed the highest resistance to the PEF treatment. The sensitivity of yeast cells is owed to larger cell sizes. The inactivation effect of the transmembrane potential created by the electric field is highly influenced by the size of the cell. The smaller the size of the cell, the smaller the membrane potential and the higher the resistance. The influence of cell size on permeabilization was studied by Wouters et al.[126] The *Lactobacillus plantarum* cells of different sizes and shapes were separated using flow cytometry, and permeabilization was studied; the larger cells showed higher permeabilization compared to smaller cells. However, the increase in time and energy of PEF treatment reduced the difference in permeabilization among different cells.

3.4.3 MICROBIAL GROWTH STAGE

The effect of PEF on inactivation can also be influenced by the growth phase of microorganisms. The exponential-phased bacteria are more sensitive to PEF when compared to stationary-phased cells.[123] The lower resistance during the exponential phase might be due to electromechanical instability, larger cell size, and continuous cell division. The impact of PEF on the inactivating process of *Listeria innocua* NCTC 11289 with electrical

TABLE 3.1 Inactivation Level of Various Organisms in Different Foods at Various Processes of PEF.

Food products	Process parameters (pulse intensity, frequency, pulse duration, number of pulses)	Microorganism	Inactivation level ($\log_{10}$)	Reference
Apple juice	23.076, 30.76 kV/cm, 0.2 kHz 2 μs 6, 9, 15, 21 pulses	*Escherichia coli*	<1	[77]
Apple juice	30.076 kV/cm 0.2 kHz 2 μs 21 pulses	*Saccharomyces cerevisiae*	5.03	[68]
Drinking water	110 kV/cm 70 Hz 130–500 ns 70 pulses	*Escherichia coli*	4	[78]
Grape juice	12–24 kV/cm 120 Hz 3–6 μs 10–60 pulses	*Saccharomyces cerevisiae,* *Escherichia coli* *Staphylococcus aureus*	6	[47]
Green tea extracts	18.1, 27.4, 38.4 kV/cm 2 μs 667 Hz 40, 80, 120, 160, 200 μs	*Escherichia coli* *Staphylococcus aureus*	5.6 4.9	[134]

TABLE 3.1 *(Continued)*

Food products	Process parameters (pulse intensity, frequency, pulse duration, number of pulses)	Microorganism	Inactivation level ($\log_{10}$)	Reference
Liquid whole egg	45, 40kV/cm	*Salmonella typhimurium*	4	[75]
	< 300 Hz	*Staphylococcus aureus*	3	
	3 µs			
	5, 10 pulses			
Phosphate-buffered saline	1–4 kV/cm	*Pseudomonas*	6	[28]
	10–20 µs	*putida*		
	100 Hz			
	200, 2000, 5000,			
	and 10,000 pulses			

conductivity of 0.51 S/m at 40 °C was investigated.[127] A 6-log reduction was observed for cells in the log phase, whereas cells in the stationary phase showed a 4.4-log reduction, indicating higher resistance to PEF treatment for cells in the exponential phase. The resistance of four enterotoxigenic *S. aureus* strains in exponential and stationary phases was studied.[16] The organisms in a stationary and mid-exponential phase were observed to be highly resistant to PEF. The resistance of the organisms was correlated with the carotenoid pigment, and the researchers observed that organisms with pigmentation were highly resistant compared to non-pigmented organisms. The impact of moderate electric field on *Lactobacillus acidophilus* at lag, log, and stationary phases were studied, and results showed that cells in the lag phase are highly susceptible to permeabilization, followed by cells in the exponential phase. No electroporation was observed in the cells in the stationary phase.[63]

3.4.4 PROPERTIES OF FOODS AND COMPOSITION

Another key parameters that impact microbial inactivation by PEF include the characteristics and the composition of food material. The inactivation of microorganisms in complex food materials is highly difficult compared to dilute buffer solution due to various ions in foods.[32] The resistivity of the microorganisms toward the electric field decreases with the increase in the ionic strength of food materials. The building of electric field strength in the medium is difficult when electrical conductivity is high. The inactivation effect of PEF on bacteria is inversely proportional to the ionic strength of food material or medium.

Jayaram et al.[55] reported the effect of PEF in the inactivation of *Lactobacillus brevis* suspended in phosphate buffer solutions with a conductivity of 17–223 mS/m. The reduction in microbial number is higher in the phosphate buffer with the lowest conductivity of 17 mS/m. The phosphate buffer with the lowest conductivity and higher resistivity showed maximum inactivation.

Vega-Mercado et al.[116] also reported the inactivation of *E. coli* ATCC11229 along with observation of a higher impact of ionic strength of the medium on microbial inactivation, and a decrement in inactivation rate as the ionic strength is enhanced. A higher ionic strength of the medium reduces the stability of the cell membrane and increases the inactivation rate.[41]

The physical structure of food materials affects the dielectric strength of the food materials. The liquid food materials with lower conductivity have a maximum reduction, which influences the permeability of membranes.[100] The particles and air bubbles present in the food materials also significantly influence the PEF treatment on microbial inactivation. The air bubbles present in foods such as sparkling products cannot tolerate higher electric field strengths. The air inside the food materials must be removed before PEF treatment for the inactivation of microorganisms.[36]

The fluctuation in dielectric characteristics of the bubbles influences treatment's efficiency. The efficiency of the treatment also gets influenced when food material has different solid particles; the difference in dielectric properties of various solid particles, such as fat globules present in the food medium, will affect the electrical conductivity. Hence, it is essential to consider the composition of food materials before selecting the processing parameters of PEF. The air bubbles must be removed completely before PEF treatment to prevent electric discharge. The foam-based food materials are completely unsuitable for PEF treatment.[125]

The divalent cations Ca^{+2} and Mg^{+2} significantly influenced the treatment of PEF. The Ca^{+2} and Mg^{+2} showed a significant influence on bacterial cell membranes as well as a protective action against PEF treatment. The effect of milk fat content on PEF treatment was investigated, and a prolongation of time needed in order to inactivate *E. coli* was reported. The effect of pH was studied well and observed to have a great influence on PEF on microbial inactivation. Gram-positive bacteria were shown to be more resistant to neutral pH than Gram-negative bacteria.[55] The products such as apple, cranberry, and orange juice with lower pH were more appropriate for PEF than the products with higher pH. This difference in resistance of organisms to PEF is mainly due to the difference in the capacity of organisms to repair the sub-lethal injuries that occur in the membrane. The products with lower pH avoid the self-repairing activity of microorganisms that are injured sub-lethally. The pH of food products such as liquid eggs and milk using organic acids enhanced the efficiency of PEF treatment. Thus, the reduction in pH before PEF treatment can effectively enhance microbial inactivation.

The water activity also significantly affects the inactivating process of microorganisms by PEF. The increase in water activity decreases the resistance of bacteria and yeast against PEF.[89] The low water activity reduces the cell volume of microorganisms, thickens the cell membrane, reduces membrane permeability and fluidity, and increases the resistance against PEF. The glycerol, when added to the medium, will passively diffuse inside the microbial cell, avoid cell shrinkage, increase cell size, reduce the resistance of microbial cells against the PEF treatment, and enhance the inactivation rate.[42]

3.5 EFFECT OF PEF ON VARIOUS MICROORGANISMS

3.5.1 EFFECT OF PEF ON VEGETATIVE CELLS AND PATHOGENS

The vegetative microbial cells are the most common contaminants of food products and are responsible for the spoilage of various foods. The effect of PEF in the inactivation of vegetative cells was most studied with the main observation of PEF inactivating them effectively. The processing parameters such as electric field strength, number of pulses, treatment time, and temperature majorly affect the inactivation of microbial cells in foods. The contamination of foods with various microorganisms is the most common cause of food spoilage, wastage, and foodborne illnesses. *E. coli* is one of the most common enterohemorrhagic pathogens responsible for outbreaks due to unpasteurized milk and juices.

The influence of PEF on the log reduction of Gram-negative *E. coli* was studied by dos Santos et al.[23] The extent of inactivation is moderate on *E. coli* inactivation with 2.5–3-log reduction at 35 kV electric field strength, 3 Hz pulse frequency, 40 °C temperature, and 0.8–1 μs pulse width. According to the US Food and Drug Administration (FDA), a 5-log reduction of microorganisms is required in foods during processing. Various kinds of food materials are treated with PEF, and various researchers studied the inactivation rate.

The liquid foods were most studied; the milk treated with PEF at 26.1 kV/cm for 101 μs showed an 8.3 log reduction,[105] and at 42.5 kV/cm for 40 μs showed a 6-log reduction in *E. coli.*[122] The efficiency of PEF treatment to inactivate *L. innocua* in 2 and 10% whey protein formulations was investigated. A maximum of 6.5-log reduction in *L. innocua* was observed without showing any effect on bioactive compounds.[102]

The increase in pH and protein concentration decreased the efficiency of inactivation. The sour cherry juice inoculated with *S. aureus*, *Erwinia carotowora*, *L. monocytogenes*, *Botrytis cinerea*, *E. coli*, and *Penicillium expansum* was studied. With increasing electric field intensity and treatment time, the considerable drop in microbial count was reduced, and the amount of inactivation increased. The extent of inactivation of yeast cells was greater, followed by Gram-negative bacteria and then Gram-positive bacteria.[3]

L. monocytogenes is another common pathogen that can grow in dairy, fruit, and vegetable-based products and causes listeriosis outbreaks. *Listeria* can be able to grow in refrigerated conditions, a wide range of pH, and harsh oxygen conditions. The skim milk, whole milk, and milk with 2% fat were inoculated with *L. monocytogenes* and treated with a PEF of 400 square pulses for 1.5 μs. The 3-log reduction was achieved in all three samples, and the fat content did not have a significant effect on PEF treatment.[94]

The increase in inactivation increased to 4-log when the treatment temperature increased from 25 to 50 °C at 30 kV/cm electric field strength for 600 μs. The skim milk added with gellan gum and increased in temperature from 35 to 55 °C increased the inactivation from 1 to 4.5 log.[33] The melon and watermelon juices inoculated with *E. coli*, *L. monocytogenes,* and *Salmonella enteritidis* were treated with PEF, and about 3.41 to 3.76-log reductions were observed, and *L. monocytogenes* were observed to be highly resistance among the three organisms.

Salmonella is an enteric pathogenic bacteria responsible for deadly outbreaks and foodborne diseases. *Salmonella* is a Gram-negative psychotropic bacteria capable of thriving at low temperatures and water activity. *Salmonella* is widely found in food items such as fruit juices, meat, eggs, and dairy products, resulting in salmonellosis infections. The improper pasteurization of foods is a major reason for the survival of *Salmonella.* According to USFDA, the *Salmonella* should be completely absent per 375 g of foods. The *Salmonella* inoculated in skim milk treated at 35 kV/cm and 2000 Hz for 163.9 μs showed a 4 log reduction.[104]

The egg white treated at 35 kV/cm, 900 Hz 30 °C for 9 μs showed a 3.5-log reduction in *S. enteritidis.*[56] The influence of PEF treatment on oxidative injury and membrane lipids of *Salmonella typhimurium* was studied by Yun et al.[133] and observed a decrease in the fluidity of the membrane owed to alteration in biosynthesis of fatty acid and related gene expressions *CFA* and *fab A* genes. The increase in malondialdehyde and modulation of cytochrome oxidase gene expressions *cyoA*, *cyoB*, and *cyoC* indicate the oxidative injury caused by PEF.

S. aureus is a spherical Gram-positive pathogenic bacterium that cause food poisoning when the organism concentration goes beyond 10^5 CFU/g, according to FDA. The food products such as milk, dairy, and egg products are commonly contaminated with *Staphylococcus* and cause food poisoning. The raw milk treated with PEF at 40 kV and 3.5 Hz showed a 4-log reduction in *S. aureus,* whereas reduction of 3.7 logs was observed when skim milk was treated with PEF at 35 kV and 250 Hz in continuous and batch mode of treatment.[30] The skim milk (pH maintained at 6.8) was treated with PEF at an electric field of 35 kV/cm, using 600 square pulses in the bipolar mode for 4 μs, only 1-log reduction in *S. aureus* was observed, and when nisin of 20 ppm was added, 6-log reduction was obtained.[108]

The PEF was used to inactivate *S. aureus* in Thai orange juice, and 5-log reduction was obtained using 10 pulses at 30 kV/cm field strength. The mechanism of inactivation of *S. aureus* was studied by Freire et al.[35]—the *S.* aureus suspended in McIlvaine buffer at pH 7 and 2 mS/cm was treated with PEF at two different electric fields of 18 and 25 kV/cm for 20 to 400 μs. The study revealed that most compounds inside the cell were released except for some proteins, and a good correlation was obtained between the release of compounds and membrane permeabilization. The increase in PEF treatment avoids the self-repairing of cells leading to the inactivation of *S. aureus*.

The apple juice inoculated with *S. cerevisiae* was treated at electric field strengths of 10–28 kV/cm. The maximum of 4-log reduction was obtained at a 20 kV/cm electric field and 10.4 pulses.[20] The influence of PEF treatment on wine spoilage organisms *Dekkera bruxellensis*, *Lactobacillus hilgardii, L. plantarum,* and *Dekkera anomalai*, was studied at 16–31 kV/cm electric field strength at 24 °C. The optimum treatment of 29 kV/cm sufficiently inactivated 99.9% spoilage microflora.[90] The apple juice inoculated with *S. cerevisiae* was treated with PEF to study the structural changes.[43]

The transmission electron microscopy revealed that the organelle of cells was disrupted, and ribosomes were completely absent, which led to the destruction of microorganisms. The study reveals that organelle damage due to PEF treatment is responsible for microbial inactivation along with electroporation.

3.5.2 EFFECT OF PEF ON SPORES

The spores were observed to be highly resistant to the processing compared to vegetative cells owing to the presence of protective layers around the spores, particularly the cortex layer around the cytoplasmic membrane. The cortex layer prevented the permeabilization of the cell membrane by PEF. The spores of yeast and mold, such as ascospores and conidiospores, were highly sensitive to PEF compared to bacterial spores.

The conidiospores of *Byssochlamys fulva* and the ascospores of *Neosartoria fischeri* were suspended in grape, cranberry, orange, pineapple, tomato, and apple juices followed by treatment of PEF with varying electric field (30–51 kV/cm) and temperature (19–34°C).[93] At 30 kV/cm, the inactivation of conidiospore is higher in cranberry juice with a 6-log reduction and the least in tomato juice with less than 1-log reduction. The ascospores in cranberry juice were not inactivated by PEF treatment even at 40 pulses of 51 kV/cm electric field. The pH is one of the major factors influencing PEF processing, and the difference in pH values of the juices is accountable for the difference in the level of inactivation. The spores of *Bacillus cereus* and *Bacillus polymyxa* were resistant to PEF treatment up to 30 kV/cm electric field strength. The *Bacillus subtilis* spores treated with PEF at 30 kV/cm for 0.5–3 ms showed 1.3-log reduction, and 4-log reduction was obtained at 60 kV/cm at 60°C.[10] The *B. subtilis* and *B. cereus* spores suspended in 0.15% NaCl were treated with PEF at 35 kV/cm, 25°C, and 3.4-log reduction was obtained for *B. subtilis* spores with 30 pulses and 5-log reduction for *B. cereus* with 50 pulses. The enhancing the number of pulses and level of temperature increased the inactivation efficiency.

The effect of the temperature, electric field frequency, width, and number of pulses on *B. cereus* spores was studied, and results showed that spores are highly resistant when PEF alone was used. The increase in temperature showed an effect on spores, and a maximum of 3.6-log reduction was obtained at 40 kV/cm, 144 pulses, and 65 °C with the addition of 50 IU/mL nisin. Skim milk was observed to have a higher reduction compared to whole milk.[14] The *B. subtilis* spores suspended in 0.01% NaCl at 10^8 cell/mL.

The cherry juice, peach, and apricot nectar were inoculated with 10^5–10^6 CFU/mL of *P. expansum* spores.[31] The electric field of 13–34 kV/cm for 62–218 μs was applied, and the result showed that an increase in field strength and processing time inhibited the germination tube elongation and germination of spores. The scanning electron microscope observation revealed coagulation of cytoplasm, shrinkage, and leakage of protoplast and vacuolation. The inhibition rate was higher in cherry juice, followed by apricot and peach nectar. Overall, the experiments show that spores are more resistant than vegetative cells, and that pulsed electric field (PEF) treatment solely is insufficient for successful spore destruction in foods.

3.6 COMBINATION OF PEF WITH OTHER PROCESSING TECHNIQUES FOR INACTIVATION OF MICROORGANISMS

The PEF processing is known for its efficiency in inactivating microorganisms, particularly vegetative cells. Although efficient in the destruction of microorganisms alone, reaching the appropriate amount of inactivation necessitates intensive treatment with a high electric field intensity as well as a longer duration of processing.[22] The use of high-intensity PEF has several technical limitations and affects the food properties.

The application of electric field strength above 30 kV/cm and scaling up the treatment chamber has technical and economic limitations. The spore cells also cannot be destroyed effectively when PEF treatment is used alone.[11] The PEF can be combined with other preservation techniques for effective inhibition of microorganisms and to protect the food properties. The combination of various preservation techniques for improving food safety and extending the shelf-life of foods is known as hurdle technology.[66]

The PEF can be combined with mild thermal treatment and can effectively destroy the microorganisms without affecting the properties of food. Enhancing the temperature lessens the thickness of the lipid bilayer owing to the conversion of phospholipids from gel to liquid, and the membrane loses its elasticity. The thermal treatment sensitizes the microbial cells toward PEF treatment and enhances microbial death. Enhancement of electric field strength (24–34 kV/cm) and preheating temperature (45–55°C) resulted in more than the 2-log reduction of *E. coli,* in smoothie of tropical fruit.[121] Inoculation of the apple cider with *E. coli* followed by ten pulses of 90 kV/cm electric field strength and temperature of 42°C reduced *E. coli* by 5.9 log.[52] The inoculation of apple juice with *E. coli* followed by treatment with 16, 800 pulses of 43 kV/cm electric field at 35 °C showed a 6.3-log reduction.[27]

The orange juice processed at a condition of 20 pulses of 90 kV/cm electric field and 45°C temperature did not affect the *S. typhimurium*; however, when the temperature increased to 55°C and pulses to 50, a 5.9-log reduction was observed.[62] The milk preheated for 60 s at 50°C followed by PEF treatment at 50 kV/cm for 50 μs effectively reduced the microorganisms by 6.5 log.[118] The 3-log reduction in microorganisms in red grape juice was obtained when the temperature of the treatment increased from 42 to 52°C and treated with 20 pulses of 80 kV/cm electric field strength.[130]

The reduction of the pH of food material can reduce the thermal resistance of microorganisms and inhibit spore germination. The PEF, when combined with lower pH, mixed results were observed, some organisms are effectively inhibited at acidic pH, some at neutral, and some organisms are unaffected by the change in pH. PEF, in combination with antimicrobial agents such as lysozyme, nisin, organic acids, sorbic acid, and ozone was observed to be highly effective in inactivating spoilage and pathogenic microorganisms. The combination of PEF with nisin was studied mostly, and it is observed that this combination is effective against both Gram-positive and Gram-negative bacteria. The simultaneous addition of nisin, lysozyme, and essential oil effectively inhibited *S. typhimurium* and *B. cereus*. Nisin is an effective bacteriocin known to have strong antimicrobial activity. The nisin, when combined with PEF treatment, the bacteria can be inactivated by two different mechanisms. The PEF causes sublethal injury to the bacteria by disrupting the cell envelope, and nisin will act on the cell membrane. The orange juice treated with nisin at 2.5 IU/mL concentration and 40 kV/cm electric field for 100 μs reduced *L. innocua* by 5.6 logs and *E. coli* by 7.9 logs.[70]

The rate of reduction increased from 0.8 to 2.1 times in combination when compared to individual PEF treatment. The Enterocin AS-48, a circular bacteriocin used at 0.613 AU/mL in apple juice along with PEF at 35 kV/cm for 1000 μs, reduced *Pediococcus parvulus* by 6.6 log.[79] The Enterocin AS-48 has prolonged antimicrobial activity and has activity during storage also after PEF treatment. The treating orange juice with 20 pulses of 80 kV/cm electric field along with 4, 250 IU/mL lysozyme at 20 °C didn't show any reduction in microorganisms, but a 3-log reduction was obtained at 44°C temperature.[46]

The milk added with 0.2% acetic acid and treated at 33–33.9 kV/cm achieved a 5.1-log reduction of *L. innocua*, while in the range of 33.8–34.9 kV/cm, milk added with 0.3% acetic acid showed 5.7-log reduction. The fruit juices treated with PEF were highly safe owing to their rich organic acid content and low pH. The malic acid and citric acid combined with PEF treatment for fresh fruit juices showed more than a 5-log reduction of *E. coli*, *S. enteritidis*, and *L. monocytogenes.*[117]

The anions from organic acids interact with cytoplasm at near-neutral pH, reduce the intracellular pH, and affect the genetic material and cell physiology leading to cell death. The plant extracts rich in phytochemicals such as polyphenols, tannins, coumarins, alkaloids, flavanols, flavones, and flavonoids have strong antimicrobial activity. The apple cider treated with clove oil alone reduced the microbial count by 0.07 log, and when combined, it increased to 3.1 log.

The PEF, combined with other non-thermal technologies, also effectively inhibits the microorganisms. The PEF at 30–40 kV/cm combined with 2000 Hz sonication showed a 4-log reduction in *B. subtilis* spores.[81] The milk inoculated with *L. innocua* was treated with 400 w sonication for 160 s followed by PEF at 40 kV/cm for 50 μs reducing the cell count by 6.8 log.[81] The *B. subtilis* spores were suspended in buffer solution treated with PEF and high hydrostatic pressure and observed a 7.1-log reduction in spores. The *S. cerevisiae* suspended in buffer treated with 6 kV/cm electric field and 8 MPa high-pressure carbon dioxide (HPCD) for 5 min at 35°C completely inactivated the organisms. The combination of HPCD and PEF completely inhibited *S. aureus*, *E. coli,* and *B. cereus* up to 4-log.[85] The apple juice suspended with *S. aureus was* treated with a combination of Ultraviolet (30 w for 30 min) and PEF (40 kV/cm for 100 μs) at 46–58 °C, and the result showed a 9.5-log reduction of the organism.[120]

3.7 KINETIC MODELS FOR INACTIVATION OF MICROORGANISMS

3.7.1 PARAMETERS AFFECTING MICROBIAL INACTIVATION KINETICS

Microbial death kinetics explains the lethality of any treatment or processing technique for microorganisms for a particular time. The microorganisms reduced by treatment of PEF are represented as the Log_{10} number of survivors at electric field strength against processing duration.[65] The first-order kinetics explains the relation in which the microorganisms decrease exponentially with treatment time.

The kinetic models help in investigating the resistance of microorganisms toward PEF using decimal reduction time (*D* value) and predicting the inactivation rate under different conditions of processing by developing secondary models. Most of the studies observed linear relations between the number of microorganisms and process time at low electric field strengths and process time. The enhancement of field strength and process time deviated from the linear relation, and the curve changed to upward concave indicating rapid inactivation of organisms at the initial time and a decrease in the rate of inactivation later.[98] The model's alternative to the first-order model was suggested by various investigators to describe the inactivation of microbes using PEF. The Weibull model is most widely used and observed as the best fit to describe microbial death using PEF. Various models were developed using quadratic equations with varied factors like medium, electric field, pH, and temperature.[99]

The deviation of microbial death from first-order kinetics is explained by various theories. In the mechanistic theory, it is believed that the microbial population exhibits similar resistance to the lethal agent; the vitalistic theory, on the other hand, implies the existence of a distribution of microbial sensitivity to the procedure of inactivation. The combination of subpopulations with varying degrees of resistance toward PEF may also be responsible for concave curves of survival. The initial part of the curve explains the destruction of less-resistant organisms, and the secondary portion defines the inactivation of highly resilient organisms.[49]

The Weibull model is based on the hypothesis that the microbial population in a suspension has the same sensitivity toward PEF, and the difference in the sensitivity among microbial populations cannot be explained.[50]

Numerous factors such as microbial strain, culture conditions, and PEF process parameters affect microbial kinetics. The inoculum preparation and culture conditions must be controlled carefully, and thorough data need to be collected.[97] The resistance of microorganisms varies with microbial strain and within the same microbial species. The electroporation mechanism is governed by the size as well as orientation of the cells of microbes. The kinetics study must be conducted with a single microbial strain with uniform morphology and growth phase. The difference between microbial strains majorly affects the results of inactivation kinetics, and conclusions drawn will be incorrect.

The process parameters of PEF treatment are extremely critical in determining the inactivation kinetics. The process parameters play a crucial role, and every parameter is interdependent on each other in inactivating the microorganisms.[44,91] Altering the temperature affects the chamber's electrical conductivity and resistance. The change in intensity of the electric field is required to apply the same amount of voltage between the electrodes. Furthermore, temperature plays a crucial role in inactivating microorganisms, and should be controlled carefully to achieve uniform inactivation and avoid false results.[67]

The electric field strength, intensity, and time of processing should be monitored continuously and regulated to achieve the correct results of inactivation and evaluation of microbes resisting PEF. The non-uniform distribution of the electric field creates dead spaces in the treatment chamber, and the process will be ineffective in inactivating microorganisms. The food materials, when mixed uniformly and homogenized, enhanced the rate of inactivation.

The non-uniform distribution of the electric field in a continuous flow chamber can be minimized using turbulent flow, which will homogenize the samples and achieve uniform microbial inactivation.[65] In a system that has a continuous flow, temperature alterations occur at various parts of the treatment chamber. Temperature is a major factor that affects the resistance of PEF, and the individual influence of process parameters on microbial inactivation is difficult to determine in continuous flow systems. Therefore, it is strongly recommended to use static treatment chambers to study the microbial death kinetics, and process parameters should be strictly controlled. The treatment medium also greatly affects the microbial death kinetics. The properties of the treatment medium, such as pH and water activity, affect the electrical conductivity affecting the electrical parameters. The introduction of toxic contaminants into the medium also affects the microbes and affects the survival curves. The combination of treatments used along with the PEF will have a great effect on the sensitivity of microorganisms toward processing and impact the kinetics and rate of microbial inactivation.[91]

3.7.2 MATHEMATICAL MODELS DESCRIBING MICROBIAL INACTIVATION BY PEF

The mathematical models are required to describe the level of inactivation of microorganisms at various process parameters of PEF that efficiently inactivate the spoilage microorganisms and pathogens to produce safe and shelf-stable food products. The models should effectively describe the inactivation artifacts during processing should be prevented. The models should be based on the mechanism of microbial inactivation and simple. The steady treatment system should be selected for investigating and developing mathematical models to avoid complexity in continuous treatment, and process parameters and critical factors should be strictly controlled.[1,38,40] The pioneering work on developing the first mathematical model describing the inactivation of microbial cells was done by Hfilsheger et al.[45] In this model, a linear relationship was established between the logarithm of survival fraction and the electric field strength (eq 3.1), as well as the logarithm of survival fraction and treatment time (eq 3.2).

$$Ln\,S = -b_t\,ln\left(\frac{t}{t_c}\right) \tag{3.1}$$

$$Ln\ S = -b_E\,ln\left(\frac{E}{E_c}\right) \tag{3.2}$$

where S: survival fraction, t: time of treatment, b_t and b_E: constants, E: electric field strength; t_c: critical treatment time, E_c: the critical electric field strength, T_c: intercept of the regression line (describes the relationship between the natural logarithm of the survival fraction and the natural logarithm of the treatment time for 100% survivors), E_c: intercept of the regression line (describes the relationship between the natural logarithm of the survivor fraction and electric field strength for 100% of survivors)

E_c and t_c values are governed by electric field strength and time of treatment. However, after the electric field intensity and treatment duration have been determined, these parameters may be assumed constant, and the survival fraction can be computed using eq 3.3.

$$S = \left(\frac{t}{t_c}\right)^{-\frac{E-E_c}{k}} \tag{3.3}$$

where E_c denotes the threshold values of the electric field strength, and t_c is the threshold values of the treatment time, respectively, and k is the constant characteristic factor of each organism. The critical electric field strength (E_c) along with and time of treatment can be calculated using this model, and these are characteristic values for each organism and help in analyzing the behavior of microorganisms during PEF treatment.

Peleg[87] proposed the double logarithmic equation based on Fermi's equation. The percentage of survivors is represented in the equation as a sigmoid function of several pulses and electric fields (eq. 3.4). The electric field strength (E_c) corresponds to the inflection point and represents the electric field strength required to inactivate 50% of the microbial population, and a_c represents the slope of the sigmoid curve.

$$S = \frac{1}{1 + exp\left(\frac{E - E_C}{a_c}\right)} \tag{3.4}$$

Researchers have offered many models to describe upward concave rising survival curves associated with the inactivation of microbes.[19,54,135] The curves are empirical equations and imagine the presence of two subpopulations with the difference in resistance toward PEF processing and distribution of resistance within the microbial population.

The Weibull distribution model gained popularity among all the equations owing to its simplicity and flexibility. The model assumes that the microbial population exhibits varying levels of resistance, and the survival graph represents a cumulative distribution of fatal effects. The microbial inactivation was estimated through two parameters, namely, scale and shape parameters.

Peleg and Cole[88] proposed the Weibull model, which was subsequently improved by Mafart et al.[64] The Weibull model fits well with a concave upward survival curve that can cover up to a 4-log reduction of Gram-positive, and Gram-negative bacteria and yeast. The inactivation of eight different organisms—*L. monocytogens*, *Enterococcus faecium*, *Yersinia enterocolitica*, *E. coli*, *S. enteritidis*, *S. typhimurium*, *Salmonella softenberg*, and *S. cerevisiae*—was carried out at electric field strength of 9–28 kV/cm.[101,109]

The survival curves fit well in the Weibull equation, and the shape factor <1 indicates upward concave survival curves and scale factors decreased with an increase in electric field strength. When microbial inactivation is modeled using kinetics of the first order, the value of D can potentially be utilized to contrast microbial resistance and death rate constant that is independent of treatment time. But, according to the Weibull model, the survival fraction depends on both scale and shape parameters.

The comparison between different organisms under different treatment conditions and times is to be established to compare the resistance of microorganisms against PEF. The inactivation of a minimum 5-log microbial population is required to compare the microbial resistance against PEF. The *L. monocytogenes* was observed as a highly resistant organism, and 28 kV/cm electric field strength is required for 5-log inactivation. *Salmonella enterica* is sensitive, and a 5-log reduction can be achieved at 15 kV/cm. The environmental factors influencing PEF in inactivating microorganisms can also be modeled using the Weibull model. The effect of electric field strength and pH on the inactivation of *L. plantarum* and *L. monocytogenes* was modeled using Weibull parameters. The relationship between pH and shape factor was modeled, and an empirical equation was developed based on the Gompertz equation (eq 3.5).

$$\beta = A + C_e^{-e^{(-B(pH-M))}} \quad (3.5)$$

where A, B, C, and M are empirical constants.

The multiple regression equations (eqs 3.6 and 3.7) developed in these studies predicted the scale parameters (α) of *L. monocytogenes* and *L. plantarum* for any combinations of electric field strength (15–28 kV/cm) and pH (3.5–7.0)

$$\text{Log}_{10}^{\alpha} = 4.82 - 0.37\,E + 0.00056E^2 + 0.63pH^2 \quad \textit{(Listeria monocytogenes)} \quad (3.6)$$

$$\text{Log}_{10}^{\alpha} = 6.32 - 0.12E - 1.09pH + 0.13pH^2 \quad \textit{(Lactobacillus plantarum)} \quad (3.7)$$

3.8 APPLICATIONS OF PEF IN FOOD PROCESSING

Thermal pasteurization is conventionally used for the pasteurization of food to inactivate spoilage and pathogenic microorganisms to produce safe products.[112] The pasteurization is usually carried out at 65–95°C temperature for seconds to minutes of time. Although thermal pasteurization is highly efficient in inactivating microorganisms, it will affect the natural characteristics such as flavor, color, texture, and nutritional value of the products. The studies conducted on the impact of PEF on microorganisms revealed that a combination of PEF with mild to moderate heating effectively pasteurized the foods. The PEF creates electroporation in microbial cells and inactivates, and when PEF is combined with heat treatment, various hurdles are created to enhance microbial inactivation. The pasteurization process includes mild heating of food material that affects cell membrane fluidity and enhances pore formation.[53] The food material was then exposed to PEF, cooled, and packed aseptically to avoid contamination. The high-intensity electric field of short, multiple pulses applied to foods reduces microbial contamination. The fruit juices with highly acidic ($pH < 4.6$) are suitable for the PEF process as the PEF can show effect only on vegetative organisms; low pH combined with chilled storage can effectively prevent spore germination and development.[113]

The apple and orange juice treated with PEF at 20 kV/cm for 2–3 μs at 60°C showed to 4–6-log reduction in *L. monocytogenes*, *E. coli*, *S. cerevisiae,* and *Bacillus megaterium*. The reduction in the electric field to 10 kV/cm and temperature to 36°C showed only ~1-log reduction. The orange juice treated at 23 kV/cm for 2 μs at 58°C did not affect the composition, such as sugar profile, carotenoid, and vitamin C content, both immediately after processing and during storage at 4°C for 58 days.[114] The microorganisms in highly acidic fruit juices can be effectively controlled with the combination of moderate heat and PEF treatment, and the quality is similar to untreated juice without any effect. The pathogenic microorganisms can be able to grow in low-acid juices of fruit with a value of pH greater than 4.6, and the treatment should be sufficient to inactivate all the organisms. When the watermelon juice with a pH of 5.3 was treated with PEF at 20 kV/cm of electric field, it showed only about 1-log reduction in *E. coli* and *L. monocytogenes*. An increase in the electric field to 35 kV/cm proves effective and led to the inactivation of 3–4 log units of *Salmonella enteritis*, *L. monocytogenes,* and *E. coli*. However, this treatment also results in a reduction of antioxidant activity and vitamin C to 78% and 46%, respectively. PEF of high strength and energy are required to effectively inactivate the pathogens in low-acid juices.[114] The schematic of pasteurization of liquid foods is shown in Figure 3.2.

The eggs and egg products are most contaminated by *Salmonella*. Indeed, *Salmonella* demonstrates thermal resistance within the pH range of 5–6, and it exhibits even higher resistance in egg yolk at a pH of 6 when compared to whole eggs at a 7.6 value of pH and egg white at a 9.1 value of pH.[74] Typically, the whole egg undergoes pasteurization at temperatures ranging from 65 to 68°C for a duration of 2–5 min, resulting in reduced *S. enteritidis* and *L. monocytogenes* by 5–6-log reductions.[60] The treatment of the egg at >60 °C will affect the viscosity and functional properties of the egg. Treating the whole liquid egg with an electric field of 25 kV/cm at temperature <35 °C showed a 1–2-log reduction in *S. typhimurium* and *S. enteritis.*[132]

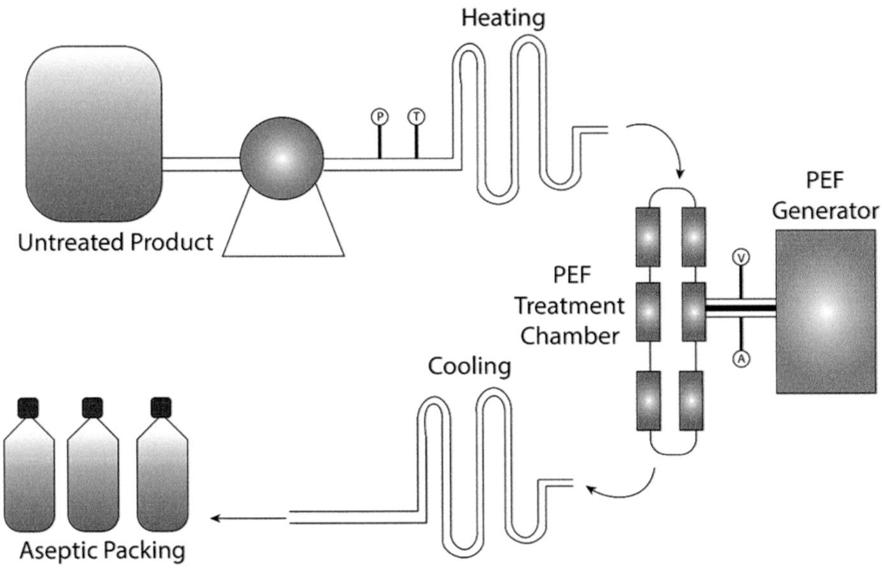

FIGURE 3.2 Schematic of liquid food pasteurization using PEF.

The increase in temperature enhances microbial inactivation, while also posing the challenge of potential protein coagulation. PEF processing at 26 kV/cm of electric field and 37°C of temperature followed by intermediate cooling resulted in a remarkable 6-log reduction in *E. coli* for the liquid egg treatment.[58] Likewise, the whole liquid egg treatment with a 60 kV/cm electric field for 800 μs at a maximum temperature of 20°C led to a significant 4.8-log reduction in *Salmonella*. The increase in temperature to 40°C showed a 6.3-log reduction in *Salmonella*. The PEF treatment of egg and egg products is highly advantageous in protecting the properties of eggs and the efficient inactivation of microorganisms.[128]

The treatment of smoothies and fruit purees with PEF is challenging due to their highly viscous nature. The flow profile of the products should be studied carefully, and the PEF treatment chamber should be designed accordingly for homogenous treatment of the product.[119] The milk-based smoothie inoculated with *L. innocua* treated at 18 kV/cm didn't show any inactivation, whereas an increase of intensity to 32 kV/cm showed a 2.7-log reduction. The tropical smoothie inoculated with *E. coli* treated at 24 kV/cm showed a 3-log reduction and a 5-log reduction at 34 kV/cm.

The fruit smoothie made of strawberry, banana, and apple was treated with the variable electric field of 13.5–24 kV/cm at a maximum temperature of 58°C; the untreated smoothie had a shelf-life of seven days and was spoiled by yeast.[113] Whereas the smoothie treated with PEF showed partial inactivation of yeast and shelf-life extended up to 14 days, and after 14 days, growth of molds was observed and which spoiled the smoothie. The PEF should be assisted along with the thermal treatment required to effectively inactivate the pathogens and to inactivate the microorganisms in highly viscous foods, a high energy of 650–900 kJ/kg is needed.[84] Overall, PEF might be a viable alternative to thermal treatment for producing food with excellent sensory and nutritional characteristics.

3.9 MICROBIAL TOLERANCE AND STRESS RESPONSE TOWARD PEF

Microorganisms, when subjected to various processing techniques apart from the physiological environment, such as osmotic shock, heat shock, and stress reactions, the metabolism of microorganisms changes partially, and resistance against the stress increases.[15] The stress response of microorganisms can be measured through flow cytometry, enzyme activity, proteomic assay, sodium dodecyl sulfate-polyacrylamide gel electrophoresis (SDS-PAGE), transcriptomic analysis, and ATPase analysis. The specific growth rate of microbial culture can be used as a direct measurement for PEF-induced stress. The growth of *E. coli* was studied by Aronsson et al.[6] at 10°C after PEF treatment. The increase in maximum specific growth rate (μ_{max}) and lag phase was observed with the increase in electric field strength representing the non-lethality of the PEF treatment.

Rivas et al.[96] investigated the change in protein concentration in E. coli and found that the protein OmpA in the outer cell membrane was absent indicating a disruption in the cell membrane. However, some enzymes such as phospheptose isomerase (GmhA), ATP-dependent Clp protease ATP-binding subunit (ClpA), ribosomal protein S6 (RS6), deoxyuridine triphosphate pyrophosphatase (DUT), and ferritin (ftnA) protein responsible for iron storage and stress adaptation were observed to be increased.[111] This indicates the response of the cell to injury caused by PEF through the expression of proteins and enzymes that can repair the cell damage. The *S. cerevisiae* exhibited oxidative stress response by upregulating genes encoding glutathione synthesis (GSH1, GLR1) and superoxide dismutase (SOF2) enzymes.[18] The sublethal PEF stress response can be used in the fermentation process for enhancing biomass production and stress-related compounds. The production of certain compounds, trehalose and glycerin, proteins, and amino acids in *S. cerevisiae* was observed to be increased after PEF treatment.[103] Sublethal injury is one of the critical factors responsible for inactivating the organisms and producing safe food. The stress response increases the resistance of cells toward PEF and reduces the inactivation.[69]

On the other hand, the stress response enhances the production of various metabolites and cell biomass during the fermentation process. An extensive study needs to be conducted further to reveal various stress responses exhibited by various strains and sublethal injuries that avoid inactivation.[124]

3.10 CHALLENGES AND FUTURE PROSPECTIVE OF PEF

The physical and chemical changes and magnitude of change that occur in food products during PEF processing are unclear, and studies must be conducted at various intensities of electric fields, treatment time, and temperature.[53] The preheating of food products is required to enhance microbial inactivation and achieve the desired levels of inactivation using PEF. The intermediate cooling of food products is required to avoid overheating and reduction in the quality of food products.[60] A longer treatment time is required to achieve higher log reduction of microorganisms which is an energy-intensive process and commercially not feasible for larger-scale processing. The scaling of PEF and the development of continuous process lines is challenging, and extensive engineering and design studies must be performed. The initial capital cost is higher and industrial-scale processing of food products using PEF results in the production of foods with higher price premium products.[72]

The PEF process must be improved for wider applications in the food industry. The PEF is ineffective in treating spores, and sterilized food products cannot be produced with this technology. The uniform food matrix is required for effective treatment using PEF; the application of PEF on non-uniform products results in non-uniform treatment and variation in microbial inactivation at various places of food. Most of the PEF designs can be used for processing only liquid food products, and a more robust and effective design is yet to develop for effective application on solid foods.[119] The solid particulates and composition of liquid foods and gas bubbles reduce the electrical conductivity and effectiveness of PEF. To effectively utilize pulsed electric field (PEF) in the field of food processing and ensure the development of safe foods, it is essential to extensively investigate and address all these limitations and challenges.[107]

3.11 SUMMARY

The food industry bears the prime responsibility of producing safe, high-quality food to instill consumer confidence in the safety of food products. In minimally processed food products, the reduction of microorganisms and improvement of shelf stability are crucial factors. Although thermal processing is highly efficient in a reduction in microorganisms, reduction in the sensory and nutritional quality of food products is a major disadvantage. Various non-thermal technologies were developed and investigated in the production of safe foods with higher quality. The pulsed electric field is one such technology effective in inactivating various vegetative microbial cells at lower temperatures and producing safe food. The electroporation causes pore formation in microbial cells, affecting osmotic equilibrium and leading to cell death. Some organisms were observed to revive after sublethal injury and increase in number, leading to food spoilage. Extensive research on sublethal injury and the mechanism of stress response and survival must be studied, and PEF design must be improved to enhance microbial inactivation. The electric field strength, specific energy, and treatment time are the critical parameters of PEF that need to be optimized and responsible for microbial inactivation. The type of microbial strain, the morphology of the cell, and culture conditions influence the inactivation rate of microorganisms. Several mathematical models were developed to explain the inactivation kinetics, and the Weibull model is the most used among them. The deviation of experimental results from theoretical inactivation was observed, and an investigation must be done to understand the mechanism of inactivation and the factors responsible for the deviation in inactivation rate. The PEF can be one of the most promising non-thermal food processing technologies for the pasteurization and production of safer foods.

KEYWORDS

- **electric field strength**
- **electroporation**
- **food safety kinetics**
- **permeabilization**
- **pulsed electric field**
- **Weibull model**

REFERENCES

1. Alkhafaji, S.; Farid, M. Modelling the Inactivation of Escherichia Coli ATCC 25922 Using Pulsed Electric Field. *Innovat. Food Sci. Emerg. Technol.* **2008,** *9* (4), 448–454.
2. Allison, D. G.; D'emanuele, A.; Eginton, P.; Williams, A. R. The Effect of Ultrasound on Escherichia Coli Viability. *J. Basic Microbiol.* **1996,** *36* (1), 1–11.
3. Altuntas, J.; Evrendilek, G. A.; Sangun, M. K.; Zhang, H. Q. Effects of Pulsed Electric Field Processing on the Quality and Microbial Inactivation of Sour Cherry Juice. *Int. J. Food Sci. Technol.* **2010,** *45* (5), 899–905.
4. Álvarez, I.; Pagán, R.; Condón, S.; Raso, J. The Influence of Process Parameters for the Inactivation of *Listeria Monocytogenes* by Pulsed Electric Fields. *Int. J. Food Microbiol.* **2003,** *87* (1–2), 87–95.
5. Amiali, M.; Ngadi, M. O.; Smith, J. P.; Raghavan, G. S. V. Synergistic Effect of Temperature and Pulsed Electric Field on Inactivation of *Escherichia Coli* O157:H7 and *Salmonella Enteritidis* in Liquid Egg Yolk. *J. Food Eng.* **2007,** *79* (2), 689–694.
6. Aronsson, K.; Borch, E.; Stenlöf, B.; Rönner, U. Growth of Pulsed Electric Field Exposed *Escherichia Coli* in Relation to Inactivation and Environmental Factors. *Int. J. Food Microbiol.* **2004,** *93* (1), 1–10.
7. Aronsson, K.; Lindgren, M.; Johansson, B. R.; Rönner, U. Inactivation of Microorganisms Using Pulsed Electric Fields: The Influence of Process Parameters on *Escherichia Coli, Listeria innocua, Leuconostoc Mesenteroides* and *Saccharomyces Cerevisiae. Innovat. Food Sci. Emerg. Technol.* **2001,** *2* (1), 41-54.
8. Aronsson, K.; Rönner, U.; Borch, E. Inactivation of *Escherichia Coli, Listeria Innocua* and *Saccharomyces Cerevisiae* in Relation to Membrane Permeabilization and Subsequent Leakage of Intracellular Compounds Due to Pulsed Electric Field Processing. *Int. J. Food Microbiol.* **2005,** *99* (1), 19–32.
9. Astráin-Redín, L.; Moya, J.; Alejandre, M.; Beitia, E.; Raso, J.; Calvo, B.; Cebrian, G.; Álvarez, I. Improving the Microbial Inactivation Uniformity of Pulsed Electric Field Ohmic Heating Treatments of Solid Products. *LWT* **2022,** *154*, 112709.
10. Barsotti, L.; Cheftel, J. C. Food Processing by Pulsed Electric Fields. II. Biological Aspects. *Food Rev. Int.* **1999,** *15* (2), 181–213.

11. Barsotti, L.; Merle, P.; Cheftel, J. C. Food Processing by Pulsed Electric Fields. I. Physical Aspects. *Food Rev. Int.* **1999,** *15* (2), 163–180.
12. Batista Napotnik, T.; Reberšek, M.; Vernier, P. T.; Mali, B.; Miklavčič, D. Effects of High Voltage Nanosecond Electric Pulses on Eukaryotic Cells (In Vitro): A Systematic Review. *Bioelectrochemistry* **2016,** *110*, 1–12.
13. Bendicho, S.; Barbosa-Cánovas, G. V.; Martín, O. Reduction of Protease Activity in Milk by Continuous Flow High-Intensity Pulsed Electric Field Treatments. *J. Dairy Sci.* **2003,** *86* (3), 697–703.
14. Bermúdez-Aguirre, D.; Dunne, C. P.; Barbosa-Cánovas, G. V. Effect of Processing Parameters on Inactivation of *Bacillus Cereus* Spores in Milk Using Pulsed Electric Fields. *Int. Dairy J.* **2012,** *24* (1), 13–21.
15. Cebrián, G.; Raso, J.; Condón, S.; Mañas, P. Acquisition of Pulsed Electric Fields Resistance in *Staphylococcus Aureus* After Exposure to Heat and Alkaline Shocks. *Food Control* **2012,** *25* (1), 407–414.
16. Cebrián, G.; Sagarzazu, N.; Pagán, R.; Condón, S.; Mañas, P. Heat and Pulsed Electric Field Resistance of Pigmented and Non-Pigmented Enterotoxigenic Strains of *Staphylococcus Aureus* in Exponential and Stationary Phase of Growth. *Int. J. Food Microbiol.* **2007,** *118* (3), 304–311.
17. Chen, C.; Smye, S. W.; Robinson, M.; Evans, J. A. Membrane Electroporation Theories: A Review. *Med. Biol. Eng. Comput.* **2006,** *44*, 5-14.
18. Chueca, B.; Pagán, R.; García-Gonzalo, D. Transcriptomic Analysis of *Escherichia Coli* MG1655 Cells Exposed to Pulsed Electric Fields. *Innovat. Food Sci. Emerg. Technol.* **2015,** *29*, 78–86.
19. Couvert, O.; Gaillard, S.; Savy, N.; Mafart, P.; Leguérinel, I. Survival Curves of Heated Bacterial Spores: Effect of Environmental Factors on Weibull Parameters. *Int. J. Food Microbiol.* **2005,** *101* (1), 73–81.
20. Cserhalmi, Z.; Vidacs, I.; Beczner, J.; Czukor, B. Inactivation of *Saccharomyces Cerevisiae* and *Bacillus Cereus* by Pulsed Electric Fields Technology. *Innovat. Food Sci. Emerg. Technol.* **2002,** *3* (1), 41-45.
21. Delemotte, L.; Tarek, M. Molecular Dynamics Simulations of Lipid Membrane Electroporation. *J. Membr. Biol.* **2012,** *245* (9), 531–543.
22. de Vito, F.; Ferrari, G.; Lebovka, N. I.; Shynkaryk, N. V.; Vorobiev, E. Pulse Duration and Efficiency of Soft Cellular Tissue Disintegration by Pulsed Electric Fields. *Food and Bioprocess Technology* **2008,** *1* (4), 307–313.
23. dos Santos Júnior, L. C. O.; Schulz, M.; Knorr, D.; Amante, E. R. Pulsed Electric Field for Escherichia Coli Inactivation in Pumpkin Juice and Nectar. *Acta Scientiarum– Technol.* **2022,** *44*, e56091.
24. Dutreux, N.; Notermans, S.; Wijtzes, T.; Gongora-Nieto, M. M.; Barbosa-Canovas, G. V.; Swanson, B. G. Pulsed Electric Fields Inactivation of Attached and Free-Living *Escherichia Coli* and *Listeria Innocua* Under Several Conditions. *Int. J. Food Microbiol.* **2000,** *54* (1–2), 91–98.
25. Dziadek, K.; Kopeć, A.; Dróżdż, T.; Kiełbasa, P.; Ostafin, M.; Bulski, K.; Oziembłowski, M. Effect of Pulsed Electric Field Treatment on Shelf Life and Nutritional Value of Apple Juice. *J. Food Sci. Technol.* **2019,** *56* (3), 1184–1191.
26. Edd, J. F.; Horowitz, L.; Davalos, R. V.; Mir, L. M.; Rubinsky, B. In Vivo Results of a New Focal Tissue Ablation Technique: Irreversible Electroporation. *IEEE Trans. Biomed. Eng.* **2006,** *53* (7), 1409–1415.

27. El-Hag, A. H.; Dadarwal, R.; Gonzalez, O. R.; Jayaram, S. H.; Griffiths, M. W. Survivability of Inoculated Versus Naturally Grown Bacteria in Apple Juice Under Pulsed Electric Fields. *IEEE Trans. Ind. Appl.* **2009,** *46* (1), 9–15.
28. Emanuel, E.; Roman, P.; Cahan, R. Influence of the Current Density in Moderate Pulsed Electric Fields on P. Putida F1 Eradication. *Bioelectrochemistry* **2019,** *126*, 172–179.
29. Evrendilek, G. A.; Jin, Z. T.; Ruhlman, K. T.; Qiu, X.; Zhang, Q. H.; Richter, E. R. Microbial Safety and Shelf-Life of Apple Juice and Cider Processed by Bench and Pilot Scale PEF Systems. *Innovat. Food Sci. Emerg. Technol.* **2000,** *1* (1), 77–86.
30. Evrendilek, G. A.; Zhang, Q. H.; Richter, E. R. Application of Pulsed Electric Fields to Skim Milk Inoculated with *Staphylococcus Aureus*. *Biosyst. Eng.* **2004,** *87* (2), 137–144.
31. Evrendilek, G. A.; Tok, F. M.; Soylu, E. M.; Soylu, S. Inactivation of *Penicillum Expansum* in Sour Cherry Juice, Peach and Apricot Nectars by Pulsed Electric Fields. *Food Microbiol.* **2008,** *25* (5), 662–667.
32. Fernández-Molina, J. J.; Altunakar, B.; Bermúdez-Aguirre, D.; Swanson, B. G.; Barbosa-Cánovas, G. V. Inactivation of Pseudomonas Fluorescens in Skim Milk by Combinations of Pulsed Electric Fields and Organic Acids. *J. Food Protect.* **2005,** *68* (6), 1232–1235.
33. Fleischman, G. J.; Ravishankar, S.; Balasubramaniam, V. M. The Inactivation of *Listeria Monocytogenes* by Pulsed Electric Field (PEF) Treatment in a Static Chamber. *Food Microbiol.* **2004,** *21* (1), 91–95.
34. Floury, J.; Grosset, N.; Leconte, N.; Pasco, M.; Madec, M. N.; Jeantet, R. Continuous Raw Skim Milk Processing by Pulsed Electric Field at Non-Lethal Temperature: Effect on Microbial Inactivation and Functional Properties. *Le Lait* **2006,** *86* (1), 43–57.
35. Freire, V.; Lattanzio, G.; Orera, I.; Manas, P.; Cebrián, G. Component Release After Exposure of *Staphylococcus Aureus* Cells to Pulsed Electric Fields. *Innovat. Food Sci. Emerg. Technol.* **2021,** *74*, 102838.
36. García, D.; Gómez, N.; Mañas, P.; Condón, S.; Raso, J.; Pagán, R. Occurrence of Sublethal Injury After Pulsed Electric Fields Depending on the Micro-Organism, the Treatment Medium pH and the Intensity of the Treatment Investigated. *J. Appl. Microbiol.* **2005,** *99* (1), 94–104.
37. Garner, A. L. Pulsed Electric Field Inactivation of Microorganisms: from Fundamental Biophysics to Synergistic Treatments. *Appl. Microbiol. Biotechnol.* **2019,** *103* (19), 7917–7929.
38. Gerlach, D.; Alleborn, N.; Baars, A.; Delgado, A.; Moritz, J.; Knorr, D. Numerical Simulations of Pulsed Electric Fields for Food Preservation: A Review. *Innovat. Food Sci. Emerg. Technol.* **2008,** *9* (4), 408-417.
39. Gómez, B.; Munekata, P. E.; Gavahian, M.; Barba, F. J.; Martí-Quijal, F. J.; Bolumar, T.; Campagnol, P. C. B.; Tomasevic, I.; Lorenzo, J. M. Application of Pulsed Electric Fields in Meat and Fish Processing Industries: An Overview. *Food Res. Int.* **2019,** *123*, 95–105.
40. Gómez, N.; García, D.; Álvarez, I.; Condón, S.; Raso, J. Modelling Inactivation of *Listeria Monocytogenes* by Pulsed Electric Fields in Media of Different pH. *Int. J. Food Microbiol.* **2005,** *103* (2), 199–206.
41. Grahl, T.; Markl, H. Killing of Microorganisms by Pulsed Electric Fields. *Appl. Microbiol. Biotechnol.* **1996,** *45*, 148–157.
42. Griffiths, M. W.; Walkling-Ribeiro, M. Pulsed Electric Field Processing of Liquid Foods and Beverages. In *Emerging Technologies for Food Processing*; Da-Wen Sun, ed.; Academic Press, 2014; pp 115–145.

43. Harrison, S. L.; Barbosa-Cánovas, G. V.; Swanson, B. G. *Saccharomyces Cerevisiae* Structural Changes Induced by Pulsed Electric Field Treatment. *LWT-Food Sci. Technol.* **1997,** *30* (3), 236–240.
44. Heinz, V.; Phillips, S. T.; Zenker, M.; Knorr, D. Inactivation of *Bacillus Subtilis* by High Intensity Pulsed Electric Fields Under Close to Isothermal Conditions. *Food Biotechnol.* **1999,** *13* (2), 155–168.
45. Hfilsheger, H.; Potel, J.; Niemann, E.-G. Killing of Bacteria with Electric Pulses of High Field Strength. *Radiat. Environ. Biophys.* **1981,** *20,* 53–65.
46. Hodgins, A. M.; Mittal, G. S.; Griffiths, M. W. Pasteurization of Fresh Orange Juice Using Low-Energy Pulsed Electrical Field. *J. Food Sci.* **2002,** *67* (6), 2294–2299.
47. Huang, K.; Jiang, T.; Wang, W.; Gai, L.; Wang, J. A Comparison of Pulsed Electric Field Resistance for Three Microorganisms with Different Biological Factors in Grape Juice Via Numerical Simulation. *Food Bioprocess Technol.* **2014,** *7* (7), 1981–1995.
48. Huang, K.; Tian, H.; Gai, L.; Wang, J. A Review of kinetic Models for Inactivating Microorganisms and Enzymes by Pulsed Electric Field Processing. *J. Food Eng.* **2012,** *111* (2), 191–207.
49. Huang, K.; Yu, L.; Liu, D.; Gai, L.; Wang, J. Modeling of Yeast Inactivation of PEF-Treated Chinese Rice Wine: Effects of Electric Field Intensity, Treatment Time and Initial Temperature. *Food Res. Int.* **2013,** *54* (1), 456–467.
50. Huang, K.; Yu, L.; Wang, W.; Gai, L.; Wang, J. Comparing the Pulsed Electric Field Resistance of the Microorganisms in Grape Juice: Application of the Weibull Model. *Food Control* **2014,** *35* (1), 241–251.
51. Hu, Q.; Joshi, R. P. Transmembrane Voltage Analyses in Spheroidal Cells in Response to an Intense Ultrashort Electrical Pulse. *Phys. Rev. E – Stat., Nonlinear, Soft Matter Phys.* **2009,** *79* (1), 011901.
52. Iu, J.; Mittal, G. S.; Griffiths, M. W. Reduction in Levels of Escherichia Coli O157: H7 in Apple Cider by Pulsed Electric Fields. *J. Food Protect.* **2001,** *64* (7), 964–969.
53. Jaeger, H.; Meneses, N.; Knorr, D. Impact of PEF Treatment Inhomogeneity Such as Electric Field Distribution, Flow Characteristics and Temperature Effects on the Inactivation of *E. Coli* and Milk Alkaline Phosphatase. *Innovat. Food Sci. Emerg. Technol.* **2009,** *10* (4), 470–480.
54. Jaeger, H.; Meneses, N.; Moritz, J.; Knorr, D. Model for the Differentiation of Temperature and Electric Field Effects During Thermal Assisted PEF Processing. *J. Food Eng.* **2010,** *100* (1), 109–118.
55. Jayaram, S.; Castle, G. S. P.; Margaritis, A. The Effects of High Field DC Pulse and Liquid Medium Conductivity on Survivability of *Lactobacillus Brevis*. *Appl. Microbiol. Biotechnol.* **1993,** *40* (1), 117–122.
56. Jeantet, R.; Baron, F.; Nau, F.; Roignant, M.; Brulé, G. High Intensity Pulsed Electric Fields Applied to Egg White: Effect on *Salmonella Enteritidis* Inactivation and Protein Denaturation. *J. Food Protect.* **1999,** *62* (12), 1381–1386.
57. Jiang, C.; Davalos, R. V.; Bischof, J. C. A Review of Basic to Clinical Studies of Irreversible Electroporation Therapy. *IEEE Trans. Biomed. Eng.* **2014,** *62* (1), 4–20.
58. Jin, T.; Zhang, H.; Hermawan, N.; Dantzer, W. Effects of pH and Temperature on Inactivation of *Salmonella Typhimurium* DT104 in Liquid Whole Egg by Pulsed Electric Fields. *Int. J. Food Sci. Technol.* **2009,** *44* (2), 367–372.

59. Kantala, C.; Supasin, S.; Intra, P.; Rattanadecho, P. Evaluation of Pulsed Electric Field and Conventional Thermal Processing for Microbial Inactivation in Thai Orange Juice. *Foods* **2022,** *11* (8), 1102.
60. Lechevalier, V.; Guérin-Dubiard, C.; Anton, M.; Beaumal, V.; Briand, E. D.; Gillard, A.; Le Gouar, Y.; Musikaphun, N.; Tanguy, G.; Pasco, M.; Dupont, D. Pasteurisation of Liquid Whole Egg: Optimal Heat Treatments in Relation to Its Functional, Nutritional and Allergenic Properties. *J. Food Eng.* **2017,** *195*, 137–149.
61. Lee, H.; Choi, S.; Kim, E.; Kim, Y. N.; Lee, J.; Lee, D. U. Effects of Pulsed Electric Field and Thermal Treatments on Microbial Reduction, Volatile Composition, and Sensory Properties of Orange Juice, and their Characterization by a Principal Component Analysis. *Appl. Sci.* **2020,** *11* (1), 186.
62. Liang, Z.; Mittal, G. S.; Griffiths, M. W. Inactivation of *Salmonella Typhimurium* in Orange Juice Containing Antimicrobial Agents by Pulsed Electric Field. *J. Food Protect.* **2002,** *65* (7), 1081–1087.
63. Loghavi, L.; Sastry, S. K.; Yousef, A. E. Effect of Moderate Electric Field Frequency and Growth Stage on the Cell Membrane Permeability of *Lactobacillus Acidophilus*. *Biotechnol. Progress* **2009,** *25* (1), 85–94.
64. Mafart, P.; Couvert, O.; Gaillard, S.; Leguerinel, I. On Calculating Sterility in Thermal Preservation Methods: Application of the Weibull Frequency Distribution Model. *Int. J. Food Microbiol.* **2002,** *72*, 107–113.
65. Manas, P.; Pagán, R. Microbial Inactivation by New Technologies of Food Preservation. *J. Appl. Microbiol.* **2005,** *98* (6), 1387–1399.
66. Martín-Belloso, O.; Sobrino-López, A. Combination of Pulsed Electric Fields with Other Preservation Techniques. *Food Bioprocess Technol.* **2011,** *4*, 954–968.
67. Martínez, J. M.; Cebrián, G.; Álvarez, I.; Raso, J. Release of Mannoproteins During Saccharomyces Cerevisiae Autolysis Induced by Pulsed Electric Field. *Front. Microbiol.* **2016,** *7*, 1435.
68. Marx, G.; Moody, A.; Bermúdez-Aguirre, D. A Comparative Study on the Structure of Saccharomyces Cerevisiae Under Nonthermal Technologies: High Hydrostatic Pressure, Pulsed Electric Fields and Thermo-Sonication. *Int. J. Food Microbiol.* **2011,** *151* (3), 327–337.
69. Mattar, J. R.; Turk, M. F.; Nonus, M.; Lebovka, N. I.; El Zakhem, H.; Vorobiev, E. *S. Cerevisiae* Fermentation Activity After Moderate Pulsed Electric Field Pre-Treatments. *Bioelectrochemistry* **2015,** *103*, 92–97.
70. McNamee, C.; Noci, F.; Cronin, D. A.; Lyng, J. G.; Morgan, D. J.; Scannell, A. G. M. PEF Based Hurdle Strategy to Control *Pichia Fermentans, Listeria Innocua* and *Escherichia Coli* k12 in Orange Juice. *Int. J. Food Microbiol.* **2010,** *138* (1–2), 13–18.
71. Mendes-Oliveira, G.; Jin, T. Z.; Campanella, O. H. Modeling the Inactivation of *Escherichia Coli* O157:H7 and *Salmonella Typhimurium* in Juices by Pulsed Electric Fields: The Role of the Energy Density. *J. Food Eng.* **2020,** *282*, 110001.
72. Meneses, N.; Jaeger, H.; Moritz, J.; Knorr, D. Impact of Insulator Shape, Flow Rate and Electrical Parameters on Inactivation of *E. Coli* Using a Continuous Co-Linear PEF System. *Innovat. Food Sci. Emerg. Technol.* **2011,** *12* (1), 6–12.
73. Min, S.; Jin, Z. T.; Min, S. K.; Yeom, H.; Zhang, Q. H. Commercial-Scale Pulsed Electric Field Processing of Orange Juice. *J. Food Sci.* **2003,** *68* (4), 1265–1271.
74. Monfort, S.; Gayán, E.; Raso, J.; Condón, S.; Álvarez, I. Evaluation of Pulsed Electric Fields Technology for Liquid Whole Egg Pasteurization. *Food Microbiol.* **2010,** *27* (7), 845–852.

75. Monfort, S.; Gayán, E.; Saldaña, G.; Puértolas, E.; Condón, S.; Raso, J.; Álvarez, I. Inactivation of *Salmonella Typhimurium* and *Staphylococcus Aureus* by Pulsed Electric Fields in Liquid Whole Egg. *Innovat. Food Sci. Emerg. Technol.* **2010,** *11* (2), 306–313.
76. Moody, A.; Marx, G.; Swanson, B. G.; Bermúdez-Aguirre, D. A Comprehensive Study on the Inactivation of *Escherichia Coli* Under Nonthermal Technologies: High Hydrostatic Pressure, Pulsed Electric Fields and Ultrasound. *Food Control* **2014,** *37* (1), 305–314.
77. Munekata, P. E.; Domínguez, R.; Budaraju, S.; Roselló-Soto, E.; Barba, F. J.; Mallikarjunan, K.; Roohinejad, S.; Lorenzo, J. M. Effect of Innovative Food Processing Technologies on the Physicochemical and Nutritional Properties and Quality of Non-Dairy Plant-Based Beverages. *Foods* **2020,** *9* (3), 288.
78. Narsetti, R.; Curry, R. D.; McDonald, K. F.; Clevenger, T. E.; Nichols, L. M. Microbial Inactivation in Water Using Pulsed Electric Fields and Magnetic Pulse Compressor Technology. *IEEE Trans. Plasma Sci.* **2006,** *34* (4 II), 1386–1393.
79. Nguyen, P.; Mittal, G. S. Inactivation of Naturally Occurring Microorganisms in Tomato Juice Using Pulsed Electric Field (PEF) with and Without Antimicrobials. *Chem. Eng. Proc.: Process Intensificat.* **2007,** *46* (4), 360–365.
80. Niu, D.; Zeng, X. A.; Ren, E. F.; Xu, F. Y.; Li, J.; Wang, M. S.; Wang, R. Review of the Application of Pulsed Electric Fields (PEF) Technology for Food Processing in China. *Food Res. Int.* **2020,** *137*, 109715.
81. Noci, F.; Walkling-Ribeiro, M.; Cronin, D. A.; Morgan, D. J.; Lyng, J. G. Effect of Thermosonication, Pulsed Electric Field and their Combination on Inactivation of Listeria Innocua in Milk. *Int. Dairy J.* **2009,** *19* (1), 30–35.
82. Ohshima, T.; Tanino, T.; Guionet, A.; Takahashi, K.; Takaki, K. Mechanism of Pulsed Electric Field Enzyme Activity Change and Pulsed Discharge Permeabilization of Agricultural Products. *Jpn. J. Appl. Phys.* **2021,** *60* (6), 060501.
83. Pagán, R.; Jordan, S.; Benito, A.; Mackey, B. Enhanced Acid Sensitivity of Pressure-Damaged Escherichia Coli O157 Cells. *Appl. Environ. Microbiol.* **2001,** *67* (4), 1983–1985.
84. Palgan, I.; Muñoz, A.; Noci, F.; Whyte, P.; Morgan, D. J.; Cronin, D. A.; Lyng, J. G. Effectiveness of Combined Pulsed Electric Field (PEF) and Manothermosonication (MTS) for the Control of *Listeria Innocua* in a Smoothie Type Beverage. *Food Control* **2012,** *25* (2), 621–625.
85. Pataro, G.; Ferrentino, G.; Ricciardi, C.; Ferrari, G. Pulsed Electric Fields Assisted Microbial Inactivation of *S. Cerevisiae* Cells by High Pressure Carbon Dioxide. *J. Supercrit. Fluids* **2010,** *54* (1), 120–128.
86. Pataro, G.; Senatore, B.; Donsì, G.; Ferrari, G. Effect of Electric and Flow Parameters on PEF Treatment Efficiency. *J. Food Eng.* **2011,** *105* (1), 79–88.
87. Peleg, M. A Model of Microbial Survival After Exposure to Pulsed Electric Fields. *J. Sci. Food Agric.* **1995,** *67* (1), 93–99.
88. Peleg, M.; Cole, M. B. Reinterpretation of Microbial Survival Curves. *Crit. Rev. Food Sci. Nutr.* **1998,** *38* (5), 353–380.
89. Picart, L.; Dumay, E.; Cheftel, J. C. Inactivation of *Listeria Innocua* in Dairy Fluids by Pulsed Electric Fields: Influence of Electric Parameters and Food Composition. *Innovat. Food Sci. Emerg. Technol.* **2002,** *3* (4), 357–369.
90. Puértolas, E.; López, N.; Condón, S.; Raso, J.; Álvarez, I. Pulsed Electric Fields Inactivation of Wine Spoilage Yeast and Bacteria. *Int. J. Food Microbiol.* **2009,** *130* (1), 49–55.
91. Qin, S.; Timoshkin, I. V.; Maclean, M.; Wilson, M. P.; Given, M. J.; Wang, T.; Anderson, J. G.; Macgregor, S. J. Pulsed Electric Field Treatment of *Saccharomyces Cerevisiae* Using Different Waveforms. *IEEE Trans. Dielectr. Electr. Insulat.* **2015,** *22* (4), 1841–1848.

92. Raso, J.; Barbosa-Cánovas, G. V. Nonthermal Preservation of Foods Using Combined Processing Techniques. *Crit. Rev. Food Sci. Nutr.* **2003,** *43* (3), 265–285.
93. Raso, J.; Calderón, M. L.; Góngora, M.; Barbosa-Cánovas, G.; Swanson, B. G. Inactivation of Mold Ascospores and Conidiospores Suspended in Fruit Juices by Pulsed Electric Fields. *LWT-Food Sci. Technol.* **1998,** *31* (7–8), 668–672.
94. Reina, L. D.; Jin, Z. T.; Zhang, Q. H.; Yousef, A. E. Inactivation of *Listeria Monocytogenes* in Milk by Pulsed Electric Field. *J. Food Protect.* **1998,** *61* (9), 1203–1206.
95. Riener, J.; Noci, F.; Cronin, D. A.; Morgan, D. J.; Lyng, J. G. Combined Effect of Temperature and Pulsed Electric Fields on Apple Juice Peroxidase and Polyphenoloxidase Inactivation. *Food Chem.* **2008,** *109* (2), 402–407.
96. Rivas, A.; Pina-Pérez, M. C.; Rodriguez-Vargas, S.; Zuñiga, M.; Martinez, A.; Rodrigo, D. Sublethally Damaged Cells of Escherichia Coli by Pulsed Electric Fields: The Chance of Transformation and Proteomic Assays. *Food Res. Int.* **2013,** *54* (1), 1120–1127.
97. Saldaña, G.; Puértolas, E.; Álvarez, I.; Meneses, N.; Knorr, D.; Raso, J. Evaluation of a Static Treatment Chamber to Investigate Kinetics of Microbial Inactivation by Pulsed Electric Fields at Different Temperatures at Quasi-Isothermal Conditions. *J. Food Eng.* **2010,** *100* (2), 349–356.
98. Saldaña, G.; Puértolas, E.; Condón, S.; Álvarez, I.; Raso, J. Modeling Inactivation kinetics and Occurrence of Sublethal Injury of a Pulsed Electric Field-Resistant Strain of *Escherichia Coli* and *Salmonella Typhimurium* in Media of Z *Innovat. Food Sci. Emerg. Technol.* **2010,** *11* (2), 290–298.
99. Saldaña, G.; Puértolas, E.; Condón, S.; Álvarez, I.; Raso, J. Inactivation kinetics of Pulsed Electric Field-Resistant Strains of *Listeria Monocytogenes* and *Staphylococcus Aureus* in Media of Different pH. *Food Microbiol.* **2010,** *27* (4), 550–558.
100. Saldaña, G.; Puértolas, E.; Monfort, S.; Raso, J.; Álvarez, I. Defining Treatment Conditions for Pulsed Electric Field Pasteurization of Apple Juice. *Int. J. Food Microbiol.* **2011,** *151* (1), 29–35.
101. San Martín, M. F.; Sepúlveda, D. R.; Altunakar, B.; Góngora-Nieto, M. M.; Swanson, B. G.; Barbosa-Cánovas, G. V. Evaluation of Selected Mathematical Models to Predict the Inactivation of *Listeria Innocua* by Pulsed Electric Fields. *LWT - Food Sci. Technol.* **2007,** *40* (7), 1271–1279.
102. Schottroff, F.; Gratz, M.; Krottenthaler, A.; Johnson, N. B.; Bedard, M. F.; Jaeger, H. Pulsed Electric Field Preservation of Liquid Whey Protein Formulations–Influence of Process Parameters, pH, and Protein Content on the Inactivation of *Listeria Innocua* and the Retention of Bioactive Ingredients. *J. Food Eng.* **2019,** *243*, 142–152.
103. Schottroff, F.; Krottenthaler, A.; Jaeger, H. Stress Induction and Response, Inactivation, and Recovery of Vegetative Microorganisms by Pulsed Electric Fields. In *Handbook of Electroporation*; D. Miklavcic, ed.; Springer, Cham, 2017.
104. Sensoy, I.; Zhang, Q. H.; Sastry, S. K. Inactivation kinetics *of Salmonella Dublin* by Pulsed Electric Field. *J. Food Process Eng.* **1997,** *20* (5), 367–381.
105. Sharma, P.; Oey, I.; Bremer, P.; Everett, D. W. Reduction of Bacterial Counts and Inactivation of Enzymes in Bovine Whole Milk Using Pulsed Electric Fields. *Int. Dairy J.* **2014,** *39* (1), 146–156.
106. Sherry, A. E.; Patterson, M. F.; Madden, R. H. Comparison of 40 *Salmonella Enterica* Serovars Injured by Thermal, High-Pressure and Irradiation Stress. *J. Appl. Microbiol.* **2004,** *96* (4), 887–893.
107. Silva, F. V. M.; Gibbs, P. Target Selection in Designing Pasteurization Processes for Shelf-Stable High-Acid Fruit Products. *Crit. Rev. Food Sci. Nutr.* **2004,** *44* (5), 353–360.

108. Sobrino-L´opez, A.; Mart´ın-Belloso, O. Enhancing Inactivation of Staphylococcus Aureus in Skim Milk by Combining High-Intensity Pulsed Electric Fields and Nisin. *J. Food Protect.* **2006,** *69*, 345–353.
109. Sobrino-López, Á.; Raybaudi-Massilia, R.; Martín-Belloso, O. High-Intensity Pulsed Electric Field Variables Affecting Staphylococcus Aureus Inoculated in Milk. *J. Dairy Sci.* **2006,** *89* (10), 3739–3748.
110. Soltanzadeh, M.; Peighambardoust, S. H.; Gullon, P.; Hesari, J.; Gullón, B.; Alirezalu, K.; Lorenzo, J. Quality Aspects and Safety of Pulsed Electric Field (PEF) Processing on Dairy Products: A Comprehensive Review. *Food Rev. Int.* **2022,** *38* (sup1), 96–117.
111. Tanino, T.; Sato, S.; Oshige, M.; Ohshima, T. Analysis of the Stress Response of Yeast Saccharomyces Cerevisiae Toward Pulsed Electric Field. *J. Electrostat.* **2012,** *70* (2), 212–216.
112. Vervoort, L.; Van der Plancken, I.; Grauwet, T.; Timmermans, R. A.; Mastwijk, H. C.; Matser, A. M.; Hendrickx, M. E.; Van Loey, A. Comparing Equivalent Thermal, High Pressure and Pulsed Electric Field Processes for Mild Pasteurization of Orange Juice. Part I: Impact on Overall Quality Attributes. *Innovat. Food Sci. Emerg. Technol.* **2011,** *12* (3), 235–243.
113. Timmermans, R. A. H.; Nederhoff, A. L.; Nierop Groot, M. N.; Van Boekel, M. A. J. S.; Mastwijk, H. C. Effect of Electrical Field Strength Applied by PEF Processing and Storage Temperature on the Outgrowth of Yeasts and Moulds Naturally Present in a Fresh Fruit Smoothie. *Int. J. Food Microbiol.* **2016,** *230*, 21–30.
114. Timmermans, R. A. H.; Nierop Groot, M. N.; Nederhoff, A. L.; Van Boekel, M. A. J. S.; Matser, A. M.; Mastwijk, H. C. Pulsed Electric Field Processing of Different Fruit Juices: Impact of pH and Temperature on Inactivation of Spoilage and Pathogenic Micro-Organisms. *Int. J. Food Microbiol.* **2014,** *173*, 105–111.
115. Toepfl, S.; Siemer, C.; Saldaña-Navarro, G.; Heinz, V. Overview of Pulsed Electric Fields Processing for Food. In *Emerging Technologies for Food Processing*, Da-Wen Sun, ed.; Academic Press, 2014; pp 93–114.
116. Vega-Mercado, H.; Pothakamury, U. R.; Chang, F. J.; Barbosa-Cánovas, G. V.; Swanson, B. G. Inactivation of *Escherichia Coli* by Combining pH, Ionic Strength and Pulsed Electric Fields Hurdles. *Food Res. Int.* **1996,** *29* (2), 117–121.
117. Venkitanarayanan, K. S.; Lin, C. M.; Bailey, H.; Doyle, M. P. Inactivation of *Escherichia Coli* O157: H7, *Salmonella Enteritidis*, and *Listeria Monocytogenes* on Apples, Oranges, and Tomatoes by Lactic Acid with Hydrogen Peroxide. *J. Food Protect.* **2002,** *65* (1), 100–105.
118. Walkling-Ribeiro, M.; Noci, F.; Cronin, D. A.; Lyng, J. G.; Morgan, D. J. Antimicrobial Effect and Shelf-Life Extension by Combined Thermal and Pulsed Electric Field Treatment of Milk. *J. Appl. Microbiol.* **2009,** *106* (1), 241–248.
119. Walkling-Ribeiro, M.; Noci, F.; Cronin, D. A.; Lyng, J. G.; Morgan, D. J. Inactivation of *Escherichia Coli* in a Tropical Fruit Smoothie by a Combination of Heat and Pulsed Electric Fields. *J. Food Sci.* **2008,** *73* (8), M395–M399.
120. Walkling-Ribeiro, M.; Noci, F.; Cronin, D. A.; Riener, J.; Lyng, J. G.; Morgan, D. J. Reduction of *Staphylococcus Aureus* and Quality Changes in Apple Juice Processed by Ultraviolet Irradiation, Pre-Heating and Pulsed Electric Fields. *J. Food Eng.* **2008,** *89* (3), 267–273.
121. Walkling-Ribeiro, M.; Noci, F.; Riener, J.; Cronin, D. A.; Lyng, J. G.; Morgan, D. J. The Impact of Thermosonication and Pulsed Electric Fields on *Staphylococcus Aureus* Inactivation and Selected Quality Parameters in Orange Juice. *Food Bioprocess Technol.* **2009,** *2* (4), 422–430.

122. Walter, L.; Knight, G.; Ng, S. Y.; Buckow, R. Kinetic Models for Pulsed Electric Field and Thermal Inactivation of *Escherichia Coli* and *Pseudomonas Fluorescens* in Whole Milk. *Int. Dairy J.* **2016,** *57*, 7–14.
123. Wang, M. S.; Wang, L. H.; Bekhit, A. E. D. A.; Yang, J.; Hou, Z. P.; Wang, Y. Z.; Dai, Q. Z.; Zeng, X. A. A Review of Sublethal Effects of Pulsed Electric Field on Cells in Food Processing. *J. Food Eng.* **2018,** *223*, 32–41.
124. Wang, M. S.; Zeng, X. A.; Sun, D. W.; Han, Z. Quantitative Analysis of Sublethally Injured *Saccharomyces Cerevisiae* Cells Induced by Pulsed Electric Fields. *LWT* **2015,** *60* (2), 672–677.
125. Wouters, P. C.; Alvarez, I.; Raso, J. Critical Factors Determining Inactivation Kinetics by Pulsed Electric Field Food Processing. *Trends Food Sci. Technol.* **2001,** *12*, 112–121.
126. Wouters, P. C.; Bos, A. P.; Ueckert, J. Membrane Permeabilization in Relation to Inactivation Kinetics of *Lactobacillus Species* Due to Pulsed Electric Fields. *Appl. Environ. Microbiol.* **2001,** *67* (7), 3092–3101.
127. Wouters, P. C.; Dutreux, N.; Smelt, J. P.; Lelieveld, H. L. Effects of Pulsed Electric Fields on Inactivation kinetics of *Listeria Innocua. Appl. Environ. Microbiol.* **1999,** *65* (12), 5364–5371.
128. Wu, L.; Zhao, W.; Yang, R.; Yan, W.; Sun, Q. Aggregation of Egg White Proteins with Pulsed Electric Fields and Thermal Processes. *J. Sci. Food Agric.* **2016,** *96* (10), 3334–3341.
129. Wu, W. J.; Chang, J. Inactivation of Vegetative Cells, Germinated Spores, and Dormant Spores of *Bacillus Atrophaeus* by Pulsed Electric Field with Fixed Energy Input. *J. Food Process Eng.* **2022,** *45* (2), e13959.
130. Wu, Y.; Mittal, G. S.; Griffiths, M. W. Effect of Pulsed Electric Field on the Inactivation of Microorganisms in Grape Juices with and Without Antimicrobials. *Biosyst. Eng.* **2005,** *90* (1), 1–7.
131. Yildiz, S.; Pokhrel, P. R.; Unluturk, S.; Barbosa-Cánovas, G. V. Identification of Equivalent Processing Conditions for Pasteurization of Strawberry Juice by High Pressure, Ultrasound, and Pulsed Electric Fields Processing. *Innovat. Food Sci. Emerg. Technol.* **2019,** *57*.
132. Yogesh, K. Pulsed Electric Field Processing of Egg Products: A Review. *J. Food Sci. Technol.* **2016,** *53*, 934–945.
133. Yun, O.; Zeng, X. A.; Brennan, C. S.; Han, Z. Effect of Pulsed Electric Field on Membrane Lipids and Oxidative Injury of *Salmonella Typhimurium. Int. J. Mol. Sci.* **2016,** *17* (8), 1374.
134. Zhao, W.; Yang, R.; Lu, R.; Wang, M.; Qian, P.; Yang, W. Effect of PEF on Microbial Inactivation and Physical-Chemical Properties of Green Tea Extracts. *LWT - Food Sci. Technol.* **2008,** *41* (3), 425–431.
135. Zhong, K.; Chen, F.; Wang, Z.; Wu, J.; Liao, X.; Hu, X. Inactivation and kinetic Model for the *Escherichia Coli* Treated by a Co-Axial Pulsed Electric Field. *Eur. Food Res. Technol.* **2005,** *221* (6), 752–758.

CHAPTER 4

PULSED ELECTRIC FIELD IN HURDLE TECHNOLOGY

PINTU CHOUDHARY, SANDHYA KUMAR, SAWINDER KAUR, and YUVRAJ K. BHOSALE

ABSTRACT

Perishable foods tend to spoil in a short period of time, leading to qualitative and quantitative losses. Conventionally, various processing techniques are applied for the preservation of such perishable commodities. Conventional techniques mainly include thermal preservation, preservation with low temperature, and application of chemical preservatives. These techniques work on single principles for preservation. However, intense processing with conventional techniques appears with respective disadvantages, which mainly include nutrient degradation, change in sensory parameters, and reduction of commercial value. Taking these issues into account, a combination of different preservation techniques to preserve the food with different approaches at the same time is being applied, and the technique is termed as hurdle technology. The pulsed electric field (PEF) is a novel nonthermal technique that can inactivate the microbes as well as enzymes. PEF is having large commercialization ability and can also be used as a technique for food preservation in combination with other techniques, giving a hurdle effect. This chapter is focused on the applications of PEF in hurdle technology and discusses various advancements in the area.

4.1 INTRODUCTION

Food is among the basic needs of humans and is consumed to perform basic functions such as providing nutrients to humans. Attaining self-sufficiency

Nonthermal Pulsed Electric Field Technology for the Food Industry.
Preeti Birwal, Megh R. Goyal, Baldev Singh Kalsi, & Ankita Kataria (Eds.)

DOI: 10.1201/9781003605904-6

in terms of quantitative aspects is an important aspect while food is being produced. To achieve this, several technological advances are being made in the fields of agriculture and food production. According to FAO statistics, the worldwide production of primary crops has reached to the record high quantity of 9.2 billion tons in the year 2018. This production figure is 50% higher in comparison with the production of the same commodities in the year 2000. This high quantity of food produced is mainly dominated by cereals, which share nearly 33% of total production. Further distinguishing revealed a higher share of sugar-producing crops. Considering the production of perishable crops, vegetables were most dominating, beholding 12% of world food production, followed by fruits (10%) and root tubers (9%) ("World Food Agric. - Stat. Yearbook 2020," 2020).

A large chunk of harvested food is lost every year in different food groups, and the major losses are observed in fruits, vegetables, and other perishable commodities. Higher levels of moisture content and water activity in perishable commodities make them prone to spoilage. The spoilage of these perishable commodities is leading to qualitative and quantitative losses of the food, which are considered postharvest losses.[8] To arrest the impoverishment of food due to spoilage, several processing techniques have been evolved. The prime aim of these processing techniques is to preserve the food from self-decay, microbial decay, and decay from external factors such as insects and rodents.

In ancient times, the preservation of food was done by basic available ingredients such as keeping under ice, drying, and the addition of salts and sugar. Furthermore, the preservation techniques have evolved, leading to processes like fermentation, followed by thermal processing, cold preservation, and preservation with the help of chemicals. Though these techniques have evolved from time to time with advancements in technological aspects, there are some drawbacks that are of major concern while applying these techniques. Simple preservation techniques such as fermentation tend to produce gases such as CO_2, which are considered greenhouse gases. Additionally, fermentation is a time-consuming process that limits the bulk processing of several commodities. Processing techniques such as thermal processing and low-temperature processing are highly energy-intensive processes. Though these techniques can be implemented for very large quantities of processing, the nutritional quality of the food is reported to degrade with a higher degree of processing, leading to inferior availability of the heat-sensitive nutrients.

The current scenario reveals that energy-intensive processes such as thermal processing not only expel CO_2 but also degrade quality, and a higher need for consumption increases the carbon footprints of the process.

The issues like the greenhouse gas effect, CO_2 emissions, and higher carbon footprints hinder the sustainability of these processes. Other manufacturing sectors are turning toward sustainability by adopting novel technologies for operations. However, considering the food processing sector, it is lagging behind when it comes to the adoption of sustainable technologies.

Lately, some technological advancements have been made for the application of novel technologies such as power ultrasound, high-pressure processing, pulsed electric field, and ozone treatment, to name a few. The prime ascendancy of these techniques is to retain the nutrients present in the food while achieving a higher degree of preservation with an optimum level of processing. Moreover, many of these techniques do not produce effluent or unwanted waste that further facilitates lowering the carbon expelling that is happening during the treatment of effluent. Thus, these techniques are now being considered for a futuristic approach with sustainable process design in the food processing sector.

Pulsed electric field (PEF) is a novel processing technique that is applied for the preservation of food as a nonthermal technique. It involves suspending the food material between two electrodes, followed by pulsating electric current in subsequent processing. In the technique of pulsed electric field, microbial inactivation of food is achieved with minimal degradation as the technique is nonthermal. Moreover, PEF is also known to inactivate the enzymes. This makes PEF a novel alternative to conventional techniques which can prevent food from self and microbial degradation. This enables the application of PEF into the food industry with great economic value and for the foods that contain several active components with multiple health benefits.

Being a nonthermal food preservation technique, apart from conserving bioactive compounds, PEF is also known to preserve the physicochemical properties such as color, flavor, texture, and nutritional value in terms of macro- and micronutrients. This enables the application of PEF for high-quality consumer products, proving its advantages over other preservation techniques, for instance, conventional thermal preservation. Consequently, PEF is viewed as the most promising technique for future applications for food preservation by microbial and enzymatic inactivation. However, the lower effectiveness of PEF on microbial spores encourages the researchers to conduct studies related to the application of it in combination with other techniques, leading toward the development of hurdle protocols.

Apart from being highly effective for food preservation, the direct application of electric current in PEF for food preservation avoids the energy losses by converting them into other forms, such as electric heaters. Though, in current times, the major energy source for electricity generation is coal and other nonrenewable forms, a futuristic green approach for electricity generation will lead toward a reduction of carbon footprint and sustainable preservation of food. Considering these rationales, this chapter is focused on the application of PEF for reduction of self-degradation and microbial degradation as a tool in hurdle protocols.

4.2 SHELF-LIFE EXTENSION USING PEF

The process of PEF involves the application of a very high voltage of electric current (10 to 80 kV/cm) with a very short impulse rate varying from microseconds to milliseconds.[34] The prime application of PEF enables the destruction of microbial cells, which consequently leads to the extended shelf-life of the food material, but if exposed to a certain level, it can also improve the growth rate of microbial cells.[42] This technology is extensively studied for the pasteurization process, which is known as the application of preservation techniques for the inactivation of pathogenic microorganisms.[6,39] Considering the heterogeneous nature of food microflora, it is of prime importance to understand the native microflora on the food material, as Gomez-Gomez et al.15 reported variable effects of PEF on different microbes. The study specifically differentiated a variety of microbes.

4.3 MICROBIAL INACTIVATION

Exposure of food and subsequently microbes present on the food material to a very high intensity of electric current passing through microbial cells at very high pulsation will lead to microbial destruction. The microbial cell wall and plasma membrane are mainly made of specific proteins, peptidoglycans, and some phospholipids.[7] Calzia et al.[10] reported that many of the proteins, when exposed to electric current, undergo the release of ATP, which further leads to irreversible structural and functional changes. Likewise, exposure of microbial cells to PEF conditions will lead to a change in the fundamental structure of proteins and peptidoglycans present in the cell wall and plasma membrane.[32] This deterioration in structural moieties consequently results in degradation of the cell wall as well as the plasma membrane.

The degradation of the cell wall and plasma membrane will lead to leaching out of the cytoplasm, which is also known as electroporation. At an extensive processing of the food commodities, slight leaching of genetic materials such as DNA is also observed.[32] However, microbial inactivation is mainly dependent on the type of microorganisms, the composition of the cell wall and plasma membrane, and its sensitivity toward the pulse of an electric field.[14] Moreover, the physical state of food also plays an important role in the effectiveness of PEF.[5]

4.4 EFFECT ON SPORES

Microbial spores are significantly different from respective vegetative cells in comparison with morphology, composition, and resistance toward processing. The morphology of spores mainly includes the spore coat, cortex, spore wall, and core sequentially from outside to inside.[35] Unlike vegetative cells, the genetic material (DNA) of spores is in condensed form, and the complete structure is nearly dehydrated. This is the reason for the resistance of spores to different preservation techniques. When exposed to PEF, spores show similar resistance. Specific analysis of spores revealed several deformations on the spore coat due to changes in native proteins. Moreover, the deformations are also observed in the core and cortex. However, the dehydrated nature of spores does not show any form of leaching of electroporation and subsequently removal of DNA from the basic structure of the core. Pillet et al.[32] reported that these changes are due to protein changes from it.

4.5 MICROBIAL INACTIVATION USING HURDLE

Application of PEF is a greener technique that is known for its ecofriendly nature and low carbon footprint in comparison with conventional techniques. However, there are some limitations when applied for food preservation applications and may lead to microbial activity by slight improvement in cell permeability.[42] To overcome these drawbacks, PEF is applied in combination with different conventional and novel technologies and further discussed in the application section.

4.6 ENZYME INACTIVATION

Enzymes are biological catalysts that are playing their role specifically in improving the rate of reactions based on their functions. Considering the basic nature of enzymes, all the enzymes are proteins, and their functionality may vary based on different factors such as basic protein structure, degree of polymerization, and specific size and sequence of peptides. Enzymes play a vital role in the self-degradation of food commodities, making their inactivation a subject of prime importance for shelf-life extension. Application of PEF precisely attacks the stabilizing forces of enzyme structure, which, namely, include electrostatic forces, ion pairing, Van der Waals forces, hydrogen bonding,[18] and at higher voltage, it is also potent to break the disulfide bonds.[46] Changes in these forces modify the basic three-dimensional structure of the enzymes and consequently diminish enzyme activity. However, different enzymes react differently toward PEF processing, changing their stability and substrate conversion.[3]

4.7 APPLICATIONS OF PEF FOR HURDLE PRESERVATION

PEF results in microbial reduction without much loss in quality parameters. When it is combined with conventional (temperature-based, preservative-based, and pH-based) or novel (HPP, CP, US, and IR) techniques, it produces a synergistic effect in reducing the microbial load and thus preserving the food products. The sections below discuss the effects of such treatments in detail.

4.7.1 APPLICATION WITH CONVENTIONAL PRESERVATION

Conventional preservation includes treating the food samples at high temperatures for a longer duration to preserve the quality attributes of the product in order to increase the shelf-life. Although the shelf-life is increased, high-temperature treatment often leads to minor changes in quality attributes such as taste, flavor, and aroma. Thus, nonthermal techniques in combination with thermal techniques might help in reducing the microbial load with little or no quality degradation. The section below discusses the effects of hurdles offered by PEF and conventional food preservation techniques in order to reduce microbial load and preserve the quality of food products.

4.7.1.1 PEF WITH THERMAL PRESERVATION

PEF is one of the nonthermal preservation techniques used in reducing the microbial proliferation in foods. However, the efficacy of inactivation is increased when the technique is combined with other processing methods.

Heat treatment-assisted PEF is one of the novel combined techniques carried out to reduce the microbial population.[28] The synergistic effect of PEF and mild temperature helps in preserving the food sample by inactivating the spoilage organisms. In one experiment, cantaloupe juice was treated using mild temperatures (50–55°C), and it was observed that the tea polyphenols and natamycin acted as natural preservatives along with heat and resulted in better preservation of the quality of the juice with microbial inactivation of *Saccharomyces cerevisiae*.[23] The use of combined treatment resulted in preserving the juice for 15 days under refrigerated conditions. PEF when used on any sample aids in reducing spoilage organisms without much quality deterioration, whereas PEF when used synergistically with thermal treatment reduces the microbial growth and provides a pasteurized effect with little or minimal quality degradation. The effect of the two treatments led to the preservation of vitamin C after the treatment. With the increase in temperature from 50 to 55°C, the inactivation rate also increases. The use of mild temperature for inactive spoilage organisms also results in reducing the sublethally injured organisms due to the damaging of cell membranes by the tea polyphenols. This leads to metabolic disorders and targets the enzymes present in them. The use of mild thermal treatment with PEF produces a similar degree of spoilage reduction as seen using a conventional thermal pasteurization method. An increase in inactivation can be achieved by maintaining a low electric field and increasing the temperature to 55°C in the case of milk samples.[4] Electroporation occurs when the samples are preheated and treated with PEF due to an alteration in the fluidity of the cell membrane.[24] The heat economy can be used to increase the efficiency of microbial reduction and maintain the overall energy usage. As the temperature of the sample medium increases upon PEF, the sample can be used to preheat the fresh sample, thus reducing the cost of cooling requirements.[25]

4.7.1.2 PEF WITH LOW TEMPERATURE

Low-temperature preservation of foods involves freezing the foods below their freezing point in order to maintain their shelf-life for a longer duration. When PEF is used in combination with such techniques, the kinetics of freezing is increased, which in turn leads to enhanced quality.[12] This section describes the phenomena of low-temperature treatment and PEF in preserving food products. PEF involves electroporation of the microbial cell walls, which leads to membrane disruption. When the freezing technique is applied to such products, the small fragments of the ruptured cells serve as sites for crystallization, thus reducing the time required for freezing.[43]

The samples treated with PEF prior to freezing exhibit a lower shrinkage and good visual appearance. The phase transition time is also reduced, leading to the formation of smaller ice crystals during freezing.[9] In one experiment, strawberries were treated with PEF prior to freezing, and it was observed that strawberries retained a fresh red color, and the time required to freeze also reduced.[38] In the case of meat samples, PEF and freezing often lead to lipid oxidation. Therefore, the electric field strength during PEF should be maintained at lower levels. In a study using Atlantic salmon, PEF was used at 1.0 kV/cm followed by freezing. The thawing time was reduced by 20 min due to the effect of PEF, and the freshness of the samples was maintained for up to 8 days.[22] In another study, to preserve the quality of lamb meat, PEF was used in combination with freezing. PEF increases the conductivity of samples, which indicates a greater movement of ions across the membrane. This in turn justifies the effective electroporation achieved during PEF treatment.[20]. Freezing of samples also reduces the water movement, which disrupts the homeostasis of the meat sample.

4.7.1.3 PEF WITH pH

The pH of the feed also has a profound impact on providing the necessary hurdle with PEF treatment. The studies conducted so far involve treating the samples at pH 4 and 6. An increase in the inactivation of microorganisms was found in samples at pH 6 due to an increase in death kinetics, whereas the sample at pH 4 leads to cell injury, which will recover later. Therefore, in order to inactivate the cells of the sample at pH 4, natural antimicrobials or heat treatment is further added to increase the death of spoilage organisms.[28] The conductivity of the sample also plays a crucial role in microbial elimination in PEF. The conductivity, temperature, and pH of the sample determine the efficacy of inactivation. At higher temperatures, the conductivity increases and pulse width decreases.

The increase in electrical energy also affects the pH of the food system. When samples are PEF-treated at high frequency, electroporation changes the conductivity, thus leading to a change in pH, whereas low-frequency PEF treatment does not have any significant changes in pH.[20]

4.7.1.4 PEF WITH DRYING AND OSMOTIC DEHYDRATION

PEF treatment prior to drying often leads to a reduction in the drying time and also preserves the quality aspects of food samples. PEF pretreatment leads to a 20.4% reduction in the drying time of red pepper.[36] When the time required for drying is reduced, the color degradation due to dying is also minimized. Cell disruption takes place during PEF treatment that disrupts the cell membrane, which in turn leads to increased mass transfer of water molecules across the membrane, thereby reducing the time required during drying operations. However, short electric field strength PEF as pretreatment helps in conserving the quality and energy utilization.

In another study, PEF with osmotic drying of berries led to a decrease of 33% drying time with good color retention, antioxidant activity, and total phenolic content.[13] The harder the peel of fruits used for treatment, the higher the electric field strength required. But lower field strength of 0.9 to 2.8 kV/cm is commonly used for fruits and vegetables prior to drying. However, at lower field strength, the number of pulsations can be increased to increase the water diffusion across the membrane during drying.

4.7.1.5 PEF WITH CHEMICAL PRESERVATIVES

The growing need to obtain a combination treatment to preserve the quality of foods and also reduce microbial load is recently being demonstrated by conducting various scientific experiments. The natural preservatives present in foods such as tea polyphenols and natamycin help in preserving the vitamin C content during PEF treatment.[23] The microbial cells become sublethally injured on exposure to PEF, and the tea polyphenols tend to kill the injured cells. On the other hand, natamycin in PEF had a profound impact on intact cells as more than 5 $\log_{10}$ reduction was achieved with a 30 min incubation time. Natamycin has better antimicrobial properties compared to tea polyphenols as the latter serves as a good antioxidant and is used as a natural additive, whereas the former is added as a biopreservative in various foods. Citral is another natural preservative that, when added at 50% of the minimum inhibiting concentration, produces a lethal effect.[28] Thus, citral, when used at pH 6, provided strong inactivation compared to acidic pH, which contradicts the fact of inactivation achieved by essential oils due to their acidic pH. This was due to the low chemical stability of citral at acidic conditions as it gets broken down into a variety of byproducts.[28]

The antimicrobial activity is enhanced upon treating samples at high temperatures in PEF due to an increase in vapor pressure, which in turn results in the solubilization of active antimicrobial compounds into the bacterial or fungal cell membrane.[17] At higher electric fields, such as 50 kV/cm, the citral enters the bacterial cell due to increased electroporation, causing lethal effects.

Another common good-compatible antimicrobial agent is nisin, which also leads to electroporation of microbial cells at an electric density range of 50 kJ/L to 90 kJ/L. Electroporation occurs when the cell reaches a threshold transmembrane polarization due to the effect of PEF.[31] The efficacy also increases when a mild heat treatment is applied to the food samples before treating with PEF and nisin. Apart from the electric field or electric density, the duration of electric pulses also plays a major role in producing the desired antimicrobial effect. The increase in pulse duration leads to increased permeabilization with enhanced cell electroporation.[11,21] With the increase in electroporation, the mass transfer across the cell membrane increases with the transfer of antimicrobials and provides the necessary lethal effect.

4.7.2 APPLICATION WITH NOVEL TECHNIQUES

Preservation of foods using hurdle technology involving novel techniques leads to synergistic log reduction of microbial organisms. The treatment time in such cases is reduced, and the preservation of food products is also achieved with little or no quality changes. The section below discusses major novel techniques that are used in combination with hurdles in PEF for food preservation.

4.7.2.1 PEF WITH POWER ULTRASOUND

PEF with high-power ultrasound proves to be synergistic toward microbial log reduction in the case of *E. coli*, *A. niger*, and *B. pumilus*. Although 6.6 log reduction was achieved for *A. niger* under PEF (152.3–176.3 kJ/kg) with high power ultrasound of 3 min, the efficacy of microbial reduction was not suitable enough for pasteurization of the food sample as reported by Gomez-Gomez et al.[15] This type of hurdle technique is used in the inactivation of vegetative bacterial and fungal spores with increased efficiency toward destruction of fungal spores.

The hurdle effect with US and PEF helps in little or no quality loss with enhanced microbial reduction. Thus, such techniques aid in preserving foods without losing their integrity.[37] Both the techniques discussed above result in electroporation and cavitation. The use of PEF leads to electrofusion and electroporation, which inhibit the proliferation of yeast and molds present in the sample, while the cavitation process reduces both TPC and yeast and molds.[1] Cavitation results in the free-radical production, which leads to localized heating that produces the required effect for microbial reduction.[2]

The combined effect of PEF and US in the microbial inactivation was studied in milk processing, where thermosonication was done to reduce the *Listeria innocua* species.[30] Thermosonication at 400 W and 160 s led to 0.6 $\log_{10}$ cfu mL^{-1} microbial inactivation of *L.innocua* as lower power and retention time do not have any significant impact on the inactivation process. With the combined treatment of thermosonication followed by PEF, the microbial reduction increased up to 3.8 $\log_{10}$ cfu mL^{-1}. The reduction further increased with the increase in power value of PEF from 30 to 40 kV/cm. The above study suggests the possible utilization of both novel treatments in the microbial reduction to achieve a longer storage life for milk samples.

The above-mentioned combined technology also aids in reducing the contamination of oil-field reinjection water by acting on saprophytic bacteria, iron bacteria, and sulfate-reducing bacteria.[45] The combined treatment greatly increased the inactivation with D values of 17.5, 4.27, and 7.3 min for saprophytic, iron, and sulfate-reducing bacteria, respectively, thus proving to be an efficient technique in their reduction.

In another study conducted by Aadil et al.,[1] the combined effect of PEF and US aided in reducing the microbial load of grapefruit juice with no or little effect on the bioactive compounds present in them. The time required to obtain the desired microbial reduction is less in combined effect compared to individual processing techniques. By reducing process time and input energy, a highly beneficial result is obtained with no detrimental effect on the environment. Thus, such novel hurdle techniques are ecofriendly processes.

Though many studies reported the energy efficiency and reduced process times of combined techniques, it is expensive compared to individual ones.[14] Apart from microbial reduction, PEF and US result in reducing the enzyme activity in fruit juices, which will further increase their storage life.[37] A 6.8 log reduction of *Staphylococcus aureus* was achieved in the hurdle technique with PEF and thermosonication. The hurdle technique was compared to the traditional thermal process as PEF targeted the resistive cells of *S. aureus* while TS targeted the sensitive cells to inactivate through electroporation and cavitation, respectively, producing bactericidal effects.[41]

4.7.2.2 PEF WITH HIGH PRESSURE PROCESSING

The synergistic effect obtained by HPP followed by PEF resulted in more than a 3 log reduction of the microbial population. The kinetic energy changes in membrane components, such as changes in the state of lipids and proteins due to the action of pressure, can increase their sensitivity toward PEF, thereby providing a hurdle effect.[16] However, there are very limited studies available that analyze the combined effect of PEF in combination with HPP, limiting the knowledge and understanding of their application as elements of a hurdle.

4.7.2.3 PEF WITH COLD PLASMA

This part discusses the preservation effect obtained using PEF and cold plasma treatment on foods. Some indigenous varieties of plants possess antimicrobial activities. Such leaf extracts, together with combined nonthermal treatments, provide the required hurdle effect to prevent microbial attack. Chamuang leaf extract, which is believed to possess the compounds Chrysoeriol 6-C-glucoside-8-C-arabinopyranoside and 2-feruloyl-1-sinapoylgentiobiose, which on penetration into the food results in antioxidant and antimicrobial activity. PEF followed by vacuum impregnation of the leaf extract at 2% w/v to shrimp samples with subsequent high-voltage cold plasma results in extending the shelf-life of shrimps to 18 days at 4°C.[33] The hurdle effect provided by PEF and CP with the presence of leaf extract resulted in reducing the microbial contamination and unacceptable oxidative effects in shrimp samples.

4.7.2.4 PEF WITH IRRADIATION

This section discusses the hurdle effect produced by PEF and the irradiation process in food materials. PEF and the irradiation-combined hurdle technique are advantageous as they result in increased log reduction compared to individual processes. In a study, apple juice was processed using the hurdle technique, where the combined effect was taken as PEF+UV irradiation and UV irradiation + PEF.[29] It was deduced from their experiment that 6.2 and 7.1 log reductions were observed for PEF+UV and UV+PEF, respectively. The combined effect has a profound impact on microbial reduction, whereas the changes in quality attributes are similar to those processed under mild heat treatment (72°C).

The incorporation of UV irradiation treatment and preheating of food samples prior to PEF treatment led to an increase in log reduction of microbial counts compared to the HTST process. The preheating effect prior to PEF had a profound impact in increasing the microbial reduction values, as interpreted from an experiment conducted by a study.[40] With the increase in temperature of preheating effect, a higher load of inactivation is possible. As the preheating temperature increases from 43°C to 46°C, the log reduction increases to 9.5 $\log_{10}$ from 8.3 $\log_{10}$ of *S. aureus*. When the temperature is increased in a PEF and IR hurdle, the inactivation achieved will be comparable to the HTST pasteurization process.

4.7.2.5 PEF WITH MAGNETIC FIELD

This section throws light on the preservation effects achieved using PEF and magnetic fields as hurdles. A study conducted by Mok et al.[26] on food freezing using the above combination technique in preserving the quality of food samples was validated in the production of small ice crystals. A synergistic approach is developed between PEF and repulsive SMF in ice crystal formation where the former produces uniform-sized crystals while the latter produces ice crystals under short phase transition. At 1.78 V/cm of PEF and 0.5 duty phase of SMF, ice crystal formation occurs at 20 kHz, reducing the phase transition time. Due to the presence of a magnetic field, the water molecules lead to bond breaking and reformation, resulting in vibration and collision that reduce the phase transition time and produce ordered and uniformly structured crystals.

An oscillating magnetic field coupled with PEF is widely used in the preservation of frozen chicken breasts to prevent microbial contamination during their transport and storage. Supercooling was achieved on sequential changes in the cycles of OMF, which maintains the frozen state of chicken meat for an increased duration.[27] The supercooled meat maintains the quality and physical state of the meat with no visible damages. The combined effect led to supercooling, thus inhibiting sudden nucleation in samples. The formation of sudden ice crystals in the sample can be ascertained by the changes in electrical conductivity of the samples as the growth of ice crystals causes protonic defects in the sample at an OMF duty phase of 0.2. Thus, supercooling helps in preserving the fresh-like nature of meat samples without any physical or chemical damages. Similar supercooling effects in beef samples were studied and reported by Kang et al.[19]

One of the main hurdles in meat samples is low-temperature freezing that maintains the fresh-like nature without any microbial attack. Thus, the combined effect results in freezing and supercooling, thereby reducing the onset of microbial spoilage and preserving the food samples.

4.8 SUMMARY

Perishable foods are prone to spoilage within a short span of time due to the onset of microbial growth feeding on the highly nutritive substrate. The primary approach for extending the shelf-life of such foods is through microbial and enzyme inactivation. Thus, processing methods are considered to reduce the spoilage by providing the necessary lethal effect. Processing methods include thermal and nonthermal treatments, of which a combination of processing techniques proves to be more significant in reducing the microbial population than a single method. PEF is one such technique that reduces the microbial contamination by damaging its cell membrane and DNA under high electric fields. PEF with various conventional and novel techniques is discussed in the chapter on providing the necessary hurdle effect to extend the shelf-life of food stuffs.

KEYWORDS

- **death kinetics**
- **electroporation**
- **hurdle**
- **microbial inactivation**
- **preservation**
- **synergistic effect**

REFERENCES

1. Aadil, R. M.; Zeng, X. A.; Han, Z.; Sahar, A.; Khalil, A. A.; Rahman, U. U.; Khan, M.; Mehmood, T. Combined Effects of Pulsed Electric Field and Ultrasound on Bioactive Compounds and Microbial Quality of Grapefruit Juice. *J. Food Process. Preserv.* **2018,** *42* (2), e13507.
2. Abid, M.; Jabbar, S.; Hu, B.; Hashim, M. M.; Wu, T.; Wu, Z.; Khan, M. A.; Zeng, X. Synergistic Impact of Sonication and High Hydrostatic Pressure on Microbial and Enzymatic Inactivation of Apple Juice. *LWT - Food Sci. Technol.* **2014,** *59* (1), 70–76.
3. Ahmad, T.; Butt, M. Z.; Aadil, R. M.; Inam-ur-Raheem, M.; Bekhit, A. E. D.; Guimarães, J. T.; Balthazar, C. F.; Rocha, R. S.; Esmerino, E. A.; Freitas, M. Q.; Silva, M.; Sameen, A.; Cruz, A. Z. Impact of Nonthermal Processing on Different Milk Enzymes. *Int. J. Dairy Technol.* **2019,** *72* (4), 481–495.

4. Alirezalu, K.; Munekata, P. E. S.; Parniakov, O.; Barba, F. J.; Witt, J.; Toepfl, S.; Wiktor, A.; Lorenzo, J. M. Pulsed Electric Field and Mild Heating for Milk Processing: A Review on Recent Advances. *J. Sci. Food Agric.* **2020,** *100* (1), 16–24.
5. Ariza-Gracia, M. Á.; Cabello, M. P.; Cebrian, G.; Calvo, B.; Alvarez, I. Experimental and Computational Analysis of Microbial Inactivation in a Solid by Ohmic Heating Using Pulsed Electric Fields. *Innov. Food Sci. Emerg. Technol.* **2020,** *65,* 102440.
6. Astráin-Red'in, L.; Moya, J.; Alejandre, M.; Beitia, E.; Raso, J.; Calvo, B.; Cebrián, G.; Álvarez, I. Improving the Microbial Inactivation Uniformity of Pulsed Electric Field Ohmic Heating Treatments of Solid Products. *LWT* 2022 *154,* 112709.
7. Baron, S., Ed. *Medical Microbiology*, 4th ed.; University of Texas Medical Branch at Galveston: Galveston, TX, 1996.
8. Bechoff, A.; Shee, A.; Mvumi, B. M.; Ngwenyama, P.; Debelo, H.; Ferruzzi, M. G.; Nyanga, L. K.; Mayanja, S.; Tomlins, K. I. Estimation of Nutritional Postharvest Losses Along Food Value Chains: A Case Study of Three Key Food Security Commodities in Sub-Saharan Africa. *Food Sec.* **2022,** *14* (3), 571–590.
9. Ben Ammar, J.; Lanoisellé, J. L.; Lebovka, N. I.; Van Hecke, E.; Vorobiev, E.; Effect of a Pulsed Electric Field and Osmotic Treatment on Freezing of Potato Tissue. *Food Biophys.* **2010i,** *5* (3), 247–254.
10. Calzia, D.; Panfoli, I.; Ravera, S.; Dazzi, E.; Gandolfo, S.; Pepe, I. M.; Vergani, L.; Morelli, A. M. Structural Modification of Proteins by Direct Electric Current From Low Voltage. *J. Biochem. Mol. Toxicol.* **2009,** *23* (5), 309–317.
11. Chen, C.; Smye, S. W.; Robinson, M. P.; Evans, J. A. Membrane Electroporation Theories: A Review. *Med. Biol. Eng. Comput.* **2006,** *44* (1–2), 5–14.
12. Dadan, M.; Nowacka, M.; Czyzewski, J.; Witrowa-rajchert, D. Modification of Food Structure and Improvement of Freezing Processes by Pulsed Electric Field Treatment. In Barba, F. J.; *Pulsed Electric Fields to Obtain Healthier and Sustainable Food for Tomorrow*; Parniakov, O., Wiktor, A., Eds.; Academic Press, 2020; pp 203–226.
13. Dermesonlouoglou, E.; Chalkia, A.; Dimopoulos, G.; Taoukis, P. Combined Effect of Pulsed Electric Field and Osmotic Dehydration Pre-treatments on Mass Transfer and Quality of Air Dried Goji Berry. *Innov. Food Sci. Emerg. Technol.* **2018,** *49,* 106–115.
14. Faisal Manzoor, M.; Ahmed, Z.; Ahmad, N.; Karrar, E.; Rehman, A.; Muhammad Aadil, R.; Al-Farga, A.; Waheed Iqbal, M.; Rahaman, A.; Zeng, X. A. Probing the Combined Impact of Pulsed Electric Field and Ultra-Sonication on the Quality of Spinach Juice. *J. Food Process. Preserv.* **2021,** *45* (5), 1–11.
15. Gomez-Gomez, A.; la Fuente, E.; Gallegos, C.; Garcia-Perez, J. V.; Benedito, J.; Combined Pulsed Electric Field and High-Power Ultrasound Treatments for Microbial Inactivation in Oil-in-water Emulsions. *Food Control.* **2021,** *130,* 108348.
16. Heinz, V.; Knorr, D. Effect of pH, Ethanol Addition and High Hydrostatic Pressure on the Inactivation of Bacillus Subtilis by Pulsed Electric Fields. *Innov. Food Sci. Emerg. Technol.* **2000p,** *1* (2), 151–159.
17. Hyldgaard, M.; Mygind, T.; Meyer, R. L. Essential Oils in Food Preservation: Mode of Action, Synergies, and Interactions with Food Matrix Components. *Front. Microbiol.* **2012,** *3,* 1–24.
18. Jäger, H. *Process Performance Analysis of Pulsed Electric Field (PEF) Food Applications.* Dissertation, Fakultat III—Prozesswissenschaften der Technischen Universitat, Berlin, 2013.

19. Kang, T.; Lee, D.; Ko, Y.; Jun, S. Effects of Pulsed Electric Field (PEF) and Oscillating Magnetic Field (OMF) on Supercooling Preservation of Beef at Different Fat Levels. *Int. J. Refrig.* **2022,** *136,* 36–45.
20. Kantono, K.; Hamid, N.; Ma, Q.; Oey, I.; Farouk, M. Changes in the Physicochemical Properties of Chilled and Frozen-Thawed Lamb Cuts Subjected to Pulsed Electric field processing. *Food Res. Inter.* **2021,** *141,* 110092.
21. Kotnik, T.; Frey, W.; Sack, M.; Haberl Meglič, S.; Peterka, M.; Miklavčič, D.; Electroporation-Based Applications in Biotechnology. *Trends Biotechnol.* **2015,** *33* (8), 480–488.
22. Li, J.; Shi, J.; Huang, X.; Zou, X.; Li, Z.; Zhang, D.; Zhang, W. Effects of Pulsed Electric Field on Freeze-Thaw Quality of Atlantic Salmon. *Inov. Food Sci. Emerg. Technol.* **2020,** *65,* 102454.
23. Li, L.; Yang, R.; Zhao, W. The Effect of Pulsed Electric Fields (PEF) Combined with Temperature and Natural Preservatives on the Quality and Microbiological Shelf-Life of Cantaloupe Juice. *Foods* **2021,** *10* (11), 2606.
24. Mannozzi, C.; Fauster, T.; Haas, K.; Tylewicz, U.; Romani, S.; Dalla Rosa, M.; Jaeger, H. Role of Thermal and Electric Field Effects During the Pre-Treatment of Fruit and Vegetable Mash by Pulsed Electric Fields (PEF) and Ohmic Heating (OH). *Innov. Food Sci. Emerg.Technol.* 2018 *48,* 131–137.
25. Martín-belloso, O.; Sobrino-lópez, A. Combination of Pulsed Electric Fields with Other Preservation Techniques, **2011,** *4*, 954–968.
26. Mok, J. H.; Choi, W.; Park, S. H.; Lee, S. H.; Jun, S. Emerging Pulsed Electric Field (PEF) and Static Magnetic Field (SMF) Combination Technology for Food Freezing. *Int. J. Refrig.* **2015,** *50,* 137–145.
27. Mok, J. H.; Her, J. Y.; Kang, T.; Hoptowit, R.; Jun, S.; Effects of Pulsed Electric Field (PEF) and Oscillating Magnetic Field (OMF) Combination Technology on the Extension of Supercooling for Chicken Breasts. *J. Food Eng.* **2017,** *196,* 27–35.
28. Montanari, C.; Tylewicz, U.; Tabanelli, G.; Berardinell, A.; Rocculi, P.; Ragni, L.; Gardini, F. Heat-Assisted Pulsed Electric Field Treatment for the Inactivation of Saccharomyces Cerevisiae: Effects of the Presence of Citral. *Front. Microbiol.* **2019,** *10,* 1737.
29. Noci, F.; Riener, J.; Cronin, D. A.; Morgan, D. J.; Lyng, J. G. Ultraviolet Irradiation and Pulsed Electric Fields (PEF) in a Hurdle Strategy for the Preservation of Fresh Apple Juice. **2008,** *85,* 141–146.
30. Noci, F.; Walkling-Ribeiro, M.; Cronin, D. A.; Morgan, D. J.; Lyng, J. G. Effect of Thermosonication, Pulsed Electric Field and Their Combination on Inactivation of Listeria Innocua in Milk. *Int. Dairy J.* **2009,** *19* (1), 30–35.
31. Novickij, V.; Stanevičienė, R.; Staigvila, G.; Gruškienė, R.; Sereikaitė, J.; Girkontaitė, I.; Novickij, J.; Servien, E. Effects of Pulsed Electric Fields and Mild Thermal Treatment on Antimicrobial Efficacy of Nisin-loaded Pectin Nanoparticles for Food Preservation. *LWT - Food Sci. Technol.* **2020,** *120,* 108915.
32. Pillet, F.; Formosa-Dague, C.; Baaziz, H.; Dague, E.; Rols, M. P. Cell Wall as a Target for Bacteria Inactivation by Pulsed Electric Fields. *Sci. Rep.* **2016,** *6* (1), 1–8.
33. Shiekh, K. A.; Benjakul, S.; Qi, H.; Zhang, B.; Deng, S. Combined Hurdle Effects of Pulsed Electric Field and Vacuum Impregnation of Chamuang Leaf Extract on Quality and Shelf-Life of Pacific White Shrimp Subjected to High Voltage Cold Atmospheric Plasma. *Food Packag. Shelf Life.* **2021,** *28,* 100660.

34. Takaki, K.; Takahashi, K.; Guionet, A.; Ohshima, T. Pulsed Power Applications for Protein Conformational Change and the Permeabilization of Agricultural Products. *Molecules* **2021,** *26* (20), 6288.
35. Tankeshwar, A. Bacterial Spores: Structure and Spore-Forming Bacteria [Online] 2022. https://microbeonline.com/bacterial-spores (accessed on 01 August 2023).
36. Taylor, P.; Won, Y.; Min, S. C.; Lee, D.; Won, Y.; Min, S. C.; Lee, D. Accelerated Drying and Improved Color Properties of Red Pepper by Pretreatment of Pulsed Electric Fields Accelerated Drying and Improved Color Properties of Red Pepper by Pretreatment of Pulsed Electric Fields. *Drying Technol.* **2014,** 37–41.
37. Vanga, S. K.; Wang, J.; Jayaram, S.; Raghavan, V.; Effects of Pulsed Electric Fields and Ultrasound Processing on Proteins and Enzymes: A Review. *Processes* **2021,** *9* (4), 1–16.
38. Velickova, E.; Tylewicz, U.; Dalla, M.; Winkelhausen, E.; Kuzmanova, S.; Romani, S. Effect of Pulsed Electric FiEld Coupled with Vacuum Infusion on Quality Parameters of Frozen/ Thawed Strawberries. *J. Food Eng.* **2018,** *233,* 57–64.
39. Verma, M. K.; Shakya, S.; Kumar, P.; Madhavi, J.; Murugaiyan, J.; Rao, M. V. R. Trends in Packaging Material for Food Products: Historical Background, Current Scenario, and Future Prospects. *J. Food Sci. Technol.* **2021,** *58* (11), 4069–4082.
40. Walkling-Ribeiro, M.; Noci, F.; Cronin, D. A.; Riener, J.; Lyng, J. G.; Morgan, D. J. Reduction of Staphylococcus Aureus and Quality Changes in Apple Juice Processed by Ultraviolet Irradiation, Pre-heating and Pulsed Electric Fields. **2008,** *89,* 267–273.
41. Walkling-Ribeiro, M.; Noci, F.; Riener, J.; Cronin, D. A.; Lyng, J. G.; Morgan, D. J. The Impact of Thermosonication and Pulsed Electric Fields on Staphylococcus Aureus Inactivation and Selected Quality Parameters in Orange Juice. *Food Bioproc. Technol.* **2009,** *2* (4), 422–430.
42. Wang, F.; Li, L.; Li, X.; Hu, X.; Zhang, B. Pulsed Electric Field Promotes the Growth Metabolism of Aerobic Denitrifying Bacteria Pseudomonas Putida W207-14 by Improving Cell Membrane Permeability. *Environ. Technol.* **2022,** *44* (15), 2327–2340.
43. Wiktor, A.; Witrowa-Rajchert, D. Pulsed Electric Fields as Pretreatment for Subsequent Food Process Operations. In *Handbook of Electroporation*; Miklavcic, D., Ed.; Springer International Publishing: Cham, Switzerland, 2016; pp 1–16.
44. FAO (Food and Agriculture Organization of the United Nations). *World Food and Agriculture—Statistical Yearbook 2020*; FAO: Rome, Italy, 2020.
45. Xin, Q.; Zhang, X.; Li, Z.; Lei, L.; Sterilization of Oil-Field Re-injection Water Using Combination Treatment of Pulsed Electric Field and Ultrasound. *Ultrason. Sonochem.* **2009,** *16* (1), 1–3.
46. Yeom, H. W.; Zhang, Q. H. Enzymatic Inactivation by Pulsed Electric Fields: A Review. *Pulsed Electric Fields Food Proc.* **2019,** 57–63.

CHAPTER 5

APPLICATIONS AND FUTURE ASPECTS OF PULSED ELECTRIC FIELD ON FRUIT PRODUCTS

PRIYA DANGI MALIK, PREETI KHANNA, SAUMYA JAIN, and VANDANA SABLANIA

ABSTRACT

The development of nonthermal technologies such as pulsed electric field (PEF)/high-intensity pulsed electric fields (HIPEFs) has been encouraged by increased customer concern for wholesome and fresh food products, as well as the hunt for ecologically benign processing solutions. Consumer awareness has increased over a while for the consumption of preserved food as fresh food cannot be available all the time. Thermal treatment might reduce their nutritional value sometime after treatment. However, it becomes essential to introduce nonthermal preservation techniques to prevent the commodity from adverse effects under high-temperature range. Hence, PEF is a nonthermal processing technique that has recently been investigated as a potential replacement for traditional heat treatments. PEF includes the use of very short electric pulses (ms or s) with high electric field strengths (kV/cm), and enough temperature is the basis of PEF technology. During the last decade, the applications of PEF/HIPEF technology have increased exponentially. The pulsed electric field is an innovative, nonthermal technique that serves a crucial role in the extraction of biopolymers from plant sources, inactivation of spoilage-causing microorganisms and enzymes, etc. PEF is used to improve the permeabilization of fruits or vegetable tissues to improve the mass transfer of vital compounds. This chapter extensively deals with the applications of PEF in various fruits. PEF might be used as

Nonthermal Pulsed Electric Field Technology for the Food Industry.
Preeti Birwal, Megh R. Goyal, Baldev Singh Kalsi, & Ankita Kataria (Eds.)

DOI: 10.1201/9781003605904-7

an alternative processing technique for the destruction of microorganisms and inactivation of enzymes with maximum retention of physicochemical, sensorial, and nutritional characteristics of the food.

5.1 INTRODUCTION

Over a while, pulsed electric field (PEF) has become one of the most important nonthermal processing techniques for numerous fruits and vegetables in terms of maintaining the qualitative properties and sensorial attributes. PEF primarily focuses on the mobilization and electroporation of cell lamina and hence contributes to the mass transfer of the cells. The PEF technology is frequently used to treat liquid items like fruit/fruit juices, milk and milk products, liquid eggs, and alcoholic beverages to inhibit the growth of undesirable microorganisms. Food dehydration, sterilization, extraction, pesticide residue reduction, and enzyme inactivation are some of the other applications of this technique. The rate of microbial and enzyme inactivation by PEF is a determinant of the electric field strength and processing time, the nature of the pulse, its duration, and frequency. The electric pulses can range between 1 and 2000 pulses per second with a period ranging from microseconds to milliseconds.[97] Sometimes, PEF also includes the usage of a high electric field for a short duration.[1,117] Electric conductivity shows significant characteristics for maintaining the conductivity between the microbial cell and the juice extracts of fruits and vegetables. Subsequently, it weakens the cell membrane and helps in the enhancement of PEF effectiveness for the better shelf-life of the food.[68]

PEF usually helps in improving the product yield, diffusion rate, dehydration rate, and reducing the microbial load to ensure the consumption of safe food without altering the nutritional composition of the products.[15,53] Electroporation is generally responsible for the pore formation over the cell membrane due to the prompted threshold potential at a shorter transition period. The process is dynamic and not instantaneous. When the potato was initially exposed to PEF, it resulted in the discharge of liquid from the tissues, which is due to the cellular damage of the membrane.[14] A large number of studies showed a significant effect of PEF on liquid products. However, some studies have been reported on meat, seafood, and some of the plant-based commodities for improving their quality. PEF treatment results in reducing the activity of infective microorganisms such as *E. coli, L. innocua, S. aureus, P. expansum*, and enzymes such as pectin methylesterase, polyphenol oxidase, peroxidase, etc.[37,145] Amiali et al.[13] reported that exposure to cells under high

PEF momentarily destabilized the phospholipid bilayer and proteins. The pores regain their shape as soon as the electric field is removed. On the other side, if the intensity is very high, then it leads to a complete breakdown of the cells, and they undergo lysis. Several factors, like the shape and size of the cell, characteristics, and growth rate of the microorganisms, cause lysis of cells and tissues. It can be used as one of the important pretreatments to develop qualitative products concerning taste, texture, yield, antioxidant activity, flavors, etc.[27]

Charles-Rodriguez et al.[34] observed a drastic increase in the pH of the apple juice when treated by using thermal processing methods such as pasteurization, sterilization, etc., whereas the change was not reported when the same juice was processed under the influence of PEF even with increasing temperature. PEF has resulted in an enhanced release of anthocyanins and bioactive substances in grapes. PEF is considered to be an alternative method for the production, processing, and better shelf-life of fruit juices under low-temperature usage. Fruit juices are largely known for their high nutritive content, sensory attributes, and healthy and functional properties.[61] The polyphenolic content of grape pomace was found to be increased significantly when treated with PEF during the degradation of grapes during wine preparation, thus improving not only the sensory quality but also the beneficial antioxidant activity as well. The utilization of PEF for the production of healthy and new products from fruits and vegetables has been increased expressively. The quality and freshness of fruit and vegetable juices might be protected if PEF treatment is used optimally. When orange juice was treated with PEF at the rate of 100–120 kJ/kg energy, it resulted in the reduction of microbial growth by 5 times. A combination of multiple techniques, such as high or low temperature, in association with PEF can minimize the energy requirement for preservation.[85] The cells of the fruit or vegetable tissue maintain the ability to repair the pores that are formed on the application of PEF treatment and thus remain viable and active. High-quality food or food ingredients can be produced with the execution of appropriate conditions during PEF treatment. This effect has been employed in many commodities, especially fruits, which will be discussed further. Implementing PEF treatment in the food industry for plant or cell cultures results in enhancing the production of useful metabolites.[104] In comparison to other extraction techniques, such as solvent extraction, ohmic extraction technique, maceration, etc., PEF is a favorable technique for supplementing the abstraction of bioactive components from numerous plant sources. A large number of pulses can be used to improve the extraction rate over a shorter period. In addition, the selective and optimum PEF-based extraction permits the appropriate extraction of various nutrients such as carbohydrates, proteins, lipids, and vitamins.[43] However, PEF efficiency is presently lesser than other traditional methods in terms of total yield and energy input.[103]

The purpose of this chapter is mainly to understand the possible use of the pulsed electric fields to maintain the quality and shelf-life of various fruits and fruit juices without compromising the nutritional and sensory quality of the products.

5.2 FOOD PRESERVATION BY PEF

Contrary to conventional techniques such as pasteurization and sterilization, PEF is one such novel processing technology that works by lowering the temperature and reducing the exposure time to ensure good quality and safer products. PEF technology makes use of electric field strength ranging from 15 to 40 kV/cm for a duration of 20 to 2000 μs depending upon the target cell.[10,133] Usually, an electric field strength for industrial applications (ranging from 10 to 20 kV/cm)[132] is of a lower order than that utilized in research studies to equilibrate the efficacy and costs of the treatment.[69] PEF is coming across as one of the gentle novel processing technologies owing to its shorter exposure time, minimal heating effects, and environmentally friendly nature. On a commercial scale, PEF-assisted pasteurization has been achieved; however, PEF-assisted sterilization is still under testing room.[64,111,112,136] Despite several advantages, Aganovic et al.[3] speculated that PEF technology is energy-driven, which leads to inflation in the price of the commodity.

5.2.1 MICROBIAL INACTIVATION BY PEF

In the course of PEF processing, the food sample is planted amid two electrodes, and an electric current is supplied in the form of pulses of varying duration.[28,63] PEF works by creating pores within the membranes of target cells on exposure to the field intensity, as reported first by Sale and Hamilton.[119] It is now well-established that exposing the target cell to a high-magnitude external electrical field for a short duration of time leads to the development of charges on the cell membrane, which later induces a corresponding membrane potential.[72] The extent of transmembrane potential generated on the membrane depends upon the electric field (concerning its strength and orientation) and the nature of cells.[62] The various modes in which the electric field can be applied are exponential decay, bipolar or oscillatory, triangular, unipolar pulses, and the square waves. The bottom-line value of the electrical field needed to attain the transmembrane potential is referred to as critical field intensity, and its exceeding values induce electroporation in the cell.[10] This minimum electric field value stretches from 4 to 14 kV/cm.[33] Often, a value of >10 kV/cm electric field strength is a prerequisite to induce pore formation in microbial cells of size 1–10 μm, and <5 kV/cm is sufficient for eukaryotic plant cells of size varying between 40–200 μm.[110] As a result of electroporation, there is a reversible or irreversible disruption in the integrity of the cell membrane, which allows the molecular exchange between the cellular material and external medium. Subsequently, the microbial cells fail to sustain homeostasis and ultimately inactivate at temperatures lower than those used in conventional heat processing.[10,42]

Contemporary reports have indicated that determinants like amplitude of the electric field, numbers, shape of pulses applied, treatment temperature and time, type and growth stage of microorganisms, an initial count of microorganisms, concentration of ions, and conductivity of the solution have a major impact on the inactivation of microorganisms during PEF treatment.[62,100,134,140] Furthermore, it was reported that the rate of membrane permeabilization and microbial inactivation rises rapidly with an escalation in the electric field strength only after the cell has reached its critical transmembrane potential.[83] Moreover, the effects of PEF on microbial inactivation were more pronounced with an increase in exposure time, specifically the numbers and width of pulses.[21,66] Consecutively, a linear dependency was highlighted among the log numbers of inactivated cells and the permeabilized cells, typically to levels of 3.6 log reduction on the application of the electric field of 12 or 15 kV/cm in date palm fruits.[147]

Fruits and fruit-based beverages are characterized by low pH values and higher sugar levels, which provides a suitable environment for the growth and proliferation of yeast than other microbial entities.[23,39] *Saccharomyces cerevisiae* is one of the species commonly involved in the degradation of such products by carrying out the fermentation process and evolving carbon dioxide and ethanol. This carbon dioxide accumulation may cause package blowing, and therefore, it is essential to avoid the proliferation of yeast in these products.[127] Henceforth, PEF, by its ability to kill yeast cells at lower temperatures and ensuring heat-sensitive compounds are well-maintained in the product, is of high potential to the fruit industry.[84] By virtue of their larger cell size, yeast cells are often more sensitive to cellular damage than cells of bacteria.[106] Jeyamkondan et al.[67,148] reported that subjecting molds and yeasts to PEF for 39 and 49 ms, respectively, brings about a reduction of 3.3 log in their total counts. In most of the treated samples of dates, it was inferred that subjecting the samples to 120 pulses of 10.82 kV/cm of electric field strength brought down bacterial count to unnoticeable levels from the initial count of 1.17×104 CFU/g. In addition, a significant reduction from the initial count of 3.27×103 CFU/g was found in the yeast and mold population when exposed to 90 pulses of 8.84 kV/cm of electric field. Table 5.1 presents an account of the effects of PEF treatment on various fruit/fruit juices in terms of microbial inactivation.

PEFs alone or in combination with various parameters such as mild heating, lowering of pH, use of antimicrobial compounds, essential oils, and their constituents, etc. have been explored so far to achieve higher antimicrobial effects.[20,29,74] In a study carried out by Aronsson and Rönner,[16] the inactivation of *S. cerevisiae* cells was compared at two different pHs. It was reported that the sample with water activity near pH 1 and pH 4, when subjected to PEF treatments at 25 kV/cm, underwent a reduction in the microbial count by 3.8 log units, whilst the sample with higher pH recorded a decrease in cell count of about 1.8–2.9 log units. Conclusively, PEF is a promising nonthermal technique that can be utilized successfully for reducing the growth of microorganisms in a number of fruits.

5.2.2 ENZYME INACTIVATION BY PEF

Apart from microorganisms, enzymes that are naturally present in fruits may cause several objectionable changes that may severely impact the color, flavor, and texture of the product. Thus, it is necessary to inhibit endogenous enzymes like polygalacturonase, pectin methylesterase, peroxidase, lipoxygenase, polyphenol oxidase, and many others for effective processing and preservation of food. Presently, PEF is under exploitation for inactivating enzymes from laboratory to industrial scale.[160] Given that the precise mechanism of the inactivation of enzymes is not clear, it is suggested that the beginning of the process is marked by the unfolding of the protein chains as a result of the imbalance of forces existing in a polypeptide chain. Any alterations in the covalent or noncovalent linkages brought about structural/conformational changes in the protein structure and resulted in reshaping the primary/secondary/tertiary/quaternary structure of the enzymes, respectively.[159] In a molecular study carried out by Ohshima et al.,[94] SDS-PAGE analysis of enzymes like peroxidase, β-galactosidase, and lactate dehydrogenase was carried out before and after PEF treatment. No apparent change in molecular size of each enzyme was recorded after PEF treatment, reflecting that no proteolysis occurs during PEF processing. Furthermore, the secondary

TABLE 5.1 Role of PEF for Microbial Inactivation in Various Fruits and Fruit Juices.

Food product	Experimental conditions	Major findings	References
Apple juice	E = 10.2 and 30 kV/cm Number of pulses: 100	2 log cycle reduction at 10.2 kV/cm and 5 log reduction at 30 kV/cm in yeast population.	[52]
Pineapple juice	E = 9–13 kV/cm Pulse type: square waveform pulse Frequency: 1000 Hz	At 9 kV/cm electric field strength, a decline in TPC was observed from an initial 150×10^3 CFU/ml to 120×10^3 CFU/ml, which further dropped to 35×10^3 CFU/ml with further increase of intensity to 13 kV/cm.	[154]
Orange juice	E = 35 kV/cm Pulse type: bipolar Frequency: 200 Hz P = 4 micros	*Saccharomyces cerevisiae* population was reduced by 5.1 log cycles.	[48]
	E = 10, 15, and 20 kV/cm Pulse type: bipolar P = 25 μs	1.4, 3, and 6.4 decimal reductions in the *E. coli* population were observed at 10 kV/cm and specific energy inputs of 50, 100, and 150 kJ/L, respectively. 20 kV/cm completely inactivated *E. coli* at fastest rate.	[75]
Strawberry juice	E = 18.5 kV/cm P = 500 μs Frequency: 100 Hz	Significant reductions in the total bacterial load, psychrophilic bacteria, and mold and yeast counts were observed as an effect of PEF.	[44]
Date palm	E = 10.82 kV/cm Number of pulses: 120	In most treated samples, total mesophilic aerobic bacterial counts were reduced to <10 CFU/g from the initial count of 1.18×10^4 CFU/g.	[87]
	E = 8.84 kV/cm Number of pulses: 90	The yeast and mold count declined to < 10 CFU/g from the initial count of 3.27×10^3 CFU/g.	
	E = 10.82 kV/cm Number of pulses: 60	Absence of yeasts and molds.	

(Here, E = electric field strength, P = pulse width, and CFU = colony-forming unit.)

structure is comprised of α-helix and β-sheet. Owing to the arrangement of amino acids in the α-helical structure, a strong dipole moment is generated with a positive pole aligned at the N-terminus and a negative pole toward the C-terminus in the structure under the effect of an electric field.[160] This has caused an upsurge in the dielectric constant of protein chains and ultimately leads to their uncoiling. Consequently, the loosening of protein structure may further cause dissociation of the quaternary structure. Concurrently, too many thiol and hydrophobic groups are exposed, which are associated with the conformational changes in the active site of enzymes, thereby preventing the substrate–enzyme reaction and reducing the enzyme's activity.[95] Several researchers have firmly established the effects of PEF on the α-helix structure of the enzyme. One such study was carried out on papain enzyme, which, when exposed to 50 kV/cm for 2 ms, undergoes an irreversible disruption in the α-helical region and loses its activity.[151] Subsequently, the same approach was applied to the horseradish peroxidase enzyme by Zhong et al.,[161] and it was revealed that there is a decrease in α-helix relative amount by 35.1% on being subjected to 22 kV/cm for 130.5 μs. Later on, Zhong et al.[162] compared the secondary structure and inactivation kinetics of PEF-treated enzymes. It was noteworthy that on treatment with PEF, peroxidase was more stable than the polyphenol oxidase enzyme owing to the higher stability of its α-helical structure. Table 5.2 summarizes the effects of PEF on various degradative enzymes in fruit/fruit juices.

Enzyme inactivation by PEF depends on treatment conditions (like strength, duration of an electric field, pulse width and frequency, wave shape), enzyme characteristics, and the suspended medium.[82] Aguiló-Aguayo et al.[8] and Marsellés-Fontanet and Martín-Belloso[81] reported an enhancement in enzyme inactivation on the application of greater electric field strength for a longer period. Giner et al.[59] studied the effects of varying pulse width on the inactivation of the polyphenol-oxidase enzyme in peach. It was reported that applying pulse width for about 20 ms and 80 ms led to the inactivation of PPO enzymes by less than 5% and 10%, respectively. Similar results were seen for the peroxidase enzyme, where the exposure to the pulse width of 10 ms and field intensity of 35 kV/cm for 600 ms leads to maximum inactivation of 97%. Furthermore, the enzyme inactivation boosts up with a surge in pulse rate. For example, the residual peroxidase activity in orange juice was cut down to 6.9% when subjected to a pulse frequency of 450 Hz (35 kV/cm, 600 ms).[47] Besides these factors, pulse polarity is an ambiguous factor that influences the efficiency of PEF in inactivating enzymes. For different enzymes, varied results have been reported as in orange juice, when treated with a monopolar and bipolar pulse, showcased greater reduction

TABLE 5.2 Utilization of PEF Processing for Inactivation of Enzyme Activity in Fruits and Fruit Juices.

Fruit/fruit juices	Treatment conditions	Effect of PEF on quality	References
Apple juice	Preheating to 40°C E = 30 kV/cm Energy: 100 kJ/kg	Deactivation of polyphenol oxidase (PPO).	[123]
	E = 12.5 kV/cm Energy: 76.4 kJ/kg Frequency: 62 Hz	Maximum deactivation of polyphenol oxidase and peroxidase activity.	[145]
	E = 24.6 kV/cm Frequency: 10 Hz P = 40 μs No. of pulses: 1000	Decrease in polyphenol oxidase activity by 32%.	[143]
Orange juice	E = 35 kV/cm P = 59 μs	88% reduction in pectin methyl esterase (PME) activity.	[148]
	E = 25 kV/cm P = 330 μs Outlet temperature: 72°C	90% reduction in PME activity.	[149]
	E = 25 kV/cm P = 250 μs Outlet temperature: 64°C	90% inactivation of PME activity.	[124]
	E = 35 kV/cm P = 1500 μs Outlet temperature: 37.5°C	78.1% reduction in PME activity.	[47]

TABLE 5.2 *(Continued)*

Fruit/fruit juices	Treatment conditions	Effect of PEF on quality	References
Orange and carrot juice	E = 25 kV/cm P = 340 μs Temperature: 63°C	81.38% inactivation of pectin methyl esterase activity.	[114]
	E = 24 kV/cm P = 89 μs Temperature: 49°C	14% inactivation of PME activity.	[30]
Tomato juice	E = 24 kV/cm P = 0.02 ms No. of pulses: 400	93.8% inactivation of pectin methyl esterase activity.	[58]
Pear juice	E = 22.3 kV/cm P = 0.02 ms	62% decrease in polyphenol oxidase (PPO) activity.	[57]
Peach	E = 24 kV/cm P = 8 ms	Reduction in PPO activity by 30%.	[59]
Carrot juice	E = 15 kV/cm P = 600 μs Frequency: 5 Hz Energy: 309 kJ/kg	83% inactivation of pectin methyl esterase activity.	[150]
Watermelon juice	E = 35 kV/cm P = 1727 μs Frequency: 188 Hz	Reduction in polygalacturonase (PG) activity by 86.4%.	[7]

TABLE 5.2 *(Continued)*

Fruit/fruit juices	Treatment conditions	Effect of PEF on quality	References
Mixed juice (orange, kiwi, mango, and pineapple) along with whole milk	E = 35 kV/cm Pulse width: 4 μs Frequency: 200 Hz	More effective in reducing PME activity at the rate of 87.33%.	[121]
Strawberry juice	E = 35 kV/cm P = 1700 μs Frequency: 100 Hz Temperature: 90°C Process time: 30 s	Reduction in PME and PG activity by 48.8% and 96.8 %, respectively.	[5]
Tomato juice	E = 35 kV/cm P = 1500 μs Frequency: 100 Hz Temperature: 90°C Process time: 30 s, 60 s	78% and 56% reduction in residual PG activity for a period of 30 s and 60 s, respectively.	[6]
Grapefruit juice	E = 20 kV/cm P = 25 μs Temperature: 25°C	59% reduction in PME activity.	[113]
	E = 30 kV/cm P = 100 μs Temperature: 60°C	90% reduction in PME activity.	

(Here, E = electric field strength and P = pulse width.)

of pectin methylesterase activity in the latter type.[49] Similarly, Giner et al.[59] documented the better effectiveness of bipolar pulse in comparison to monopolar pulse in inactivating the polyphenol oxidase enzyme in peach. However, orange juice containing peroxidase enzyme, when exposed at 25 kV/cm in monopolar and bipolar mode, indicated that a monopolar wave is more effective than a bipolar wave. If the intensity applied is increased to 35 kV/cm, it leads to higher levels of enzyme inactivation in bipolar mode.[47] Another peer factor that is of great importance is the nature of the enzyme. Although there is a lack of knowledge in this regard, it has been observed that certain enzymes like peroxidase, polyphenoloxidase, and pectin methylesterase exhibited higher degrees of inactivation (>80%), and enzymes like polygalacturonase, lipoxygenase, and glucosidase may undergo maximum inactivation until 50%.[160]

5.3 PEF AS A PREPROCESSING STEP

5.3.1 PEF AS A PRETREATMENT TO DRYING

Drying is a commonly used preservation method for lengthening the stability of food products by reducing their water activity due to which microorganisms may not get favorable conditions for their growth. A large number of drying techniques have been included to increase the production of dried or dehydrated products. Sometimes drying results in loss of nutritional composition due to high-temperature processing. In the current scenario, PEF has been dragged into drying or dehydration techniques to retain the utmost quality of the products.[88] PEF, by improving the permeabilization of cells of various fruits, results in an increased rate of mass and heat transfer among the cells and external environment and therefore is used as a preprocessing method before drying any fruits. Ostermeier et al.[98] observed disintegration in the cell membrane of onions with an increase in the electric field up to 1.07 kV/cm. Thus, it facilitated moisture removal at a faster rate and reduced the drying time by 30%. It was also reported that PEF-treated basil leaves took a lesser period in drying when compared with the untreated sample.[129] Tylewicz et al.[139] have observed in their studies that the permeabilization of the cell membrane of strawberries was improved on exposure to a high-intensity (1 kV/cm) electric field. This change might have occurred due to the active participation of various enzymes such as polyphenol oxidase, peroxidase, etc. Moreover, Orikasa et al.[96] stated that a significant change in the color of the kiwifruit was observed when treated with PEF at 60°C, which could be due to suppressed nonenzymatic activity during drying. Further, plums treated with PEF showed a decrease in lightness value with increasing intensity of electric pulses. Some studies have also reported a change in the rate of the water diffusion coefficient in samples treated with PEF before drying. Rahaman et al.[109] reported that plums treated with PEF showed an increased water diffusion rate of up to 60% as compared to untreated plums. A study conducted by Yu et al.[155] on dehydrated blueberries also suggested the change in water diffusion rate from 0.94 to 1.07 10^{-9} m^2/s by changing the intensity of PEF. The water diffusion rate depends on several parameters such as the designing tool, the intensity of pulses, dehydration time–temperature combination, composition, and porosity of the commodity. Goji berry pretreated with PEF was followed by osmotic dehydration and freeze-drying. In both the drying methods, goji berries experienced a large amount of cell disintegration, resulting in high solid gain and more water removal.[41] In addition to this, the antioxidant activity of goji berries was also increased when treated with 750 and 1500 pulses as compared to untreated or fresh samples. This could be presumably due to the modification in its structure in terms of moisture loss and increased solid uptake. The penetration of PEF through the waxy layer of goji berry has reduced its surface hardness and improved the drying rate.[41] Alongside, PEF-treated fruits showed lower volume reduction when accompanied by the freeze-drying process. However, limited benefits from PEF treatment can be acquired when freezing of the fruit is done at a higher temperature. This itself is a limitation, but the fact that PEF-treated samples frozen at high temperature showed better rehydration in comparison to the frozen samples not treated with PEF. After a mere 60 min immersion of PEF-treated samples, the rehydration rate improved from 22.1% to 54.0% higher as compared to the untreated one. This is because PEF induces a larger number of pores in the membranes of the treated fruit and thus favors the capillary diffusion, which helps in increasing the rehydration capacity of the PEF-treated samples. Interestingly, PEF-treated samples had a softer texture due to increased pore formation that caused less shrinkage and higher porosity of the dried product.[35] Also, PEF-treated products remained firm during the storage period. In the case of drying, generally, the efficiency was improved in the range of 40–70% by PEF. Samples treated with PEF have improved water removal characteristics during air drying or osmotic drying for numerous fruits and vegetables like bananas, pears, pumpkin, etc.[2] Toepfl and Knorr[134] have observed that implying PEF as a pretreatment leads to reduced usage of total energy due to improved porosity of cell walls on exposure to PEF, which reduces the drying time. Table 5.3 shows the PEF processing conditions and their effects on various fruits and fruit juices.

TABLE 5.3 Effects of PEF on Polyphenolic and Bioactive Components of Various Fruits.

Fruits	PEF process conditions	Observations	References
Grapes	E = 0.75 kV/cm Energy: 20 kJ/kg Duration: 45 min	Increased and improved quality of juice, Improved antioxidant activity.	[70]
Apple	E = 1 to 2 kV/cm No. of pulses: 25–75 Blanching temperature: 60 to 80°C	Optimized conditions for better permeabilization were obtained at 1.25 kV/cm with 50 pulses and a blanching temperature of 50°C.	[35]
	E = 0.5 to 2 kV/cm Energy: 0.6–9.6 kJ/kg	Increased permeabilization of the cell membrane with an increase in electric strength and pulse.	[128]
	E = - 0.90 kV/cm No. of pulses: 750 P = 100 μs Energy: 13.5 kJ/kg	Induced cell permeability, acceleration of pore formation for easy transportation of solutes and water.	[12]
Plum	E = 1 to 3 kV/cm Energy: 10–90 kJ/kg	Higher cell disintegration, increase in drying rate, reduced drying period, and increased water diffusion rate from 0.27 to 16.47×10^{-9} m^2/s.	[109]
Goji berry	E = 0.9 to 2.8 kV/cm Up to 7500 pulses	Large tissue permeabilization, improved mass transfer during osmotic dehydration	[41]
Citrus fruits (orange and lemon)	E = 3 kV/cm P = 230 μs	Increase in yield of orange juice by 25%, 59% for lemon juice, with improved extraction of polyphenolic content.	[46]
Blueberry	E = 1, 3, and 5 kV/cm Energy: 10 kJ/kg	Increase in yield by 28%, improvement in antioxidant activity, anthocyanin content, and phenolic content.	[26]
	E = 2 kV/cm	Reduction in drying time from 6 to 4 h at 75°C, 10 to 7 h at 60°C, and 70 to 40 h at 45°C, retention of nutritional composition.	[156]
Strawberry	E >0.1 kV/cm	Reduction in the lightness of the product, activation of peroxidase and polyphenol oxidase during the electroporation of cell membrane.	[139]

(Here, E = electric field strength and P = pulse width.)

5.3.2 PEF AS A PRETREATMENT TO FREEZING

Freezing is also one such imperative processing method, commonly used for the preservation of various food commodities. Nucleation during freezing is a major disadvantage responsible for causing destruction to the cell membrane and tissues and subsequently affects the shape, structure, texture, and sensory attributes of the products.[104] Being a nonthermal technique, PEF helps to keep food products at a lower temperature to control excessive deterioration. It was reported that the inclusion of PEF before freezing has augmented an increase in product yield, improved quality of the product, retention of nutritional value, reduced the losses during processing, etc. PEF reduces the formation of large-sized ice crystals, thus resulting in better product quality after thawing. PEF maintains the lower temperature at the center of the commodity and prevents structural losses and the shape of the treated sample. This also helps in controlling the minimum volume loss after freeze-drying the PEF-treated products.[54]

Freezing or freeze-drying combined with the application of PEF technology affects freezing time and rate. A study by Jalté et al.[65] observed an improvement in the quality attributes of the freeze-dried potato that was given a pretreatment with PEF. This could be due to a decrease in freezing time, an improvement in the rate of freeze-drying, and thus enhanced the quality of the same. According to Wiktor et al.,[146] treatment of apple tissues with PEF before freezing resulted in a significant change in thawing, and the thawing rate was found to be reduced up to 71.5%. It was reported that as PEF induced electroporation in multicellular tissues, a greater association between intra- and intercellular content was seen, which allowed a better possibility of ice crystal nucleation and quicker ice propagation after freezing and thus shortened the freezing time.[89] Additionally, PEF treatment preserved the color of thawed fruits.[144] Nevertheless, PEF is one of the versatile techniques used for preserving volatile compounds, thus helping in allowing color and flavor to be retained when compared to HTST and LTLT pasteurization methods.[71] PEF processing in the citrus industry leads to the destruction of microorganisms and controls the development of off-flavors during storage.[126] PEF treatment also improves the quality of the processed fruit by freeze-drying. A study conducted on freeze drying of apples after subjecting them to PEF treatment signified good retention of the original shape, less shrinkage, and better rehydration capacity of the product. In contrast, control freeze-dried samples induced deformity and contraction of the macroshape apart from the homogenization of the fruit juice and the absence of visible micropores.[104]

5.4 PEF AS A PROCESSING AID FOR THE EXTRACTION OF BIOACTIVE COMPOUNDS

Bioactive compounds are the naturally occurring constituents that possess therapeutic potential and are available in minute quantities in foods such as whole grains, fruits, and vegetables. Such compounds are gaining wide popularity these days for their exploitation in the food, pharmaceutical, cosmetic industries, and many others. Numerous conventional processing techniques (boiling, grinding, heat refluxing, hydrodistillation, membrane separation, magnetic stirring, maceration, soxhlet extraction, water percolation, etc.) and advanced techniques (supercritical fluid extraction, pressurized liquid extraction, enzyme-assisted extraction, accelerated solvent extraction, microwave-assisted extraction, and ultrasonic-assisted extraction) have been utilized till date for the extraction of such compounds. However, all these methods lack one or more attributes like low yield, poor efficiency, greater solvent requirement, poor sustainability, and many others.[125] To overcome these limitations, PEF is proposed to be a highly cost- and time-effective option for the recovery of bioactive compounds.[32]

As a consequence of PEF, the intracellular molecules align themselves according to the charge by the effect of the dipole moment. Additionally, cellular lipid bilayers and proteins are impaired, and channels are constructed in the cell membrane, which may transverse the migration of bioactive components across the membrane.[72] As discussed earlier, there is a possibility of creating reversible and fixed channels in the membrane depending upon the strength of the electric field. To obtain better extraction without affecting the integrity of the cells permanently, temporary electroporation is achieved by applying an optimal electric field to the sample, which further facilitates the migration of intracellular compounds to the surrounding suspension. Recent studies have utilized PEF for extracting bioactive compounds like anthocyanins, alkaloids, betanine, betulin, carotenoids, lycopene, lutein, lactase, proteins, polysaccharides, polyphenols, essential oils, etc. from various plant products.[131]

5.4.1 POLYPHENOLIC COMPOUNDS

Polyphenols are a class of phytochemicals that are abundant in fruits. Amongst 8000 identifiable polyphenolic compounds, anthocyanins, catechins, tannins, isoflavones, and flavanols occupy sizable amounts of total polyphenols.[26,107,130] PEF has been successfully utilized for extracting polyphenols from orange peel[78] and juices,[153] red grape pomace,[119] and tea leaves.[157] The authors observed an increment in the proportion of the polyphenolic compounds by 36% from the red grape pomace sample when treated with PEF.[45] Several research studies have documented the extraction of anthocyanins from blueberries,[26] cabbage,[55] Merlot grapes[40] and their byproducts,[36] and potato[105] by PEF. It was found that the quantity of anthocyanin extracted increased by changing the numbers, duration, and intensity of the electric field. Carotenoids, specifically α-carotene, β-carotene, lycopene, lutein, and zeaxanthin, have been extracted from carrot,[118] orange juice,[137] and tomato waste.[79] González-Casado et al.[60,141] reported elevated levels of absolute and respective carotenoids in the tomato puree when the tomatoes were treated with PEF as a preprocessing step. The maximum rise in the content of carotenoids in the tomato puree when compared to untreated fruit was 52%, and it was achieved by subjecting the samples to an electric field for 30 pulses at 2 kV cm^{-1}. This transition in the levels of carotenoid concentrations was a function of specific energy input applied to the intact fruit. Furthermore, an escalation in the concentration of lycopene (4400 ± 211 $\mu g\ kg^{-c}$ to 6072 ± 292 $\mu g\ kg^{-t}$) was also reported in purees derived from PEF-treated tomatoes. Lycopene concentration was increased 1.5 fold in treated samples when exposed to an intense electrical field of 2.31 kJ kg^{-1}. The possible reason for this increase in lycopene concentration was due to the activation of the genes encoding enzymes. These enzymes, phytoene synthase (PS) and carotenoid isomerase (CARISO), are behind the biosynthesis of carotenoids in the fruits. Further enhancement in the extraction is presumably due to the electropermeabilization, which allows easy movement across the cell membranes. It was further revealed that exposing tomato fruit to varying electric fields in the range of 0.4 to 2 kV $cm^{-1.}$ affected the color and texture of the fruit considerably, apart from increasing the content of carotenoids. Recently, it has been contemplated that PEF treatment induces stress in metabolically active plants, which directs various metabolic pathways and causes the accumulation of secondary metabolites. These metabolites serve a crucial role in the self-defense mechanism of plants under biotic and abiotic stress environments.[19,142] Similarly, the concentration of carotenoids improved significantly in carrot puree when treated with an electric field intensity of up to 1 $kVcm^{-1}$ at 5 Hz.[118] In another study, betanine, a red carotenoid pigment, was extracted from beetroot by applying an electric field of 7 kV/cm with a 4.2-fold increase in betanine yield in comparison to untreated samples.[77]

PEF has shown an improved extraction rate of polyphenolic constituents, especially anthocyanin, from grape juice. The electric strength of 750 kV was applied to grape juice, which resulted in increasing the yield by 5% at a constant pressure of 5 atmospheres.[61] Similarly, the quality of pomegranate juice was also found to be improved when juice was treated with PEF, and shelf-life was reported to be increased for 84 days on a pilot plant scale. Likewise, various fruit juices were treated with PEF to understand the change in quality parameters of the end products after treatment.[53] Grapes are considered to be a good source of phenolic content. These components can be extracted partially due to resistance to mass transfer of the cell walls and cytoplasmic membranes. The extraction of these components can be increased by using appropriate maceration enzymes to facilitate the permeabilization of cells during fermentation, which again takes more time for the extraction. With the idea to improve the extraction of these bioactive components from grapes, PEF was introduced as a pretreatment before maceration. Therefore, it resulted in increased quality and yield of the juice from 50 to 80%.[70]

Carotenoids are pigments that can be found largely in various fruits and vegetables. PEF treatment has been observed to exhibit no significant effects on the extraction of the individual carotenoids, including α-carotene, β-carotene, zeaxanthin, cryptoxanthin, cryptoxanthinlutein, etc., or the total carotenoid concentration of orange juice.[50] However, as reported by Cortés et al.[37] and Odriozola-Serrano et al.,[92] the level of total carotenoids was lower in heat-pasteurized orange juice than in PEF-processed orange juice. Interestingly, when an orange carrot juice was treated with PEF, a substantial upsurge in the retention of individual compounds (cis-cryptoxanthin, 9-cis-violaxanthin, neoxanthin, 9-cis-carotene, phytoene + phytofluene, -carotene, neoxanthin, mutatoxanthin, 9-cis-α-carotene, α-carotene, phytoene + phytofluene, β-carotene, 13-cis-β-carotene, ζ-carotene, and 9-cis-β-carotene) was noticed.[37] Similarly, exposing the tomato juice to PEF processing for an extended duration of time (35 kV/cm for 1500 s) increased lycopene content.[90] It was reported that giving PEF treatment to tomato juice resulted in improved extraction of lycopene content due to the electric field, frequency, treatment time, and pulse width.[90,93] Table 5.4 summarizes the effects of PEF on carotenoids.

5.4.2 BIOPOLYMERS

The PEF technique can be used for the modification of starch present in fruits. Generally, the starch content in fruits is significantly less compared to vegetables. Though fruits do not contain much starch, the modification of starch can still be done if it is present in certain fruits. The modification of starches leads to changes in structures, improving properties, an increase in the utilization of both modified polysaccharides and proteins, agglomerating

TABLE 5.4 Effect of PEF on the Nutritional Composition of Fruits and Fruit Juices.

Fruits	HIPEF treatment	Retention (%)	References
Orange juice	E = 35, t = 750, P = bipolar	Carotenoids and vitamin A content were 90.7% and 90.9%, respectively.	[38, 122]
	E = 30, t = 100	Carotenoids and vitamin A content were 93.3% and 92.5%, respectively.	
		Total carotenoids (92.2%)	
	E = 40, t = 130, T = 69	Vitamin A (90.1%)	
Orange–carrot juice	E = 35, t = 150, T = 64	Carotenoids (70%) and Vitamin A (48%)	[137]
Tomato juice	E = 40, t = 57, bipolar, T = 53	Lycopene (100%)	[86, 90, 92]
	E = 35, t = 1000, monopolar, T < 40	Lycopene (142.7%)	
	E = 35, t = 1000, bipolar	Lycopene (125%)	
Fruit juice with the addition of stevia	E = 30 kV/cm P = 230 μs	Ascorbic acid (74%)	[31]
	E = 21 kV/cm P = 300 μs Stevia = 2.5%	Retention of maximum bioactive components.	
Blueberry fruit	E = 1, 3, and 5 kV/cm, Energy = 10 kJ/kg	Total phenolic content (+43%), Total anthocyanin content (+60%), Antioxidant activity (+31%)	[26]

f = frequency of pulse (Hz); E = electric field (kV/cm); P = polarity of pulse; t = time of treatment (μs); and T = temperature of treatment (°C).

algae and sugar complexes, etc. During PEF treatment, the produced electric potential difference triggers the mobility rate of reacting ions and switches their movement in unidirectional motion for effective collision. Moreover, the effective reaction between the targeted groups and water molecules in starch granules is based on electroporation theory. At the same time, the inclusion of PEF for the starch modification sufficiently causes the structural orientation of numerous molecules. Also, deformation of the particle with enhanced properties in starch molecules can be considerably perceived with high pulse width. The PEF is responsible for affecting the texture, secondary, and tertiary structures for the various starch and protein complexes. PEF technology may be referring to different polar molecules for causing various chemical/physical reactions. This has resulted in increased efficiency, transforming the functional performance and enhancing potential applications in an improved manner.[158]

5.4.3 WASTE VALORIZATION

Fruit waste valorization is essentially helpful in converting fruit waste such as peels, seeds, inedible parts, etc. to value-added substances. The various inedible portions of fruits, such as peel, skin, seed, stone, etc., are constituted of several bioactive compounds that, if recovered efficiently, could be utilized as nutraceuticals or supplements to provide health benefits. PEF is more selective toward the extraction of specific biocompounds, is sustainable, requires low energy, and reduces temperature rise. PEF constructively extracts value-added substances from food wastes. Pulp is often used or devoured up in a fruit; what is left behind is peel, seed, unused pulp, and bagasse, which are also used as fertilizers on farms. However, substantial quantities of phytochemicals will be derived from the leaves, seeds, peels, and stems of the fruits. These recovered bioactive compounds can be used widely in pharmaceutical applications, for creating food and natural supplements. PEF is an essential technique to improve the bioactive compounds from the waste of grapes and is additionally beneficial in decreasing microbial growth.[115] Studies have shown that the recovered bioactive compounds from the peels and other waste from fruits have a great possibility to be used as food supplements in the food processing industry as dietary enhancements, cosmetics, and pharmaceuticals. Perhaps it also decreases the microbial growth besides recovering the value-added compounds from the food. PEF treatment of prickly pear peels and pulp is directed to the natural colorant outputs. PEF was used to extract phenolic components from frozen European blueberries and apple pomace. PEF treatment resulted in a 1.59-fold increase in polyphenol content from orange peel. Recovery of polyphenols from fresh peels using the PEF technique could be a cost-effective and environment-friendly substitute as compared to other conventional methods.[115]

The citrus industry is one of the leading producers of waste in the form of peels, which are abundant in valuable compounds such as dietary fibers, polyphenolic compounds, and essential oils. These peels provide an ample opportunity for their valorization by exploiting PEF technology. Limonene, which is a lipophilic monoterpene, is known for its aroma and antioxidant properties and can be extracted from orange peels.[46] Furthermore, cocoa bean residue is rich in alkaloids, especially theobromine and caffeine, and exhibits stimulative effects on the nervous, gastrointestinal, vascular, and respiratory systems of the body.[22, 80] Phenolic compounds extracted from the food waste using the PEF technique help in increasing the resistance to radical injury, circulatory sicknesses, and cancer. Pectin is used as a thickener in many food confectionaries, like jams, jelly, etc., and is recovered within the fruit pomaces. Pomace also provides food additives along with nutrients like fibers, cellulose, vinegar, and edible dyestuffs. Some tropical fruits like papaya and pineapple comprise the enzymes papain or bromelain. These components are used as meat tenderizing agents, which promote the degradation of protein metabolites. Phenols obtained from plant extracts may be used as antibacterial agents in various food products.[17]

5.5 EFFECTS OF PEF ON THE QUALITY CHARACTERISTICS OF FRUITS

5.5.1 NUTRITIVE VALUE OF FRUITS

The effects of PEF on the nutritional quality of fruits and vegetables, especially their extracts and juices, have been documented by many researchers, and it is largely evident that PEF induced better retention of carotenoids, polyphenolic compounds, vitamins, etc. Various studies discussing these aspects are detailed below.

5.5.1.1 VITAMINS

Amongst the various studies, retention of vitamin C in juices is one of the keynote research elements of high-intensity PEF. Compared to conventional thermal methods like pasteurization, sterilization, etc., high-intensity PEF has shown better retention of vitamin C in the treated product.[108] In a study carried out by Yeom et al.,[149] it was observed that heat-treated orange juice (94.6°C, 30 s) has shown a reduction of 7–15% in ascorbic acid levels than high-intensity PEF-treated juices (4–5%). A similar study was carried out by Sánchez-Moreno et al. (2005), where the effects of PEF and pasteurization treatment were compared in terms of retention of total vitamin C content and L-ascorbic acid in orange juice. Further, higher vitamin C possession was reported when orange juice was treated at 35 kV/cm having a pulse width of 1000 s. Vitamin C retention was observed to be 91.2% higher when compared with thermally pasteurized orange juice (90°C, 1 min, 82.8%).[50] Min et al..[86] reported no significant difference in the levels of ascorbic acid between fresh and HIPEF-treated orange juices; however, they stated that heat processing (90°C, 90 s) has substantially reduced the ascorbic acid levels. The varied PEF conditions employed in the juices could explain the differences in results published in various articles regarding the influence of HIPEF on vitamin C preservation. In a similar study carried out on apple juice by Evrendilek,[53] it was observed that PEF treatment (35 kV/cm, 94 s) was substantially successful in preserving vitamin C in juice samples. Moreover, the mixture of orange and grape juice added with bacteriocin showed a nonsignificant change in ascorbic content when treated at 80 kV/cm. Strawberry juice showed more than 87% retention in ascorbic acid content when treated at 35 kV/cm up to 2000 s.[91] Min et al.[86] studied and compared the effects of PEF (40 kV/cm, 57 s) and thermal treatment on the levels of ascorbic acid in the juice of tomato (92°C, 90 s). According to these studies, ascorbic acid levels declined by roughly 10% under the influence of high-temperature treatment but not significantly after HIPEF processing. It was also reported that HIPEF-treated tomato juice at 35 kV/cm for 1500 μs at 100 Hz with an energy density of 8269 kJ/L showed higher lycopene retention as compared to vitamin C content. Immediately after processing, no significant change was reported in the antioxidant activity and vitamin C content of treated and untreated juice, whereas the values decreased during the storage period.[93] A handful of studies have been conducted to investigate the impact caused by PEF's important factors on vitamins. A bipolar electric pulse with lowered field strength, duration, pulse frequency, and width resulted in the retention of higher vitamin C content in orange juice as well as "gazpacho" (a Spanish cold vegetable beverage). Conclusively, vitamin C retention is always higher in PEF-treated juices than in heat-pasteurized juices.[49]

Apart from vitamin C, few researchers have also documented the effect of PEF processing on vitamin A content in juices. In accordance to the results of Torregrosa et al.,[138] fresh orange juice contained the maximum levels of vitamin A, followed by PEF-treated juice (30 kV/cm, 100 s), and the least levels were found in heat-treated juice (90°C, 20 s). Cortés et al.[38] further discovered that as the electric field intensity and treatment time increased, the vitamin A concentration of orange juice decreased slightly. PEF treatments at 25 and 30 kV/cm, on the other hand, produced higher vitamin A concentrations in orange–carrot juices than heat-pasteurized juices.

5.5.1.2 PROTEINS

PEF's impacts on plant food proteins have gotten little attention thus far. The underlying structure of proteins appears to be unaffected by PEF treatments (peptides and amino acids). After rigorous PEF treatments with 4 s pulses at 35 kV/cm for 1 ms, Garde-Cerdán et al.[56] reported no observable variations in the amino acid composition of grapes. Li et al.[76] considered the effect of PEF on physicochemical attributes of soybean protein isolates (SPIs) at the rate of 0–40 kV/cm with pulse width ranging between 0–547 s. They discovered that PEF therapy can change physicochemical attributes. They also discovered that a PEF treatment could alter SPI's secondary structure in subtle ways.

5.5.1.3 FATTY ACIDS

Strong treatments of grape juice with 35 kV/cm for 1 ms were applied, which resulted in grape juices with equal total fatty acid content before and after the treatment according to Garde-Cerdán et al.[56] There were no changes in fatty acid composition except for lauric acid, which appeared to be negatively influenced by PEF activity. PEF did not affect the content of linolenic and oleic acids. In the same manner, the treatment did not disturb long-chain saturated fatty acids (C14 and C16).

5.5.1.4 PIGMENTS

Dietary chlorophyll derivatives are known for their chemopreventive, antioxidant, and antimutagenic activity. Components such as pheophytins, pyropheophytins, and metal-free pheophorbides play an important role in these activities.[138] However, the effect of PEF on these compounds is uncertain.

Chlorophyll degradation in spinach puree was controlled by combining PEF treatment with water-soluble Zn^{2+}[152] by keeping the electric field up to 60 kV/cm. This may be responsible for the death of microorganisms and enzymes involved in chlorophyll degradation.

Anthocyanin pigment content in cranberry juices did not alter significantly after treating at 40 kV/cm for 150 s. After a heat treatment (90°C for 90 seconds), the anthocyanin level in the juices was drastically reduced.[4] The amount of anthocyanin retention in strawberry juice varied from 96.1 to 100% under the influence of electric field strengths varying from 20 to 35 kV/cm and a processing period of 2000 seconds.[91] Furthermore, there have been limited qualitative data on the effect of PEF on phenolic content. Sánchez-Moreno et al.[122] looked at the flavanone content of orange juice following an electric strength of 35 kV/cm at 800 Hz. Total flavanones, particularly flavanone glycosides, and their aglycons, as well as hesperetin and naringenin, did not alter.

5.5.1.5 PHENOLIC COMPOUNDS

PEF treatments did not affect the amount of phenolic compounds in plant-based diets. Pasteurized juice of an apple after PEF treatment at 35 kV/cm with 4 s bipolar pulses (1200 pulses per second) was evaluated in relation to one heat pasteurized at 90°C for a time period of 30 s (Aguilar-Rosas et al., 2007). The heat treatment resulted in a large loss of total phenols compared to the PEF treatment, which only reduced total phenols by 14.5%. Furthermore, tomato juice showed no significant differences in total phenolic content after processing at 35 kV/cm for 1500 s with 4 s bipolar pulses.[90]

5.5.2 RHEOLOGICAL BEHAVIOR OF FRUITS

For any food commodity, rheological properties play a vital role in sustaining the quality and wholesomeness of the product. Any fundamental rheological test helps in the identification of the nature of the products. The rheology of any fruit or fruit products describes its flow behavior and texture before and after the application of processing and preservation methods. Protein plays an essential role while preparing emulsions and gels, which can be designated as time-dependent and time-independent viscoelastic behaviors. These rheological changes are also affected by processing parameters that include pressure, temperature, concentration, and dispersion state.[9,101,120]

The rheological behavior of apple juice was ascribed by using a power law model. The apparent viscosity of the apple juice was reported to decrease with increased shear rate, which corresponds to the pseudoplastic flow behavior. The consistency index (K) decreased substantially as electric field strength increased, but the flow behavior index (n) increased.[25] Enzymatically treated mango pulp showed a significant impact on its rheological properties, and it was discovered that all process variables, including enzyme concentration, incubation period, and incubation temperature, had constant involvement in these changes.[24] The majority of research has been dedicated to understanding the effects of PEF treatment on watermelon juice and its flavor composites,[18] inactivation of pectin methyl esterase and microbial growth,[116] vitamin C and antioxidant activity,[102] and the physiochemical characteristics of blended orange and carrot juice.[114] The rheological properties of PEF-processed and thermally pasteurized mango nectar (*Mangifera indica*) were studied by Kumar et al.[73] The Herschel-Bulkley model was used to determine the behavior of mango nectar (PEF and thermally treated) like a pseudoplastic fluid under yield stress ($0.9780 < r < 0.9999$, $p < 0.01$). Temperature and other processing parameters modified the consistency coefficient (k) of mango nectar significantly ($p < 0.05$), although yield stress (0) of mango nectar did not show any significant ($p > 0.05$) variation. PEF-treated sample showed a significant change in flow behavior index (n) as compared to the nontreated sample. The temperature had a substantial impact on the consistency coefficient (k) of mango nectar ($r > 0.893$, $p < 0.05$), and processing conditions had a significant impact on flow activation energy (Ea) ($p < 0.05$). According to the findings, PEF along with pasteurization treatment caused a change in the rheological properties of mango nectar.

5.6 FUTURE PERSPECTIVES

PEF induces cell disintegration and microbial decontamination of fruits. Upcoming and ongoing research and new technology will result in increased treatment capacities and the development of more applications of PEF in various food groups. Application of PEF processes and treatments can enhance and extend the shelf-life period of protein-based drinks, dairy products, beer, dressings, soups, fruits, vegetable juices, etc.[85]

PEF technology is expected to bring a lot of changes in the food industry by focusing on improved and enhanced processing and preservation methods for food commodities. Due to ignorance of food preservation or lack of information about the mechanisms of electroporation and pore

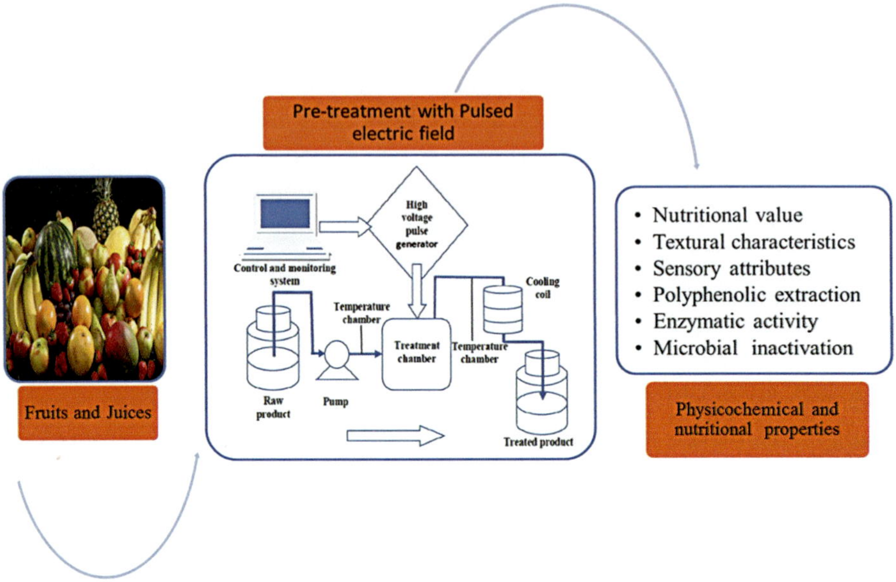

FIGURE 5.1 Effect of PEF on physicochemical properties of fruit and fruit juices.

resealing, this method has not been exploited properly yet. This technology promotes the extraction of potential nutrients or bioactive components by induction of membrane permeabilization of the plant/animal cells. PEF significantly helps in improving the preservation rate, increasing the yields in juice preparation, and achieving maximum retention of valuable cell metabolites with improved functional and chemical properties of food. Researchers should take initiatives to optimize the processing conditions such as temperature, no. of pulses, pulse rate, and strength of electric field for each commodity available in the market. Furthermore, PEF treatments can be used pragmatically to enhance mass transfer into numerous food products by permeabilization of the plant or animal cell membranes and thus enriching the quality of functional compounds or antimicrobials into foods with minimum reduction quality and organoleptic characteristics. PEF treatment also helps in the microbial inactivation of the food, and therefore bacterial endospores can also be eliminated if PEF technology combines with certain thermal methods, keeping in mind that the quality of the food does not deteriorate. As a result, PEF may be utilized for improving the quality and safety of fruits, vegetables, juices, and other food-related goods. Furthermore, PEF technology can be implied to enhance the rate of heat and mass transfer in the food environments by improving the cell permeability and enriched infusion of functional composites or antimicrobials into foods by insignificantly modifying their organoleptic characteristics.

5.7 SUMMARY AND CONCLUSIONS

PEF processing deals with the food industry to produce safe food by minimizing the processing time. Moreover, it operates the technique in such a way that the quality can be retained, such as organoleptic, nutritional, and rheological properties, in comparison to heat processing methods. PEF has also resulted in retaining color and aroma, preserving nutritional value for better sustainability in the market, and providing wholesomeness of the commodity to the consumer. From a food processing/engineering standpoint, the significant benefits of pulsed electric fields to various food groups are independent of the size and shape of the sample during processing. PEF causes the inactivation of microorganisms and enzymes and causes alteration in the structure of the cell, however, having insignificant modification in the nutritive and sensory quality of foods. The exploitation of PEF is being improved and altered by most of the food industry. Various innovative and new food products can also be developed using this technology.

KEYWORDS

- **electroporation**
- **nutritional quality**
- **microbial inactivation**
- **rheological characteristics**
- **bioactive compounds**
- **waste valorization**

REFERENCES

1. Aadil, R. M.; Zeng, X.-A.; Ali, A.; Zeng, F.; Farooq, M. A.; Han, Z.; Jabbar, S. Influence of Different Pulsed Electric Field Strengths on the Quality of the Grapefruit Juice. *Int. J. Food Sci. Technol.* **2015,** *50* (10), 2290–2296.

2. Abgail Suelen da Costa, R. Optimization of Osmotic Dehydration of Pear Followed by Conventional Drying and Their Sensory Quality. *Lebensm.-Wiss. Technol.* **2016,** *72,* 407–415.
3. Aganovic, K.; Smetana, S.; Grauwet, T.; Toepfl, S.; Mathys, A.; Van Loey, A.; Heinz, V. Pilot Scale Thermal and Alternative Pasteurization of Tomato and Watermelon Juice: An Energy Comparison and Life Cycle Assessment. *J. Clean. Prod.* **2017,** *141,*–525.
4. Aguilar-Rosas, S. F.; Ballinas-Casarrubias, M. L.; Nevarez-Moorillon, G. V.; Martin-Belloso, O.; Ortega-Rivas, E. Thermal and Pulsed Electric Fields Pasteurization of Apple Juice: Effects on Physicochemical Properties and Flavour Compounds. *J. Food Eng.* **2007,** *83* (1), 41–46.
5. Aguiló-Aguayo, I.; Oms-Oliu, G.; Soliva-Fortuny, R.; Martín-Belloso, O. Changes in Quality Attributes Throughout Storage of Strawberry Juice Processed by High-intensity Pulsed Electric Fields or Heat Treatments. *LWT - Food Sci. Technol.* **2009,** *42* (4), 813–818.
6. Aguiló-Aguayo, I.; Soliva-Fortuny, R.; Martín-Belloso, O. Comparative study on color, viscosity and related enzymes of tomato juice treated by high-intensity pulsed electric fields or heat. *Eur. Food Res. Technol.* **2008,** *227* (2), 599–606.
7. Aguiló-Aguayo, I.; Soliva-Fortuny, R.; Martín-Belloso, O.; Color and Viscosity of Watermelon Juice Treated by High-intensity Pulsed Electric Fields or Heat. *Innov. Food Sci. Emerg. Technol.* **2010a,** *11* (2), 299–305.
8. Aguiló-Aguayo, I.; Soliva-Fortuny, R.; Martín-Belloso, O.; High-intensity Pulsed Electric Fields Processing Parameters Affecting Polyphenoloxidase Activity of Strawberry Juice. *J. Food Sci.* **2010b,** *75* (7), C641–646.
9. Ahmed, J.; Ramaswamy, H. S.; Alli, I.; Ngadi, M. Effect of High Pressure on Rheological Characteristics of Liquid Egg. *LWT - Food Sci. Technol.* **2003,** *36* (5), 517–524.
10. Álvarez, I.; Condón, S.; Raso, J. Microbial Inactivation by Pulsed Electric Fields. In *Pulsed Electric Fields Technology for the Food Industry: Fundamentals and Applications*; Raso, J.; Heinz, V., Eds.; Springer US: Boston, MA, 2006; pp 97–129.
11. Álvarez, I.; Pagán, R.; Condón, S.; Raso, J. The Influence of Process Parameters for the Inactivation of *Listeria monocytogenes* by Pulsed Electric Fields. *Int. J. Food Microbiol.* **2003,** *87* (1), 87-95.
12. Amami, E.; Vorobiev, E.; Kechaou, N. Effect of Pulsed Electric Field on the Osmotic Dehydration and Mass Transfer Kinetics of Apple Tissue. *Dry. Technol.* **2005,** *23* (3), 581–595.
13. Amiali, M.; Ngadi, M. O.; Smith, J. P.; Raghavan, V. G. S. Inactivation of *Escherichia coli* O157:H7 and *Salmonella enteritidis* in Liquid Egg White Using Pulsed Electric Field. *J. Food Sci.* **2006,** *71* (3), M88-M94.
14. Angersbach, A.; Knorr, D. High Intensity Electric Field Pulses as Pretreatment for Affecting Dehydration Characteristics and Rehydration Properties of Potato Cubes. *Nahrung-Food* **1997,** *41,* 194–200.
15. Angersbach, A.; Heinz, V.; Knorr, D. Effects of Pulsed Electric Fields on Cell Membranes in Real Food Systems. *Innov. Food Sci. Emerg. Technol.* **2000,** *1* (2), 135–149.
16. Aronsson, K.; Rönner, U. Influence of pH, Water Activity and Temperature on the Inactivation of ESCHERICHIA COLI and Saccharomyces Cerevisiae by Pulsed Electric Fields. *Innov. Food Sci. Emerg. Technol.* **2001,** *2* (2), 105–112.
17. Arshad, R. N.; Abdul-Malek, Z.; Roobab, U.; Qureshi, M. I.; Khan, N.; Ahmad, M. H.; Liu, Z.W.; Aadil, R. M. Effective Valorization of Food Wastes and by-products Through Pulsed Electric Field: A Systematic Review. *J. Food Process Eng.* **2021,** *44* (3), e13629.

18. Augusto, P. E. D.; Ibarz, A.; Cristianini, M. Effect of High Pressure Homogenization (HPH) on the Rheological Properties of a Fruit Juice Serum Model. *J. Food Eng.* **2012,** *111* (2), 474–477.
19. Balasa, A.; Knorr, D. Pulsed Electric Field Processing: Food Preservation. In *Encyclopedia of Biotechnology in Agriculture and Food*, 1st ed.; Heldman, D. R.; Wheeler, M. B.; Hoover, D. G., Eds.; CRC Press, 2010.
20. Barba, F. J.; Koubaa, M.; do Prado-Silva, L.; Orlien, V.; Sant'Ana, A. de S. Mild Processing Applied to the Inactivation of the Main Foodborne Bacterial Pathogens: A Review. *Trends Food Sci. Technol.* **2017,** *66,* 20–35.
21. Barbosa-Cánovas, G. V.; Góngora-Nieto, M. M.; Pothakamury, U. R.; Swanson, B. G. Fundamentals of High-Intensity Pulsed Electric Fields (PEF). In*Preservation of Foods with Pulsed Electric Fields*; Barbosa-Cánovas, G. V., Góngora-Nieto, M. M., Pothakamury, U. R., Swanson, B. G., Eds.; Academic Press: San Diego, 1999; pp 1–19.
22. Barbosa-Pereira, L.; Guglielmetti, A.; Zeppa, G. Pulsed Electric Field Assisted Extraction of Bioactive Compounds from Cocoa Bean Shell and Coffee Silverskin. *Food Bioproc. Tech.* **2018,** *11* (4), 818-835.
23. Barth, M.; Hankinson, T. R.; Zhuang, H.; Breidt, F. Microbiological Spoilage of Fruits and Vegetables. In *Compendium of the Microbiological Spoilage of Foods and Beverages*; Sperber, W. H., Doyle, M. P., Eds.; Springer New York: New York, NY, 2009; pp 135–183.
24. Bhattacharya, S.; Rastogi, N. K. Rheological Properties of Enzyme-treated Mango Pulp. *J. Food Eng.* **1998,** *36* (3), 249-262.
25. Bi, X.; Liu, F.; Rao, L.; Li, J.; Liu, B.; Liao, X.; Wu, J. Effects of Electric Field Strength and Pulse Rise Time on Physicochemical and Sensory Properties of Apple Juice by Pulsed Electric Field. *Innov. Food Sci. Emerg. Technol.* **2013,** *17,* 85–92.
26. Bobinaitė, R.; Pataro, G.; Lamanauskas, N.; Šatkauskas, S.; Viškelis, P.; Ferrari, G. Application of Pulsed Electric Field in the Production of Juice and Extraction of Bioactive Compounds From Blueberry Fruits and Their by-products. *J. Food Sci. Technol.* **2015,** *52* (9), 5898–5905.
27. Bouzrara, H.; Vorobiev, E. Solid–liquid Expression of Cellular Materials Enhanced by Pulsed Electric Field. *Chem. Eng. Process.* **2003,** *42,* 249–257.
28. Buckow, R.; Ng, S.; Toepfl, S. Pulsed Electric Field Processing of Orange Juice: A Review on Microbial, Enzymatic, Nutritional, and Sensory Quality and Stability. *Compr. Rev. Food Sci. Food Saf.* **2013,** *12* (5), 455–467.
29. Burt, S. Essential Oils: Their Antibacterial Properties and Potential Applications in Foods—a Review. *Int. J. Food Microbiol.* **2004,** *94* (3), 223–253.
30. Caminiti, I. M.; Noci, F.; Morgan, D. J.; Cronin, D. A.; Lyng, J. G. The Effect of Pulsed Electric Fields, Ultraviolet Light or High Intensity Light Pulses in Combination with Manothermosonication on Selected Physico-chemical and Sensory Attributes of an Orange and Carrot Juice Blend. *Food Bioprod. Process.* **2012,** *90* (3), 442–448.
31. Carbonell-Capella, J. M.; Buniowska, M.; Cortes, C.; Zulueta, A.; Frigola, A.; Esteve, M. J. Influence of Pulsed Electric Field Processing on the Quality of Fruit Juice Beverages Sweetened With Stevia Rebaudiana. *Food Bioprod. Process.* **2017,** *101,* 214–222.
32. Carpentieri, S.; Režek Jambrak, A.; Ferrari, G.; Pataro, G. Pulsed Electric Field-Assisted Extraction of Aroma and Bioactive Compounds from Aromatic Plants and Food by-Products. *Front. Nutrit.* **2022,** *8.*

33. Castro, A. J.; Barbosa-Canovas, G. V.; Swanson, B. G. Microbial Inactivation of Foods by Pulsed Electric Fields. *J. Food Process. Preserv.* **1993,** *17* (1), 47–73.
34. Charles-Rodriguez, A. V.; Nevarez-Moorillon, G. V.; Zhang, Q. H.; Ortega-Rivas, E. Comparison of Thermal Processing and Pulsed Electric Fields Treatment in Pasteurization of Apple Juice. *Food Bioprod. Process.* **2007,** *85,* 93–97.
35. Chauhan, O. P.; Sayanfar, S.; Toepfl, S. Effect of Pulsed Electric Field on Texture and Drying Time of Apple Slices. *J. Food Sci. Technol.* **2018,** *55* (6), 2251–2258.
36. Corrales, M.; Toepfl, S.; Butz, P.; Knorr, D.; Tauscher, B. Extraction of Anthocyanins from Grape by-products Assisted by Ultrasonics, High Hydrostatic Pressure or Pulsed Electric Fields: A Comparison. *Innov. Food Sci. Emerg. Technol.* **2008,** *9* (1), 85–91.
37. Cortés, C.; Esteve, M. J.; Rodrigo, D.; Torregrosa, F.; Frígola, A. Changes of Colour and Carotenoids Contents During High Intensity Pulsed Electric Field Treatment in Orange Juices. *Food Chem. Toxicol.* **2006,** *44* (11), 1932–1939.
38. Cortés, C.; Torregrosa, F.; Esteve, M. J.; Frígola, A. Carotenoid Profile Modification During Refrigerated Storage in Untreated and Pasteurized Orange Juice and Orange Juice Treated with High-intensity Pulsed Electric Fields. *J. Agric. Food Chem.* **2006,** *54* (17), 6247–6254.
39. da Cruz Almeida, E. T.; de Medeiros Barbosa, I.; Tavares, J. F.; Barbosa-Filho, J. M.; Magnani, M.; de Souza, E. L. Inactivation of Spoilage Yeasts by Mentha spicata L. and M. × villosa Huds. Essential Oils in Cashew, Guava, Mango, and Pineapple Juices. *Front. Microbiol.* **2018,** *9.*
40. Delsart, C.; Ghidossi, R.; Poupot, C.; Cholet, C.; Grimi, N.; Vorobiev, E.; Milisic, V. and Mietton-Peuchot, M. Enhanced Extraction of Phenolic Compounds from Merlot Grapes by Pulsed Electric Field Treatment. *Am. J. Enol. Vitic.* **2012,** *63,* 205–211.
41. Dermesonlouoglou, E.; Chalkia, A.; Dimopoulos, G.; Taoukis, P. Combined effect of Pulsed Electric Field and Osmotic Dehydration Pre-treatments on Mass Transfer and Quality of Air Dried Goji Berry. *Innov. Food Sci. Emerg. Technol.* **2018,** *49,* 106–115.
42. Đukić-Vuković, A.; Tylewicz, U.; Mojović, L.; Gusbeth, C. Recent Advances in Pulsed Electric Field and Non-thermal Plasma Treatments for Food and Biorefinery Applications. *J. Process. Energy Agric.* **2017,** *21* (2), 61–65.
43. Eing, C.; Goettel, M.; Straessner, R.; Gusbeth, C.; Frey, W. Pulsed Electric Field Treatment of Microalgae—Benefits for Microalgae Biomass Processing. *IEEE Trans. Plasma Sci.* **2013,** *41* (10), 2901–2907.
44. El- Sayed, G. A.; Osman, M. A.; El-Hamied, A.; Ghanem, T. H. Introducing Pulsed Electric Field Treatments for Fruit Juices Industry in Egypt as an Alternative of Thermal Pasteurization. *Arch. Agric. Sci. J.* **2020,** *3,* 148–167.
45. El Darra, N.; Grimi, N.; Vorobiev, E.; Louka, N.; Maroun, R. Extraction of Polyphenols from Red Grape Pomace Assisted by Pulsed Ohmic Heating. *Food Bioproc. Tech.* **2013,** *6* (5), 1281–1289.
46. El Kantar, S.; Boussetta, N.; Lebovka, N.; Foucart, F.; Rajha, H. N.; Maroun, R. G.; &Vorobiev, E. Pulsed Electric Field Treatment of Citrus Fruits: Improvement of Juice and Polyphenols Extraction. *Innov. Food Sci. Emerg. Technol.* **2018,** *46,* 153–161.
47. Elez-Martínez, P.; Aguiló-Aguayo, I.; Martín-Belloso, O. Inactivation of Orange Juice Peroxidase by High-intensity Pulsed Electric Fields as Influenced by Process Parameters. *J. Sci. Food Agric.* **2006,** *86* (1), 71–81.
48. Elez-Martínez, P.; Escolà-Hernández, J.; Soliva-Fortuny, R. C.; Martín-Belloso, O. Inactivation of *Saccharomyces cerevisiae* Suspended in Orange Juice Using High-intensity Pulsed Electric Fields. *J. Food Protect.* **2004,** *67* (11), 2596–2602.

49. Elez-Martínez, P.; Martín-Belloso, O. Effects of high intensity pulsed electric field Processing Conditions on Vitamin C and Antioxidant Capacity of Orange Juice and Gazpacho, a Cold Vegetable Soup. *Food Chem.* **2007,** *102* (1), 201–209.
50. Elez-Martínez, P.; Soliva-Fortuny, R. C.; Martín-Belloso, O. Comparative Study on Shelf Life of Orange Juice Processed by High Intensity Pulsed Electric Fields or Heat Treatment. *Eur. Food Res. Technol.* **2005,** *222* (3), 321.
51. Elez-Martínez, P.; Suárez-Recio, M.; Martín-Belloso, O. Modeling the Reduction of Pectin Methyl Esterase Activity in Orange Juice by High Intensity Pulsed Electric Fields. *J. Food Eng.* **2007,** *78* (1), 184–193.
52. Eshtiaghi, M. N.; Nakthong, N. Application of Pulsed Electric Field for Inactivation of Yeast S. Cerevisiae in Apple Juice. *J. Phys.* **2021,** *1893* (1), 012008.
53. Evrendilek, G. A. Impacts of Pulsed Electric Field and Heat Treatment On Quality and Sensory Properties and Microbial Inactivation of Pomegranate Juice. *Food Sci. Technol. Int.* **2017,** *23* (8), 668–680.
54. Fauster, T.; Giancaterino, M.; Pittia, P.; Jaeger, H. Effect of Pulsed Electric Field Pretreatment on Shrinkage, Rehydration Capacity and Texture of Freeze-dried Plant Materials. *LWT* **2020,** *121,* 108937.
55. Gachovska, T.; Cassada, D.; Subbiah, J.; Hanna, M.; Thippareddi, H.; Snow, D. Enhanced Anthocyanin Extraction from Red Cabbage Using Pulsed Electric Field Processing. *J. Food Sci.* **2010,** *75* (6), e323–329.
56. Garde-Cerdán, T.; Arias-Gil, M.; Marsellés-Fontanet, A. R.; Ancín-Azpilicueta, C.; Martín-Belloso, O. Effects of Thermal and Non-thermal Processing Treatments on Fatty Acids and Free Amino Acids of Grape Juice. *Food Control* **2007,** *18* (5), 473–479.
57. Giner, J.; Gimeno, V.; Barbosa-Cánovas, G. V.; Martín, O. Effects of Pulsed Electric Field Processing on Apple and Pear Polyphenoloxidases. *Food Sci. Technol. Int.* **2001,** *7* (4), 339–345.
58. Giner, J.; Gimeno, V.; Espachs, A.; Elez, P.; Barbosa-Cánovas, G. V.; Martín, O. Inhibition of Tomato *(Licopersicon esculentum Mill.)* Pectin Methylesterase by Pulsed Electric Fields. *Innov. Food Sci. Emerg. Technol.* **2000,** *1* (1), 57–67.
59. Giner, J.; Ortega, M.; Mesegué, M.; Gimeno, V.; Barbosa-Cánovas, G.V.; Martín, O. Inactivation of Peach Polyphenoloxidase by Exposure to Pulsed Electric Fields. *J. Food Sci.* **2002,** *67* (4), 1467–1472.
60. González-Casado, S.; Martín-Belloso, O.; Elez-Martínez, P.; Soliva-Fortuny, R. Application of Pulsed Electric Fields to Tomato Fruit for Enhancing the Bioaccessibility of Carotenoids in Derived Products. *Food Funct.* **2018,** *9* (4), 2282–2289.
61. Grimi, N.; Lebovka, N.; Vorobiev, E.; Vaxelaire, J. Effect of a Pulsed Electric Field Treatment on Expression Behavior and Juice Quality of Chardonnay Grape. *Food Biophys.* **2009,** *4,* 191–198.
62. Heinz, V.; Álvarez, I.; Angersbach, A.; Knorr, D. Preservation of Liquid Foods by High Intensity Pulsed Electric Fields - Basic Concepts for Process Design. *Trends Food Sci. Technol.* **2001,** *12,* 103–111.
63. Jaeger, H.; Knorr, D.; Meneses, N.; Reineke, K.; Schlueter, O. Food Safety: Shelf Life Extension Technologies. In *Encyclopedia of Agriculture and Food Systems*; Van Alfen, N. K.; Ed.; Academic Press: Oxford, 2014; pp 289–303.
64. Jäger, H.; Knorr, D. Pulsed Electric Fields Treatment in Food Technology: Challenges and Opportunities. In *Handbook of Electroporation*; Miklavcic, D.; ed.; Springer International Publishing: Cham, 2017; pp 1–24.

65. Jalté, M.; Lanoisellé, J.-L.; Lebovka, N. I.; Vorobiev, E. Freezing of Potato Tissue Pre-treated by Pulsed Electric Fields. *LWT - Food Sci. Technol.* **2009,** *42* (2), 576–580.
66. Jayaram, S.; Castle, G. S. P.; Margaritis, A. The Effects of High field DC Pulse and Liquid Medium Conductivity on Survivability of Lactobacillus Brevis. *Appl. Microbiol. Biotechnol.* **1993,** *40* (1), 117–122.
67. Jeyamkondan, S.; Jayas, D. S.; Holley, R. A. Pulsed Electric Field Processing of Foods: A Review. *J. Food Protect.* **1999,** *62* (9), 1088–1096.
68. Jin Zhonglin, T.; Mendes-Oliveira, G.; Jin, T. Z.; Campanella, O. H. Modeling the Inactivation of *Escherichia coli* O157:H7 and *Salmonella Typhimurium* in Juices by Pulsed Electric Fields: The Role of the Energy Density. *J. Food Eng.* **2020,** *282,* 110001.
69. Kempkes, M. A. Pulsed electric field (PEF) Systems for Commercial Food and Juice Processing. In*Case Studies in Novel Food Processing Technologies,* Woodhead Publishing, 2010; pp 73–102.
70. Knorr, D. Impact of Non-thermal Processing on Plant Metabolites. *J. Food Eng.* **2003,** *56* (2), 131–134.
71. Korma, S. A.; Kamal-Alahmad.; Ali, A. H.; Shoaib, M.; Abed, S. M.; Yves, H. Application of Pulsed Electric Field Technology in Apple Juice Processing. *Austin J. Nutrit. Food Sci.* **2016,** *4* (2), 1080.
72. Kotnik, T.; Pucihar, G.; Miklavcic, D. Induced Transmembrane Voltage and its Correlation with Electroporation-Mediated Molecular Transport. *J. Membr. Biol.* **2010,** *236* (1), 3–13.
73. Kumar, R.; Bawa, A.; Manjunatha, S.; Tamilselvan, K.; Nadanasabapathi, S. Effect of Pulsed Electric Field and Pasteurisation Treatments on the Rheological Properties of Mango Nectar (*Mangifera indica*). *Croatian J. Food Sci. Technol.* **2015,** *7,* 22–33.
74. Lanciotti, R.; Gianotti, A.; Patrignani, F.; Belletti, N.; Guerzoni, M. E.; Gardini, F. Use of Natural Aroma Compounds to Improve Shelf-life and Safety of Minimally Processed Fruits. *Trends Food Sci. Technol.* **2004,** *15* (3), 201–208.
75. Lee, H.; Choi, S.; Kim, E.; Kim, Y.-N.; Lee, J.; Lee, D.-U. Effects of Pulsed Electric Field and Thermal Treatments on Microbial Reduction, Volatile Composition, and Sensory Properties of Orange Juice, and Their Characterization by a Principal Component Analysis. *Appl. Sci.* **2021,** *11* (1), 186.
76. Li, Y.; Chen, Z.; Mo, H. Effects of Pulsed Electric Fields on Physicochemical Properties of Soybean Protein Isolates. *LWT - Food Sci. Technol.* **2007,** *40* (7), 1167–1175.
77. López, N.; Puértolas, E.; Condón, S.; Raso, J.; Alvarez, I. Enhancement of the Extraction of Betanine From Red Beetroot by Pulsed Electric Fields. *J. Food Eng.* **2009,** *90* (1), 60–66.
78. Luengo, E.; Álvarez, I.; Raso, J. Improving the Pressing Extraction of Polyphenols of Orange Peel by Pulsed Electric Fields. *Innov. Food Sci. Emerg. Technol.* **2013,** *17,* 79–84.
79. Luengo, E.; Álvarez, I.; Raso, J. Improving Carotenoid Extraction from Tomato Waste by Pulsed Electric Fields. *Front. Nutrit.* **2014,** *1,* 12.
80. Mariatti, F.; Gunjević, V.; Boffa, L.; Cravotto, G. Process Intensification Technologies for the Recovery of Valuable Compounds from Cocoa by-products. *Innov. Food Sci. Emerg. Technol.* **2021,** *68,* 102601.
81. Marsellés-Fontanet, Á. R.; Martín-Belloso, O. Optimization and Validation of PEF Processing Conditions to Inactivate Oxidative Enzymes of Grape Juice. *J. Food Eng.* **2007,** *83* (3), 452–462.

82. Martin-Belloso, O.; Bendicho, S.; Elez-Martinez, P.; Barbosa-Canovas, G. V. Do High-Intensity Pulsed Electric Fields Induce Changes in Enzymatic Activity, Protein Conformation, and Vitamin and Flavor Stability. In *Novel Food Processing Technologies*, Barbosa-Canovas, G. V., Tapia, M. S., Cano, P., Martin, O., Martinez, A., Eds.; Marcel Dekker: New York, 2004; pp 87–104.
83. Mastwijk, H.; Gulfo, K.; Pol-Hofstad, I.; Schuten, H.; Boonman, M.; Bartels, P. Definitions and guidelines for reporting on Pulsed Electric Field experiments. In *Food preservation by Pulsed Electric Fields*; Lelieveld, H.L.M., Notermans, S., de Haan, S. W. H., Eds.; Cambridge: Woodhead, 2007; pp 320–345.
84. Mathys, A.; Reineke, K.; Jäger, H. Microbial Decontamination by Novel Technologies–Mechanisms and Application Concepts. *Front. Microbiol.* **2019,** *10,* 1476.
85. McHugh, T. Pulsed Electric Field Processing for Fruits and Vegetables. In *Food Technology Magazine*, Institute of Food Technologists: Chicago, 2016; Vol. 70, No.1.
86. Min, S.; Jin, Z. T.; Min, S. K.; Yeom, H.; Zhang, Q. H. Commercial-Scale Pulsed Electric Field Processing of Orange Juice. *J. Food Sci.* **2003,** *68* (4), 1265-1271.
87. Mohammed, M. E. A.; Eissa, A. H. A.; Aleid, S. M. Application of Pulsed Electric Field for Microorganisms Inactivation in Date palm Fruits. *J. Food Nutrit. Res.* **2016,** *4* (10), 646–652.
88. Niu, D.; Wang, L. H.; Zeng, X. A.; Wen, Q. H.; Brennan, C.; Tang, Z. S.; Wang, M. S. Effect of Ethanol Adaption on the Inactivation of Acetobacter sp. by Pulsed Electric Fields. *Innov. Food Sci. Emerg. Tech.* **2019,** *52,* 25–33.
89. Nowosad, K.; Sujka, M.; Pankiewicz, U.; Kowalski, R. The Application of PEF Technology in Food Processing and Human Nutrition. *J. Food Sci. Technol.* **2021,** *58* (2), 397–411.
90. Odriozola-Serrano, I.; Aguiló-Aguayo, I.; Soliva-Fortuny, R.; Gimeno-Añó, V.; Martín-Belloso, O. Lycopene, Vitamin C, and Antioxidant Capacity of Tomato Juice as Affected by High-Intensity Pulsed Electric Fields Critical Parameters. *J. Agric. Food Chem.* **2007,** *55* (22), 9036–9042.
91. Odriozola-Serrano, I.; Soliva-Fortuny, R.; Gimeno-Añó, V.; Martín-Belloso, O.; (2008a). Kinetic Study of Anthocyanins, Vitamin C, and Antioxidant Capacity in Strawberry Juices Treated by High-intensity Pulsed Electric Fields. *J. Agric. Food Chem., 56* (18), 8387–8393.
92. Odriozola-Serrano, I.; Soliva-Fortuny, R.; Gimeno-Añó, V.; Martín-Belloso, O.; (2008b). Modeling Changes in Health-related Compounds of Tomato Juice Treated by High-intensity Pulsed Electric Fields. *J. Food Eng., 89* (2), 210–216.
93. Odriozola-Serrano, I.; Soliva-Fortuny, R.; Martín-Belloso, O. Changes of Health-Related Compounds Throughout Cold Storage of Tomato Juice Stabilized by Thermal or High Intensity Pulsed Electric Field Treatments. *Innov. Food Sci. Emerg. Technol.* **2008,** *9* (3), 272–279.
94. Ohshima, T.; Tamura, T.; Sato, M. Influence of Pulsed Electric Field on Various Enzyme Activities. *Journal of Electrostatics* **2007,** *65* (3), 156–161.
95. Ohshima, Takayuki, Tanino, Takanori, Guionet, Alexis, Takahashi, Katsuyuki, Takaki, Koichi. **2021,** Mechanism of pulsed electric field enzyme activity change and pulsed discharge permeabilization of agricultural products. *Jpn. J. Appl. Phys.* **2007,** *60* (6), 060501.
96. Orikasa, T.; Koide, S.; Okamoto, S.; Imaizumi, T.; Muramatsu, Y.; Takeda, J.; &Tagawa, A. Impacts of Hot Air and Vacuum Drying on the Quality Attributes of Kiwifruit Slices. *J. Food Eng.* **2014,** *125,* 51–58.

97. Ortega-Rivas, E. Critical Issues Pertaining to Application of Pulsed Electric Fields in Microbial Control and Quality of Processed Fruit Juices. *Food Bioproc. Tech.* **2011,** *4* (4), 631–645.
98. Ostermeier, R.; Siemer, C., Jäger, H., Giersemehl, P., Töpfl, S. Influence of Pulsed Electric Field (PEF) Pre-treatment on the Convective Drying Kinetics of Onions. *J. Food Eng.* **2018,** *237,* 110–117.
99. Ozturk, B.; Winterburn, J.; Gonzalez-Miquel, M. Orange Peel Waste Valorisation Through Limonene Extraction Using Bio-based Solvents. *Biochem. Eng. J.* **2019,** *151,* 107298.
100. Peleg, M. A Model of Microbial Survival After Exposure to Pulsed Electric Fields. *J. Sci. Food Agric.* **1995,** *67* (1), 93–99.
101. Phillips, L. G.; Whitehead, D. M.; Kinsella, J. Protein Gelation. In *Structure–Function Properties of Food Proteins*; Phillips, L. G., Whitehead, D. M., Kinsella, J., Eds.; Academic Press: Boston, 1994; pp 179–204.
102. Plaza, L.; Sánchez-Moreno, C.; Elez-Martínez, P.; de Ancos, B.; Martín-Belloso, O.; Cano, M. P. Effect of Refrigerated Storage on Vitamin C and Antioxidant Activity of Orange Juice Processed by High-pressure or Pulsed Electric Fields with Regard to Low Pasteurization. *Eur. Food Res. Technol.* **2006,** *223* (4), 487–493.
103. Postma, P. R.; Pataro, G.; Capitoli, M.; Barbosa, M. J.; Wijffels, R. H.; Eppink, M. H. M.; Olivieri, G.; Ferrari, G. Selective Extraction of Intracellular Components from the Microalga *Chlorella vulgaris* by Combined Pulsed Electric Field–temperature Treatment. *Bioresour. Technol.* **2016,** *203,* 80–88.
104. Puértolas, E.; Barba, F. J. Electrotechnologies Applied to Valorization of by-Products From Food Industry: Main Findings, Energy and Economic Cost ofTheir Industrialization. *Food Bioprod. Process.* **2016,** *100,* 172–184.
105. Puértolas, E.; Cregenzán, O.; Luengo, E.; Álvarez, I.; Raso, J. Pulsed-electric-Field-assisted Extraction of Anthocyanins from Purple-fleshed Potato. *Food Chem.* **2013,** *136* (3), 1330–1336.
106. Qin, B. L.; Chang, F. J., Barbosa-Canovas, B. V.; Swanson, B. G. Non-thermal Inactivation of *Saccharomyces cerevisiae* in Apple Juice Using Pulsed Electric Fields. *LWT* **1995,** *28,* 564–568.
107. Qin, Bai-Lin, Zhang, Q.; Barbosa-Canovas, G. V., Swanson, B. G.; Pedrow, P. D. Inactivation of Microorganisms by Pulsed Electric Fields of Different Voltage Waveforms. *IEEE Trans. Dielectr. Electr. Insul.* **1994,** *1* (6), 1047-1057.
108. Qiu, X.; Sharma, S.; Tuhela, L.; Jia, M.; H. Zhang, Q. An integrated PEF pilot plant for continuous non theermal pasteurization of fresh orange juice. *Trans. ASAE* **1998,** *41* (4), 1069–1074.
109. Rahaman, A.; Siddeeg, A.; Manzoor, M. F.; Zeng, X. A.; Ali, S.; Baloch, Z.; Li, J. &Wen, Q. H. Impact of Pulsed Electric Field Treatment on Drying Kinetics, Mass Transfer, Colour Parameters and Microstructure of Plum. *J. Food Sci. Technol.* **2019,** *56* (5), 2670–2678.
110. Raso, J. Fundamental and Applied Aspects of Pulsed Electric Fields for Microbial Inactivation. In *1st World Congress on Electroporation and Pulsed Electric Fields in Biology, Medicine and Food & Environmental Technologies: Portorož, Slovenia,* Springer Singapore, 2016; pp 11–14.
111. Raso, J.; Calderón, M. L.; Góngora, M.; Barbosa-Cánovas, G. V.; Swanson, B. G. Inactivation of *Zygosaccharomyces Bailii* in Fruit Juices by Heat, High Hydrostatic Pressure and Pulsed Electric Fields. *J. Food Sci.* **1998,** *63* (6), 1042–1044.

112. Reineke, K.; Schottroff, F.; Meneses, N.; Knorr, D. Sterilization of Liquid Foods by Pulsed Electric Fields–an Innovative Ultra-high Temperature Process. *Front. Microbiol.* **2015,** *6,* 400.
113. Riener, J. Combined Effect of Temperature and Pulsed Electric Fields on Pectin Methyl Esterase Inactivation in Red Grapefruit Juice (Citrus paradisi). *Eur. Food Res. Technol.* **2009,** *228* (3), 373–379.
114. Rivas, A.; Rodrigo, D.; Martínez, A.; Barbosa-Cánovas, G. V.; Rodrigo, M. Effect of PEF and Heat Pasteurization on the Physical–chemical Characteristics of Blended Orange and Carrot Juice. *LWT - Food Sci. Technol.* **2006,** *39* (10), 1163–1170.
115. Rocha, C. M. R.; Genisheva, Z.; Ferreira-Santos, P.; Rodrigues, R.; Vicente, A. A.; Teixeira, J. A.; Pereira, R. N. Electric field-based Technologies for Valorization of Bioresources. *Bioresour. Technol.* **2018,** *254,* 325–339.
116. Rodrigo, D.; Barbosa-Cánovas, G. V.; Martínez, A.; Rodrigo, M. Pectin methyl Esterase and Natural Microflora OF Fresh Mixed Orange and Carrot Juice Treated with Pulsed Electric Fields. *J. Food Protect.* **2003,** *66* (12), 2336–2342.
117. Roobab, U.; Aadil, R. M.; Madni, G. M.; Bekhit, A. E. The Impact of Nonthermal Technologies on the Microbiological Quality of Juices: A Review. *Compr. Rev. Food Sci. Food Saf.* **2018,** *17* (2), 437–457.
118. Roohinejad, S.; Oey, I.; Everett, D. Effect of Pulsed Electric Field Processing on Carotenoid Extractability of Carrot Purée. *Int. J. Food Sci. Technol.* **2014,** *49,* 2120–2127.
119. Sale, A. J. H.; Hamilton, W. A. Effects of High Electric Fields on Microorganisms: I. Killing of Bacteria and Yeasts. *Biochim. Biophys. Acta (BBA)-Gen. Subj.* **1967,** *148* (3), 781–788.
120. Salehi, F. Physico-chemical Properties of Fruit and Vegetable Juices as Affected by Pulsed Electric Field: A Review. *Int. J. Food Propert.* **2020,** *23* (1), 1036–1050.
121. Salvia-Trujillo, L.; Morales-de la Peña, M.; Rojas-Graü, M. A.; Martín-Belloso, O. Microbial and Enzymatic Stability of Fruit Juice-milk Beverages Treated by High Intensity Pulsed Electric Fields or Heat During Refrigerated Storage. *Food Control* **2011,** *22* (10), 1639–1646.
122. Sánchez-Moreno, C.; Plaza, L.; Elez-Martínez, P.; De Ancos, B.; Martín-Belloso, O.; Cano, M. P. Impact of High Pressure And Pulsed Electric Fields on Bioactive Compounds and Antioxidant Activity of Orange Juice in Comparison With Traditional Thermal Processing. *J. Agric. Food Chem.* **2005,** *53* (11), 4403–4409.
123. Schilling, S.; Schmid, S.; Jäger, H.; Ludwig, M.; Dietrich, H.; Toepfl, S.; Knorr, D.; Neidhart, S.; Schieber, A.; Carle, R. Comparative Study of Pulsed Electric Field and Thermal Processing of Apple Juice with Particular Consideration of Juice Quality and Enzyme Deactivation. *J. Agric. Food Chem.* **2008,** *56* (12), 4545–4554.
124. Sentandreu, E.; Carbonell, L.; Rodrigo, D.; Carbonell, J. V. Pulsed Electric Fields Versus Thermal Treatment: Equivalent Processes To Obtain Equally Acceptable Citrus Juices. *J. Food Protect.* **2006,** *69* (8), 2016–2018.
125. Srivastava, N.; Singh, A.; Kumari, P.; Nishad, J. H.; Gautam, V. S.; Yadav, M.; Bharti, R.; Kumar, D.; Kharwar, R. N. Advances in Extraction Technologies: Isolation and Purification of Bioactive Compounds from Biological Materials. In *Natural Bioactive Compounds*; Sinha, R. P., Häder, D. P., Eds.; Academic Press, 2021; pp 409–433.
126. Syed, Q. A., Ishaq, A.; Rahman, U. U. Pulsed Electric Field Technology in Food Preservation: A Review. *J. Nutrit. Health Food Eng.* **2017,** *6* (5), 168–172.
127. Tabanelli, G.; Montanari, C.; Arioli, S.; Magnani, M.; Patrignani, F.; Lanciotti, R.; Gardini, F. Physiological Response of Saccharomyces Cerevisiae to Citral Combined with Thermal Treatment. *LWT* **2019,** *101,* 827–834.

128. Taiwo, K. A.; Angersbach, A.; Knorrr, D. Effects of Pulsed Electric Field on Quality Factors and Mass Transfer During Osmotic Dehydration of Apples. *J. Food Process. Eng.* **2003,** *26* (1), 31–48.
129. Telfser, A.; Gómez Galindo, F. Effect of Reversible Permeabilization in Combination with Different Drying Methods on the Structure and Sensorial Quality of Dried Basil *(Ocimum basilicum L.)* Leaves. *LWT- Food Sci. Technol.* **2019,** *99,* 148–155.
130. Thulasidas, J.; Sree, V. G.; Sundararajan, R. Enhancement of Bioactive Compounds from Green Grapes Extract using Pulsed Electric Field treatment. *J. Cancer Prevent. Curr. Res.* **2018,** *9,* 226–228.
131. Thulasidas, J.; Varadarajan, G.; Sundararajan, R. Pulsed Electric Field for Enhanced Extraction of Intracellular Bioactive Compounds From Plant Products: An Overview. *Novel Approaches in Drug Designing and Development* **2019,** *5* (2), 26–31.
132. Toepfl, S. Pulsed Electric Field Food Processing -industrial Equipment Design and Commercial Applications. *Stewart Postharvest Review* **2012,** *8,* 1–7.
133. Toepfl, S.; Heinz, V.; Knorr, D. High Intensity Pulsed Electric Fields Applied for Food Preservation. *Chem. Eng. Process.* **2007,** *46* (6), 537–546.
134. Toepfl, S.; Knorr, D. Pulsed Electric Fields as a Pretreatment Technique in Drying Processes. *Stewart Postharvest Review* **2006,** *2,* 1–6.
135. Toepfl, S.; Mathys, A.; Heinz, V.; Knorr, D. Review: Potential of High Hydrostatic Pressure and Pulsed Electric Fields for Energy Efficient and Environmentally Friendly Food Processing. *Food Rev. Int.* **2006,** *22* (4), 405–423.
136. Toepfl, S.; Siemer, C.; Saldaña-Navarro, G.; Heinz, V. Overview of Pulsed Electric Fields Processing for Food. In *Emerging Technologies for Food Processing*; Sun, D. W., ed.; 3rd edn. Academic Press: San Diego, 2014; pp 93–114.
137. Torregrosa, F.; Cortés, C.; Esteve, M. J.; Frígola, A. Effect of High-intensity Pulsed Electric Fields Processing and Conventional Heat Treatment on Orange-carrot Juice Carotenoids. *J. Agric. Food Chem.* **2005,** *53* (24), 9519–9525.
138. Torregrosa, F.; Esteve, M. J.; Frígola, A.; Cortés, C. Ascorbic Acid Stability During Refrigerated Storage of Orange–carrot Juice Treated by High Pulsed Electric Field and Comparison with Pasteurized Juice. *J. Food Eng.* **2006,** *73* (4), 339–345.
139. Tylewicz, U.; Tappi, S.; Mannozzi, C.; Romani, S.; Dellarosa, N.; Laghi, L.; Ragni, L.; Rocculi, P.; Dalla Rosa, M. Effect of pulsed electric field (PEF) pre-treatment Coupled With Osmotic Dehydration on Physico-chemical Characteristics of Organic Strawberries. *J. Food Eng.* **2017,** *213,* 2–9.
140. Unal, R.; Yousef, A. E.; Dunne, C. P. Spectrofluorimetric Assessment of Bacterial Cell Membrane Damage by Pulsed Electric Field. *Innov. Food Sci. Emerg. Technol.* **2002,** *3* (3), 247–254.
141. Vallverdú-Queralt, A.; Odriozola-Serrano, I.; Oms-Oliu, G.; Lamuela-Raventós, R. M.; Elez-Martínez, P.; Martín-Belloso, O. Impact of High-intensity Pulsed Electric Fields on Carotenoids Profile of Tomato Juice Made of Moderate-intensity Pulsed Electric Field-treated Tomatoes. *Food Chem.* **2013,** *141* (3), 3131–3138.
142. Vallverdú-Queralt, A.; Oms-Oliu, G.; Odriozola-Serrano, I.; Lamuela-Raventos, R. M.; Martín-Belloso, O.; Elez-Martínez, P. Effects of Pulsed Electric Fields on the Bioactive Compound Content and Antioxidant Capacity of Tomato Fruit. *J. Agric. Food Chem.* **2012,** *60* (12), 3126–3134.
143. Van Loey, A.; Verachtert, B.; Hendrickx, M. Effects of High Electric Field Pulses on Enzymes. *Trends Food Sci. Technol.* **2001,** *12* (3), 94–102.

144. Velickova, E.; Tylewicz, U.; Dalla Rosa, M.; Winkelhausen, E.; Kuzmanova, S.; Romani, S.. Effect of Pulsed Electric Field Coupled with Vacuum Infusion on Quality Parameters of Frozen/thawed Strawberries. *J. Food Eng.* **2018,** *233,* 57–64.
145. Wibowo, S.; Essel, E. A.; De Man, S.; Bernaert, N.; Droogenbroeck, B.; Grauwet, T.; Hendrickx, M. Comparing the Impact of High Pressure, Pulsed Electric Field and Thermal Pasteurization on Quality Attributes of Cloudy Apple Juice Using Targeted and Untargeted Analyses. *Innov. Food Sci. Emerg. Technol.* **2019,** *54.*
146. Wiktor, A.; Schulz, M.; Voigt, E.; Witrowa-Rajchert, D.; Knorr, D. The Effect of Pulsed Electric Field Treatment on Immersion Freezing, Thawing and Selected Properties of Apple Tissue. *J. Food Eng.* **2015,** *146,* 8–16.
147. Wouters, P. C.; Bos, A. P.; Ueckert, J. Membrane Permeabilization in Relation to Inactivation Kinetics of Lactobacillus Species due to Pulsed Electric Fields. *Appl. Environ. Microbiol.* **2001,** *67* (7), 3092–3101.
148. Yeom, H. W.; Streaker, C. B.; Zhang, Q. H.; Min, D. B. Effects of Pulsed Electric Fields on the Activities of Microorganisms and Pectin Methyl Esterase in Orange Juice. *J. Food Sci.* **2000a,** *65* (8), 1359–1363.
149. Yeom, H. W.; Streaker, C. B.; Zhang, Q. H.; Min, D. B. Effects of Pulsed Electric Fields on the Quality of Orange Juice and Comparison with Heat Pasteurization. *J. Agric. Food Chem.* **2000b,** *48* (10), 4597–4605.
150. Yeom, H. W.; Zhang, Q. H.; Chism, G. W. Inactivation of Pectin Methyl Esterase in Orange Juice by Pulsed Electric Fields. *J. Food Sci.* **2002,** *67* (6), 2154–2159.
151. Yeom, H. W.; Zhang, Q. H.; Dunne, C. P. Inactivation of Papain by Pulsed Electric Fields in a Continuous System. *Food Chem.* **1999,** *67* (1), 53–59.
152. Yin, Y.; Han, Y.; Liu, J. A Novel Protecting Method for Visual Green Color in Spinach Puree Treated by High Intensity Pulsed Electric Fields. *J. Food Eng.* **2007,** *79* (4), 1256–1260.
153. Yongguang, Y.; Yuzhu, H.; Yong, H. Pulsed Electric Field Extraction of Polysaccharide from Rana Temporaria Chensinensis David. *Int. J. Pharm.* **2006,** *312* (1), 33–36.
154. Yousuf, A.; Rahman, M. A.; Uddin, M. R.; Hoque, M. M.; Sayem, A. Pineapple Juice Preservation by Pulsed Electric Field Treatment. *Open J. Biol. Sci.* **2020,** *5* (1), 6–12.
155. Yu, Y.; Jin, T. Z.; Fan, X.; Wu, J. Biochemical Degradation and Physical Migration of Polyphenolic Compounds in Osmotic Dehydrated Blueberries with Pulsed Electric Field and Thermal Pretreatments. *Food Chem.* **2018,** *239,* 1219–1225.
156. Yu, Y.; Jin, T. Z.; Xiao, G. Effects of Pulsed Electric Fields Pretreatment and Drying Method on Drying Characteristics and Nutritive Quality of Blueberries. *J. Food Process. Preserv.* **2017,** *41* (6), e13303.
157. Zderic, A.; Zondervan, E.; Meuldijk, J. Breakage of Cellular Tissue by Pulsed Electric Field: Extraction of Polyphenols From Fresh Tea Leaves. *Chem. Eng. Transact.* **2013,** *32,* 1795–1800.
158. Zeng, X.-A.; Hong, J. Modification of Plant Biopolymers by Pulsed Electric Fields. In *Handbook of Electroporation*; Miklavčič. D.; ed.; Springer International Publishing: Cham, 2017; pp 2423–2438.
159. Zhao, W.; Yang, R. Pulsed Electric Fields for Inactivation of Endogenous Enzymes in Foods. In *Handbook of Electroporation*; Miklavčič. D., ed.; Springer International Publishing: Cham, 2017; pp 1–13.

160. Zhao, W.; Yang, R.; Zhang, H. Q. Recent Advances in the Action of Pulsed Electric Fields on Enzymes and Food Component Proteins. *Trends Food Sci. Technol.* **2012,** *27* (2), 83–96.
161. Zhong, K.; Hu, X.; Zhao, G.; Chen, F.; Liao, X. Inactivation and Conformational Change of Horseradish Peroxidase Induced by Pulsed Electric Field. *Food Chem.* **2005,** *92* (3), 473–479.
162. Zhong, K.; Wu, J.; Wang, Z.; Chen, F.; Liao, X.; Hu, X.; Zhang, Z. Inactivation Kinetics and Secondary Structural Change of PEF-treated POD and PPO. *Food Chem.* **2007,** *100* (1), 115–123.

CHAPTER 6

APPLICATIONS AND FUTURE ASPECTS OF PULSED ELECTRIC FIELD FOR STABILITY OF VEGETABLES

SURADEEP BASAK, SHUBHAM ROHILLA, RESHU RAJPUT, and VANDANA SABLANIA

ABSTRACT

The purpose of including nonthermal techniques is not only to maintain the quality but also to optimize the process conditions required for various commodities, reducing the operation time and cost, increasing the shelf life for a longer time without altering the chemical, the functional properties of the food products and inactivation of spoilage causing enzymes and microorganisms. Conversely, conventional technologies such as drying, freezing, cooking, blanching, etc., are suggested to be better techniques when used in combination with advanced nonthermal techniques, including ultrahigh temperature, cooking for a short time, high-pressure cooking, microwave heating, ultrasonication, pulsed electric field (PEF). These techniques usually highlight on utmost retention of quality characteristics of food commodities. Permeabilization of vegetables, tissues, or organic compounds could be improved by adapting PEF as a pretreatment to upsurge the efficacy of mass transfer during drying, extraction, and diffusion processes. This chapter comprises the inactivation of microorganisms and enzymes, the extension of shelf life, and the retention of bioactive components by adopting optimum operating conditions in several vegetables. Therefore, PEF can be used as an alternative technique for increasing the shelf life stability of fruits and vegetables without compromising their quality attributes.

Nonthermal Pulsed Electric Field Technology for the Food Industry.
Preeti Birwal, Megh R. Goyal, Baldev Singh Kalsi, & Ankita Kataria (Eds.)

DOI: 10.1201/9781003605904-8

6.1 INTRODUCTION

Safe and quality food is of paramount importance to humankind for growth and survival. Several traditional and nontraditional efforts have been made in the past and are being made to ensure food safety and quality to be served to the population. Food processing is a major contributor to the global trade and economy of several nations, which is ever-increasing with the escalating demand for safe and quality food products. To address the demand, several postharvest operations, namely, drying, freezing, evaporation, osmosis, extraction, mechanical disruptions, etc., are applied to ensure shelf-stable food products.[79] During such operations (mechanical and thermal), biologically active components of food often get reduced. The problems associated with thermal postharvest operations and the emergence of consumer awareness regarding the impact of thermal processing on food qualities have resulted in the development of several minimally processed technologies that retain the color, flavor, taste, and nutritional attributes of the food after processing.[13] Among such minimal processing technology, pulsed electric field (PEF) is an area of electromagnetic technologies that has gained importance in the food sector where food samples are exposed to high-energy discharge with varying electric strengths (20–80 kV/cm) for a short time (in milliseconds).[38] PEF processing involves the passage of the food products between two electrodes in a chamber wherein high-intensity pulses are discharged onto the food system. Based on the processing time, the strength of the electric field, pulse number, and refrigeration system (To control a sudden increase in temperature within the treatment chamber), the indigenous organoleptic and physicochemical properties of the food sample are mostly retained.[65]

PEF technology is rapidly substituting the usage of conventional techniques to permeabilize or disintegrate cell membranes of fruits and vegetables due to its low cost and low heat generation, resulting in a reduced loss due to thermal degradation; eventually, the basic structure of food remains unaltered.[78,97] The cause behind the disruption of cell membranes of fruits and vegetables or microorganisms under PEF treatment is the development of electrocompressive forces across the cellular membrane due to escalation in *trans*-membrane potential as well as conductivity of the cell membrane.[105,118] The dielectric breakdown results in enhanced permeability and pore formation within the cell membranes.[118] However, the requirement of electric field intensity and energy input varies for the disintegration of plant materials (0.7–3 kV/cm and 1–20 kJ/kg) and microorganisms (10–20 kV/cm and >40 kJ/kg), respectively.[22,97] A increase in energy and the intensity of electricity will result in enhancing the pore formation or dielectric breakdown of the tissues, which will eventually lead to excessive tissue softening and loss in texture or the basic structure of the food product. Therefore, the optimum electric field and energy need to be optimized for the selected food product, and the degree of cell disintegration shall routinely be evaluated using microscopic or conductivity analysis.[8] PEF technology is now widely used as a pretreatment to conventional postharvest operations to achieve the desired product with reduced cost, energy, and minimal loss of nutritional components.

The present chapter emphasizes the application of PEF technology in the processing of vegetables and the associated future perspectives.

6.2 PEF AS A PREPROCESSING STEP IN VEGETABLES

6.2.1 PEF AS A PRETREATMENT TO DRYING

Drying is one of the oldest and most extensively employed processing techniques used for the preservation of food materials. In drying, moisture is vaporized from the food sample using thermal energy ultimately providing the product with better qualities.[46] Removing the moisture helps extend the shelf life of the food materials because there is no availability of free moisture for the microorganism to attack the dried food. Most industries are commonly using connective or hot air drying methods for the drying of many fruits, vegetables, meat, spices, condiments, etc.,[66] However, connective drying is known for its numerous beneficial properties, such as food products with more extended shelf life, low storage costs, cheap, easy to transport, and user-friendly technique, but the main demerit is it requires a longer time, degradation to bioactive compounds, and change in the nutritional and physicochemical properties of the food sample.[52] Therefore, to overcome this problem, many researchers use hybrid techniques such as ultrasound, microwave, and PEF followed by drying to overcome the aforementioned issue.[69] Researchers reported that applying nonthermal technologies before the drying helps prevent the degradation of bioactive compounds and is able to maintain the optimum food quality.[14,92] Among these novel nonthermal technologies, the pulse electric field is known for its economic, environmental, and consumer-friendly technology used for improving mass transfer and inactivating the growth of unwanted microorganisms.[111]

Applying a pulse electric field before the drying process leads to easy and fast removal of moisture from the food sample because PEF leads to electroporation of cell membrane, which leads to more efficient removal of moisture, the release of cellular compounds, enhanced permeability from the sample matrix, and makes the drying process firm with minimum changes in the nutritional as well as physicochemical properties of the sample. Researchers reported that applying PEF before drying could reduce 57% of the drying time and are able to retain the shape of the product as compared with the untreated sample.[48]

Similar results have been reported in the case of carrots, where the electric field of 5 kV/cm reduced the drying time compared to the untreated samples.[53] PEF has been coupled with the drying of many food samples, and many scientific reports have been published on the physicochemical and phytochemical content of the pre- and posttreated sample.[69] PEF reportedly enhanced the dehydration process with an increase in diffusion coefficient in the potato samples.[49] Table 6.1 illustrates the effects of PEF on the drying characteristics of various vegetables.

6.2.2 PEF AS A PRETREATMENT TO FREEZING

Freezing is a significant and widely employed preservation technique used to preserve various food materials. Freezing is known to suppress the growth of microorganisms and enzyme reactions, and because of that, it can preserve the nutritional, phytochemical, flavors, and colors of the food sample for a long time.[103] In freezing, the free water is reduced by converting into frozen form; therefore, it is not available for biochemical and microbial reactions, and the shelf life of the food is enhanced.[7] The freezing time has a significant impact on the quality attributes, especially the appearance of the frozen food products. However, the application of PEF has been shown to reduce not only the freezing time but also to retain the better shape of the food sample as opposed to the untreated samples.

Jalte[47] researched the application of PEF on potatoes and it was found that applying PEF leads to reducing the freezing time by almost 65%. Similar results were reported in apples where PEF-treated frozen apples showed a reduction in the freezing time and retained better shape compared to the untreated sample after defrosting.[72] Applying PEF in the form of electric voltage to the food sample leads to electroporation in the biological cells over a period of time.[11] However, in scientific literature, it is reported that electric field strength in the range of 0.5–2 kV/cm is preferred for electroporation in animal and plant cells.[20] PEF is widely employed in various food processing industries for the inactivation of microorganisms by nonthermal means, which does not affect their volatile and phytochemical compounds

TABLE 6.1 Impact of PEF on Drying Characteristics of Various Vegetables.

Vegetables	PEF conditions	Observations	Reference
Carrots and parsnip	E_n (carrots) = 65.2 kJ/kg E_n (parsnips) = 65.8 KJ/kg P = 20 μs Frequency = 50 Hz	Reducing drying time by 13–28% in carrots and 5–28% in parsnips improved moisture diffusivity and retention of color after drying	[6]
Onion	E = 0.36–1.07 kV/cm E_n = 0.2–20.0 kJ/kg	Reduction in drying time at lower drying temperature, improved diffusion rate	[69]
Okra	E = 4 kV/cm P = NS N = NS	Better drying quality and improved rehydration ratio	[1]
Red beetroot	Time = 10 μs P = 200 μs N = 5500	Reduction in drying temperature by 20–25°C More shrinkage, improved rehydration time, no difference in textural properties	[90]
Mushroom	E = 1.61 kV/cm Pulse time = 2.5 μs	High-quality mushroom slices Reduced energy consumption Better rehydration and overall quality properties	[63]
Bell pepper	E = 1.07 kV/cm E_n = 0–4 kJ/kg N = 6–44	Retention of vitamin C reduced total phenolic, larger particle size as compared to untreated samples	[84]
Carrots	E = not specified Frequency = 0.5 Hz Time = 7 μs N = not specified	50% reduction in drying time, better retention of carotenoids	[102]

E: Electric field strength, E_n: Energy, P: Pulse width, NA: Not specified, N: Number of pulses.

and makes the freezing process overwhelming.[9] The impact of the PEF on the freezing characteristics of various vegetables is enlisted in Table 6.2.

6.3 FOOD PRESERVATION BY PULSED ELECTRIC FIELD

6.3.1 INACTIVATION OF ENZYMATIC ACTIVITY IN VEGETABLES

Ohshima[68] mentioned that enzymatic activity in protein-rich food could be inhibited under the influence of PEF as a pretreatment. It was reported that enzymes are mostly more resistant to PEF than microorganisms. Inactivation of these enzymes depends on frequency, electric field strength, energy, the membrane of a microorganism, thickness, and porosity of plant cell, which results in the instability of the enzymes, electroporation, and electropermeabilization, which corresponds to complete inactivation or resealing of the cell. Irrespective of the nature of food commodities, enzymes can cause numerous detrimental impacts. Commonly, spoilage-causing enzymes are enlisted as pectin methylesterase (PME), peroxidase (POD), polyphenol oxidase (PPO), etc.

Van Loey[100] reported that enzymes such as PPO, POD, lipoxygenase, and pectin-methylesterase are hardly affected by PEF treatment in distilled water. Still, these enzymes are very sensitive when present in complex food systems, namely, dairy products, fruits, and vegetables under escalated electric fields. Several studies showed significant changes in the inactivation of enzymes when varying PEF strengths and temperatures act upon it.[60,87] Products such as sugar beet juice (with PEF 0.6 kV/cm and 5 kJ/kg electric energy),[76] maize germ oil, and phytosterol with 0.6 kV/cm electric field and 0.62 kJ/kg electric energy),[36] betanine with 1 kV/cm electric field, 7 kJ/kg electric energy[33]; and antioxidants from fennel (0.6 kV/cm, 5 kJ/kg)[27] are treated with PEF to cause the breakdown of cell membrane for releasing endogenous enzymes from the cells of fruits and vegetables to maintain the shelf stability.

6.3.2 PECTIN METHYLESTERASE AND POLYGALACTURONASE

Enzymes present inside cell membranes, namely, PME and polygalacturonase (PG) are mainly responsible for maintaining physical and sensory attributes of several fresh crops, such as vegetable/juice and fruit/juice during processing. PME, in turn, is responsible for the de-esterification of pectin, which sequentially decreases the viscosity of vegetable and fruit juices, reduces

TABLE 6.2 Impact of PEF Pretreatment on Freezing Characteristics of Various Vegetables.

Vegetables	PEF process conditions	Observations	Reference
Potato	E = 600 kV/cm, Pulses = 10 μs, monopolar pulse	Increased in the freezing rate and thawing process depending upon the moisture content, controlled deterioration, increased shelf life, reduction in operational cost	[51]
Carrots	E = 1.85 kV/cm N = 0–100 pulses,	Reduced freezing time by 32% Reduction in thawing time significantly improved the quality of the treated carrots	[103]
Spinach	E = 580 kV/cm in combination with vacuum	Improved tolerance to trehalose sugar during freezing tolerance, improved shelf life	[73]
Potato, green bean, onion, and spinach	E = 400–600 kV/cm	Reduction in freezing time by 18%, Retention of color	[17]
Carrots	E = 1 kV/cm P = 100 Frequency = 4 Hz	Increased permeability, reduction in freezing time, and combination of PEF with cryoprotectant improved the texture, color, and overall quality of the treated samples	[89]

E: Electric field strength, P: Pulse width.

the cloudiness, etc. PME plays a crucial role in the ripening stages of fresh produce, where it is responsible for the modification of pectin for the release of other substrates during maturation. PME is naturally deliberated to cause spoilage and slightly modifies several sensory attributes throughout their shelf life. In addition, PG catalyzes α–1, 4– glycosidic bonds by hydrolytic breakdown.[21] PG usually occurs in combination with PME in fresh produce and is specifically used as a precursor of blanching. It ruptures the pectin chains once PME allows the opening of polymeric chains. The collective impact of both enzymes results in the softening of fruits and vegetables at the beginning of the senescence stage. In such cases, it becomes crucial to inactivate these enzymes. In some of the investigations, the activity of residual PG was reported as less than 25%.[91]

To produce clear juice, PME should be inactivated to reduce the cloudiness and vice versa. PME acts on the pectin present in the substance, resulting in the destabilization of cloudiness and thus reducing the turbidity of the juice. Along with PME, PG inactivation is also required as it is accountable for causing a decrease in the viscosity of the tomato juice through direct pectin solubilization.[31] Thermally treated tomato juice at 90°C for 1 min resulted in 96% inactivation of PME, whereas the activity was observed to be reduced to 82% when treated with HIPEF. On the other hand, PME activity was reported to be reduced by only 71% at 90°C for 30 s. A similar change in PME inactivation (87%) was observed at 0.5 Hz for 1600 μs at 16.8 kV/cm.[30] Maximum inactivation (93.8%) of PME in tomatoes was reported by Giner et al.[34] Several research articles have shown different PEF conditions required for microbial inactivation and enzyme inactivation, which are listed in Table 6.3.

6.3.3 *POLYPHENOL OXIDASE AND PEROXIDASE (POD)*

PPO and POD are the precursors of detrimental activity of fruits and vegetables over a while. PPO is responsible for causing oxidative and reductive catalysis of phenolic components, whereas POD is described only for oxidative catalysis. Exposure of processed food to these enzymes in the presence of oxygen generates secondary and tertiary metabolites in the form of quinones that further result in a browning reaction. They are highly conjugated complex compounds that lead to the development of off-flavor and also reduce the vitamin content in food products. Therefore, the activation rate of POD and PPO can be used as a quality indicator to assess the freshness of processed fruits and vegetables. The inactivation of enzymes is dependent on various parameters such as treatment temperature, processing

TABLE 6.3 Impact of PEF on Enzymatic and Microbial Activity of Vegetables/Vegetable Juices.

Vegetables/vegetable juice	Treatment conditions	Effect of PEF	Reference
Carrot (unblanched)	E = 1 kV/cm Frequency = 1 Hz Number of pulses = 20, monopolar	<5% reduction in peroxidase (POD) activity	[33]
Carrot (blanched)	E = 1 kV–1.5 Kv Frequency =1 Hz Number of pulses = 20, monopolar Blanching = 100°C, 3 min	30–50% reduction in POD activity	
Carrot	E = 35 kV Frequency = 50–250 Hz P = 1–7 µs, monopolar or bipolar mode	Inactivation of POD enzyme activity and color preservation	[80]
Carrot puree	E = 0.2–1.2 kV/cm P = 20 µs, square wave bipolar Frequency = 5–300 Hz E_n = 516.28 kJ/kg Number of pulse = 130–9000	Inactivation of ascorbic acid oxidase (AAO) by 61.48%	[50]
Horseradish	E = 5–25 kV/cm Number of pulses = 207–1242	35% reduction in POD activity with PEF 57.7% decrease in POD activity when treated with PEF and heated at 100°C, 5 min	[114]
Tomato	E = 35 kV/cm P = 4 µs, bipolar T_m = 1500 µs Frequency = 100 Hz	97% inactivation of hydroperoxidase 82% inactivation of pectin methylesterase (PME), Only 12% inactivation of polygalacturonase (PG)	[5]

TABLE 6.3 *(Continued)*

Vegetables/vegetable juice	Treatment conditions	Effect of PEF	Reference
	$E = 35$ kV/cm $P = 4$ μs, bipolar $T_m = 1000$ μs Frequency = 200 Hz	81.6% decrease in PME activity	[28]
	$E = 40$ kV/cm Time –57 μs $T = 92$°C, 90 sec	47% reduction in lipoxygenase activity, 6-log reduction in the growth of aerobic microflora, yeast, and molds	[61]
Broccoli juice	$E = 35$ kV/cm $P = 2000$ μs, bipolar, monopolar	68.7 and 82.1% reduction in lipoxygenase activity under bipolar and monopolar mode, respectively	[20]
	$E = 15$–35 kV/cm $P = 4$ μs, bipolar $T_m = 500$–1000 μs	Decrease in residual polyphenol oxidase (PPO) activity from 78.99% to 57.51% at 15 kV/cm and 50.78–36.11% at 35 kV/cm in bipolar mode	
Pea soup	$E = 30$ kV/cm $P = 10$ or 15 $T_m = 0.75$ μs Frequency = 6.7 Hz	4.8 D logarithmic reduction in *Escherichia coli*, and *Bacillus subtilis*	[101]
	$E = 33$ kV/cm $P = 30$ $T_m = 0.5$ μs Frequency = 4.3 Hz	6.5 D logarithmic reduction in *E. coli*, and 5.3 D in *B.subtilis*	

E: Electric field, T_m = Treatment time, E_n = Electrical pulse energy, *P*: Pulse width.

conditions, nature of the commodity, etc. There are several studies that showed the inactivation of PPO and POD under combined treatment of PEF and conventional thermal techniques.[113]

PPO and POD are known for their oxidative activity and are found abundantly in living organisms. POD is reported to be highly sensitive when treated with PEF as compared to pasteurization using conventional thermal treatment (72°C for 26 s).[82] The rate of enzymatic inactivation for PPO and POD was reported to be 71 and 68%, respectively, when treated at 50°C, having an electric field strength of 40 kV/cm at 100 μs pulse rate. On the contrary, the same enzymes have shown reduced values, which are 46 and 48% for PPO and POD, respectively, under conventional pasteurization treatment. PPO activity is reported to be decreased substantially when the pretreatment temperature achieves 80°C before PEF treatment. This activity is even more operative if used in combination with techniques such as ohmic heating, ultrasonication, microwave heating, etc.,[56] Zhong[113] stated that PEF could be successfully applied to reduce the efficiency of PPO and POD enzymes when treated with an electric field of more than 25 kV/cm.

The highest inactivation of POD activity in carrots was observed at 60°C pretreatment conditions. It could lead to the inactivation of POD from 50 to 90% if the preheating conditions are maintained between 40 and 80°C.[56] An increase in temperature results in increasing the internal energy of the enzymes and consequently leads to a decrease in POD activity by developing three-dimensional (3D) structures within the cell.[20] Moreover, the procedure of inactivation of an enzyme is likely to be encouraged by causing modification in the conformational structure of enzymes. However, it has been proved that high-intensity PEF perhaps shows more changes in the enzyme structure by polarizing the protein molecules through hydrophobic linkages or covalent bonds.

Zhong et al.[114] explained the reduction in POD activity of horseradish due to a change in its secondary structure under the influence of PEF at 25 kV/cm for 1740 μs. POD activity in horse radish was reported to decrease with the intensification of pulses (207–1242 pulses). The maximum reduction of horse radish POD activity was observed as 34.7% at 22 kV/cm and 1242 pulses.

6.3.4 LIPOXYGENASE (LOX)

LOX is known for its nonsulfur, an iron-containing enzyme that is responsible for the dioxygenation of polyunsaturated fatty acids. LOX is accountable for the rancidity and leads to the development of off-flavor. Various first-order and second-order models have been applied to study the inactivation kinetics of LOX enzyme under the influence of PEF with varying electric field strength from 0–35 kV/cm, processing time ranging from 20 to 70 μs, and temperature of the treatment varied from 10 to 50°C.[61] They demonstrated that residual LOX in tomato juice after PEF treatment has a significant impact on the flavor of the juice. The inhibition of LOX enzyme in tomato juice samples is reportedly irreversible under thermal or PEF processing of the juice. Reports on the reduction of LOX activity by 54% during the initial cold break roller, and furthermore by the action of PEF, suggested the impact of this minimal processing technology on the enzyme in the tomato juice stored under refrigerated conditions for at least 90 days.[61]

6.3.5 INACTIVATION OF MICROBIAL GROWTH IN VEGETABLES

Several studies have been explored to understand the approach of PEF in inhibiting the microbial load present in various food products. PEF is found to be an efficient method in the inactivation of vegetative bacterial pathogens, yeast, and molds.[106] However, the exact scientific route for the inactivation of microorganisms under the influence of PEF is unclear in the present scientific literature. However, it is assumed that PEF treatment disrupts the integrity of microbial cell membranes by inducing electric fields and electromechanical compression that results in the formation of porins.[12] The mechanism exhibits the combined action of electroporation and electropermeabilization under high electric voltage.[96,106]

Bacterial spores are highly resistant to PEF, inconsistently their growth cannot be destroyed by confining PEFs. Thus, this technology is primarily intended to destroy pathogenic and spoilage-causing vegetative microorganisms in food systems.[12,94] This technique is reported to be adequate for sterilizing various juices due to their low acid and protein value. However, the vegetative spores and microbial growth can be reduced under highly acidic conditions. A defensive layer of PEF may be developed on the surface of electrodes during the supply of high electric voltage, which allows the movement of microbes and proteins over the surface and successively leads to inactivation.[61] Exposure of any food product for PEF treatment can be implied in the form of oscillatory, double exponential, and square wave pulses. Of these, the most effective form of energy delivery in PEF can be performed by square-wave pulse. The bipolar square-wave pulse leads to maximum inactivation of microorganisms when compared to the monopolar pulse. However, PEF is effective against bacterial inactivation under specific operating conditions.[19]

6.4 PEF AS PRETREATMENT FOR THE EXTRACTION OF BIOACTIVE COMPOUNDS FROM VEGETABLES

6.4.1 *POLYPHENOLIC COMPOUNDS*

Several studies have been conducted for the optimum extraction of bioactive components from various crops. Compounds such as terpenes, monoterpenes, sesquiterpenes, terpenoids, phenols, flavonoids, carotenoids, caryophyllene, etc., are commonly known as bioactive compounds and possess several bioactive properties such as antimicrobial, antifungal, antitumor, anti-inflammatory, antiobesity. To obtain these compounds, it becomes important to follow adequate extraction conditions for maximum recovery. The extraction of bioactive compounds depends on several factors, such as the nature of raw material, extraction time and temperature, solute/solvent ratio, and so on.[23] Hence, the quality of these compounds has been controlled by adopting nonthermal techniques such as PEF. In this case, processing could be performed by regulating the low-temperature conditions in association with an electric field and thus reducing the activity of enzymes and microbes, responsible for the loss of bioactive components. PEF results in damage to cells or induces sublethal stress by increasing the cell permeabilization, consequently, helping in the extraction of cellular metabolites from the commodity and enhancing the extraction efficiency.[99] It was reported that PEF is responsible for producing primary metabolites from vegetables in the form of reactive oxygen species, which further releases bioactive components as secondary metabolites and combat defense mechanisms against strain.[18]

PEF-induced electroporation leads to substantial softening of tissues and reduction in turgor pressure. Consequently, PEF can improve the peeling of fruits and vegetables by weakening the cell membrane. The energy required for steam peeling is greater when compared with PEF-treated vegetable peeling. Thus, the utilization of lye peeling does not require neutralization, and thus, the pollution caused due to effluents is minimized. Moreover, the usage of somewhat expensive chemicals is minimized.[59]

Carrot is a good source of carotenoids located in the chromoplast of the cell, which shows great antioxidant activity if extracted in an appropriate amount under nonthermal processing conditions.[93] High temperature during pretreatment leads to the inactivation of enzymes, due to which degradation of phenolic components gets reduced and ultimately results in increased antioxidant activity of the commodity. It was reported that PEF processing of carrots without pretreatment did not show any change in the extractability of phenolic components.[57] A study was conducted by Puertolas[77] on the extraction of anthocyanin from purple potato. The extraction was done at a temperature ranging from 10 to 40°C, extraction time of 60–480 min. by using water and ethanol as a solvent. For improved recovery of anthocyanin, potato extract was treated with PEF conditions (3.4 kV/cm, 35 pulses of 3 μs, 8.92 kJ/kg). It was noticed that 63.9 and 65.8 mg/100 g of anthocyanin extract were obtained by using 96% ethanol without PEF and 100% water with PEF treatment, respectively. PEF is a favorable technique not only to recover the maximum quantity of anthocyanin extract but also to curtail the processing temperature, with minimum utilization of organic solvents from purple potato, without affecting the extraction yield. Therefore, PEF can likely be used with water as a solvent for a more economical and environmentally friendly solvent than ethanol.

According to Min,[62] PEF-treated tomato juice (electric field strength = 40 kV/cm, time– 57 μs, temperature = 90°C) resulted in maximum retention of the redness, flavor compounds, namely trans-2-hexanal, 2-isobutylthiazole, *cis*-3-hexanol with reduced nonenzymatic, and enzymatic browning reaction in the juice as opposed to the juice obtained after thermal processing.

Roohinejad[83] investigated the effect of PEF on the extraction of carotenoids from a plant source (carrot) and reported increased carotenoid content up to 1 kV/cm at 5 Hz, whereas further raising the frequency beyond 10 Hz at 1 kV/cm showed no significant change in carotenoid content. Usually, betanine extraction is reported as a 45–70% yield for non-PEF-treated samples, whereas the extraction is found to be higher up to 90% recovery when treated with PEF. It has been examined that the yield of bioactive components from carrot, coconut, and many other polyphenolic components, such as betanine, anthocyanin, etc., were enhanced when pretreated at 10 kV/cm electric field.[15,32,55] It was recommended that the combination of PEF with pressure (10 kg/cm^2) would assist the extraction of betanine up to 80% in 22 min and 90% in 35 min (6 times faster) when treated at 1 kV/cm at a 10 kg/cm^2 or 7 kV/cm and no pressure.[55] Min[61] showed significantly similar lycopene content of tomato juice processed under thermal and PEF treatment conditions. The possible reason could be the chemical stability of lycopene pigment due to the isomerization of *trans-cis* compounds that results in resistance to degradation under thermal conditions. Moreover, components such as tocopherols, ascorbic acid, and phenolic antioxidants assist lycopene in its improved stability during processing.[95] Table 6.4 includes different conditions required for the retention of maximum bioactive components from various vegetables.

TABLE 6.4 Impact of PEF on Bioactive Components in Vegetables/Vegetable Juices.

Vegetable juice	Treatment conditions	Effect on bioactive components	Reference
Tomato	$E = 1.2$ kV/cm $P = 30$	44% increase in total phenolic content	[99]
	$E = 1.2$ kV/cm $P = 18$	44% rise in hydrophilic antioxidant activity	
	$E = 1.2$ kV/cm $P = 5$	37% increase in lipophilic antioxidant activity, lycopene content	
	$E = 0.8$–1.2 kV/cm $P = 18$–25	99.4% retention of ascorbic acid content	
	$E = 40$ kV/cm Time– 57 μs $T = 92$°C, 90 s	Substantially no difference in lycopene content	[61]
	$E = 40$ kV/cm Time = 57 μs $T = 90$°C, 90 s	Retention of flavoring compounds, namely *trans*-2-hexanal, 2–isobutylthiazole, *cis*–3–hexanol ($p < 0.05$)	[62]
Broccoli	$E = 35$ kV/cm Time = 2000 μs	Relatively higher content of bioactive compounds and antioxidant capacity	[85]
Red bell pepper	$E = 2$ kV/cm $P = 50$ Time = 400 μs	High vitamin C ranging from 89.6 to 96.5%	[4]
Spinach	$E = 20$–100 kV/cm Addition of water- soluble zinc ion concentration = 20–200 ppm	Improved stability of chlorophyll. Spinach puree with acetic zinc below 75 ppm treated by PEF can be stored at room temperature for a longer time	[109]
Carrot	$E = 0.6$ kV/cm $T = 18$–35°C Number of pulses = 30	Increase in yield of bioactive components	[26]

TABLE 6.4 *(Continued)*

Vegetable juice	Treatment conditions	Effect on bioactive components	Reference
Carrot puree	E = 0.1–kV/cm P = 20 µs, bipolar square pulses Frequency = 5–75 Hz	Increase in carotenoids extractability	[83]
Beetroot	E = 1 kV/cm E_n = 7 kJ/kg Time = 60 min	Extraction of 90% of batanine when treated with PEF	[32]
	E = 7 kV/cm P = 2 µs E_n = 2.5 kJ/kg Time = 300 min T = 30°C	90% extraction of betanine (5-fold faster than non-PEF-treated sample)	[55]
	E = 375–1500 kV/cm P = 100 µs, monopolar Time = 0–0.2 s	Decrease in the degradation rate of betanine with the increase in extraction temperature	[54]
Maize	E = 0.6–7.3 kV/cm E_n = 0.6–90 kJ/kg	Increase in phytosterol content by 32% and isoflavanoids content by 20–21%	[37]
Purple potato	E = 3.4 kV/cm P = 105 µs, bipolar square pulses E_n = 8.92 kJ/kg Number of pulses = 35, 3 µs	Highest anthocyanin extraction	[77]
Potato peels	E = 0.75 kV/cm P = 600 µs,	99.9% higher extraction of steroid alkaloids (1856.2 µg/g)	[44]

E: Electric field, *T*: Processing temperature, E_n = Electric energy, *P*: Pulse width.

6.4.2 BIOPOLYMERS

Nowadays, consumers shift toward food products that are natural and healthy; therefore, the processing or modification of the food ingredients is done in the form of physical or nonchemical methods such as in starch and other polysaccharides. Researchers developed many modification methods for obtaining starch that is natural and healthy, including ultrasound, γ-irradiation, electric fields, annealing, ultrafine milling, extrusion, and many other techniques.[16,117] Starch is one of the major components present in food products, especially products made from cereals. It is composed of amylose and amylopectin assembled in the natural granular form (1–100 μm). However, there are some limits to natural starch; therefore, researchers modified the starches so that they can be used for wider applications and better functionalities. Pulse electric field is a nonthermal, cost-effective, and innovative method for food preservation in which the cell membrane of the food sample is permeabilized using an electric field and is widely employed in the food industries, pharmaceutical, and biotechnology industries.[35] Using PEF for the modification of the biopolymers, especially starch, will have better applicability if the electric field is applied in an optimized way in the substitution process, and hydrolysis possibly will give better results.[117]

Several scientific articles suggested the rearrangement of starch granules that resulted in alteration of physicochemical properties of the starch and also improved functionality. However, many studies have also suggested the modification of starch using PEF.[117] PEF is known to decrease the viscosity and breakdown viscosity of the starch during the pasting environment. Han[40] studied the application of PEF in maize starch given at the rate of 50 kV/cm and has been found that there is a decrease in the initial viscosity of native starch to the modified starch from 330 BU to 250 BU (Brabender unit) during pasting.[40] In the thermal analysis of modified starch done by using PEF, it was found that modified starch showed low enthalpy and gelatinization temperature range. The decrease in the gelatinization temperature is associated with the disruption in the crystalline region of a similar structure of starch.[42,110]

Han[40] reported that applying the PEF power level at 50 kV/cm on maize starch has shown a change in the molecular structure of starch, and there is a decrease in the molecular weight from 10.2×10^7 to 1.3×10^7. A similar type of study where the z-average radius and molecular weight were changed to 331–301 nm and 10×10^7 to 9×10^7 was reported for the waxy rice.[110] PEF is known to change the helical structure of the starch molecules, and confirmation was done using nuclear magnetic resonance (NMR) results in the maize starch. Although PEF is a powerful technique, it does not disrupt the granular structure of the tapioca starch; however, the crystallinity of the starch was found to decrease from 22 to 7%.[42] Applying PEF is known to increase the susceptibility of enzymes while modifying the starch molecules.[117] Although PEF is a useful and widely employed technique for the modification of starch, high pressure, ultrasound, and γ-irradiation are found to modify the starch more in comparison to PEF at 50 kV/cm.[117]

Ignat[45] studied the effect of PEF as a pretreatment on the potato and found that starch was leached by the samples because of the electroporation process in the sample; however, treated samples showed an increase in moisture and softness in the potato samples. However, some researchers stated that applying PEF does not affect the small molecule structure and durability (lipids and proteins). Through extensive PEF treatment, it has been reported changing the protein structure decreases vitamin C levels and isomerizes carotenoids but is less effective in comparison to the thermal processing methods.[108] A group of researchers worked on the PEF technology for 3D printing using cassava and wheat starch. They reported that on cassava starch, no damage on granule surface was found whereas on wheat starch fracture on granule surface was found, therefore, the starch source matters a lot in affecting the properties of the starch, and PEF acts differently.[56]

Zeng[110] studied the effect of PEF on the structure modification of waxy starch, and they reported that applying PEF caused to decrease in the slowly digestible starch from 46 to 35% and resistant starch content from 22 to 14%; however, by increasing the content of rapidly digestible starch from 32 to 55%.[110] PEF is known to increase the digestibility of the modified starch by affecting the structure of starch molecules and is also known to reduce the percentage of resistant starch (reducing the stability of the system).[117] In studies, it was reported that applying PEF at 3.75 kV/cm shows an increase in the slowly digestible starch from 6.63 to 17.53% in the PEF-treated sample on the potato starch.[43] The digestibility significantly affected the glycemic index and starch hydrolysis in the cooked rice samples. However, PEF induces the increase of acid and aldehydes. Still, a decrease in phenols and ketones was reported in the cooked rice samples, ultimately affecting the volatile compounds in the rice samples.[10] In scientific literature, it was reported that PEF acts on the outer surface of the starch granules; therefore, there is a slight change in the molecular structure, which leads to damage to the crystalline and amorphous structure. The intensity of the electric field acts to alter the digestibility and enthalpy of the starch samples reported for japonica rice.[107] Devkota[25] reported the effect of PEF and thermal treatment on the dried bean and found that it reduced hydration time and leaching of bioactive compounds from the beans. They suggested the PEF-assisted hydration is a promising technique that has a potential future in the food industry.

PEF is an excellent technology that enhances the rate of processing and produces products with higher quality. However, PEF has also been used for the enhancement of the better of tomatoes without affecting their cellulose and hemicellulose network.[3] Moreover, it was also reported that using PEF helps in the accumulation of carotenoid content enhances the extraction rate of carotenoids from the samples, and helps in reducing postharvest loss.[36] Some researchers used PEF coupled with high-pressure processing to help reduce contaminants such as aflatoxins B2 and G1 in various fruit and vegetable juices.[70] An increase in the polyphenols from the food matrix is significant from the industries' point of view and helpful for consumers in maintaining a healthy lifestyle. In vegetables, it helps in preserving the pigments such as chlorophyll a, b, and PEF-treated samples were able to prevent the loss during thermal processing in the case of spinach.[112] With its application in fruits and vegetables, PEF has been widely used for the pasteurization of various products and has very minimal impact on the nutrients and phytochemical compounds in fruits and vegetables.[75]

6.4.3 WASTE VALORIZATION

In the 21st century, food safety, along with suitable waste management, is one of the utmost requirements in upcoming future challenges handled by the agri-food sector. In scientific literature, it was reported that about one-third of the food meant for human consumption was discarded as waste, and many factors such as poor storage, postharvest loss, and transportation, were the major factors affecting the increase of waste in the environment.[39] However, there are many sectors from where food loss occurs, such as meat industries, fruit, and vegetable processing industries, edible oil industries, liquor industries, etc., although food waste is different from food loss because food waste is the term used for the waste that was generated at the last phase such as at the end of the supply chain (retailer and final consumption).[58] Waste generated from the fruits and vegetable processing industries is a matter of urgent concern because this waste is loaded with valuable bioactive compounds and could be utilized in the manufacturing of different products related to food, cosmetics, biodiesel, and bio-refinery, etc.,[74] Segre[88] and his coworkers reported that approximately 87% of the fruit and vegetables are discarded as waste before reaching the consumer.[74,88]

Vegetable waste is a potential candidate to valorize as these materials are loaded with polyphenols, antioxidants, and dietary fiber which are in turn known for value-added products if processed in the right direction. It was reported that vegetable waste is generally in the phases, such as farm to fork, food supply chain, and prepostharvest processing.[71] Vegetable waste contains high moisture; therefore, they are highly biodegradable and pose a threat to the environment. These wastes are more prone to be attacked by spoilage microorganisms, which degrades their nutrient composition and ultimately affects the economic value of the product.[24] Therefore, the utilization of waste material from fresh crops has increased continuously.

Hossain[44] recovered around 99.9% higher alkaloids as compared to nontreated PEF potato peels while using an electric field of 0.75 kV/cm and treatment time of 600 μs. In contrast, the recovery was observed to be reduced when electric field strength was used beyond 0.75 kV/cm with a treatment time of 150–1500 μs. Steroidal alkaloid content was obtained as 1856.2 μg/g of dried peels of potato. Steroids such as glycoalkaloids and aglycone alkaloids are used as precursors for the invention of hormones, antibiotics, and anticancer drugs. However, PEF can be used as a novel technique to maximize the release of steroids from potato peel as an emerging substrate for the phytopharmaceutical industry.

Roohinejad[83] extracted carotenoid from PEF-treated pomace of carrot using hexane as solvent at electric field strength (0.6 kV/cm), the pulse width of (20 μs), and frequency (5–50 Hz). The yield of carotenoids was reported to be significantly higher in treated pomace as compared to the untreated sample. Improved yield in carotenoid content is ascribed to the physical damage of cell membranes under a specific electric field and frequency for retention of higher cytoplasm.

Andreou[2] studied the effect of PEF on valorizing tomato waste from various processing industries. PEF was implemented in three different ways for the utilization of tomato waste. In the first stage, different PEF treatments were applied to tomatoes to improve the peeling rate by 72.3%, followed by cutting of tomatoes for the extraction of juice at the rate of pulsed energy ranging between 0.5–2.5 kV/cm with 15 s pulse width resulted in 20.5% higher yield. Further, extracted juice was utilized for the extraction of carotenoids from the tomato peel. The yield of carotenoid extraction improved to 56.4%. A PEF treatment at 1.0 kV/cm electric field and pulse width of 7.5 μs led to an increase in lycopene content from 98.4 μg/g to 143.1 μg/g from tomato residue. Moreover, the total phenolic content increased 4 times (56.16 mg gallic acid/kg) of the untreated tomato peel waste when treated at 2 kV/cm with 700 pulses. Carotenoid content was linked to improved antioxidant capability. Overall, using tailored PEF pretreatments in commercial tomato processing reduces energy consumption and enhances productivity.

6.5 IMPACT OF PEF ON THE QUALITY CHARACTERISTICS OF VEGETABLES

6.5.1 NUTRITIVE VALUE OF VEGETABLES

The extraction potency of carotenoid from carrot pomace under the influence of PEF was studied by Roohinejab[83] with electric field strength (0.1–1 kV cm^{-1}), and the frequency varied from 5 to 75 Hz. As per the investigations, it is noticed that the carotenoid extraction from the pomace of carrot using several vegetable oils instead of organic solvent, namely, hexane, has significantly increased when electric field intensities were raised to 1 kV cm^{-1} at 5 Hz. PEF treatment resulted in no improvement in the extraction of carotenoid content beyond 10 Hz frequency at 1 kV cm^{-1} electric field strength. This has significant implications for producing carotenoid-rich vegetable oils from agro-industrial carrot waste. This can further be used as a value-added product for numerous culinary and medicinal applications. Carrot pomace can be used as a source of fiber for the formulation and development of functional food products.

Torregrosa[98] carried out research to estimate the shelf life of orange–carrot juice treated with PEF, and the deprivation in ascorbic acid content was investigated. The researchers looked at varied electric field intensities (25–40 kV/cm) and treatment periods (30–340 μs), and the shelf stability of the juice of orange–carrot treated with pulses at 25 kV/cm for two periods (280 and 330 μs) was compared with thermally treated juice (98°C, 21 s) stored at 2 and 10°C. The residual ascorbic acid concentration in pasteurized orange–carrot juice was 83%, while it was 90% in PEF-treated juice. The rate of ascorbic acid degradation in juice stored at 2°C was lower than that in juice stored at 10°C, and the degradation was even higher in pasteurized juice. The shelf life of the PEF-treated (25 kV/cm for 280–330 μs) juice stored at 2°C was extended to 50 days.

TABLE 6.5 Impact of PEF on the Nutritional Quality of Various Vegetables.

Vegetables/juices	Process conditions	Effect of PEF	Reference
Tomato	E = 18–25 kV/cm N = 5–30	Vitamin C content retained a maximum 99.4% and reduced to 81% at 2 kV, 5 pulses	[99]
Bell pepper	E = 2 kV/cm N = 50 Time = 400 µs	Vitamin C retention from 89.6–96.5%	[4]
Carrot	E = 0.25–1 kV/cm E_n = 1.5–576 KJ/kg Time = 2–40 µs	Increased sucrose value by 43% and β– glucose by 48%	[6]
	E = 0.2–1 kV/cm E_n = 1–520 kJ/kg Temperature = 55–77°C	Vitamin C retained in the reduced form, i.e., L-Dihdroxy ascorbic acid, resulted in active participation of reduced ascorbic acid in oxidation	[50]
Tomato	E = 1 kV/cm P = 4 µs, monopolar Frequency = 0.1 Hz	High carotenoid content with an improved shelf life	[9]
	E = 40 kV/cm P = 57 µs	Improved lycopene content leading to redness in tomato, preferable flavor as compared to thermally treated juice	[6]
	E = 35 kV/cm E_n = 8269 KJ/kg P = 4 µs, bipolar Time = 1500 µs Frequency = 100 Hz	Higher lycopene content and vitamin C content	[67]

TABLE 6.5 *(Continued)*

Vegetables/juices	Process conditions	Effect of PEF	Reference
Gazpacho (a vegetable soup containing tomato, cucumber, green pepper, onion, and garlic)	$E = 25$ kV/cm $P = 4$ µs, bipolar Time = 400 µs Frequency = 200 Hz Temperature = 90°C, 1 min.	92.3% retention of vitamin C	[29]
	$E = 15$, 25,35 kV/cm $P = 2$, 4, 10 µs, monopolar, bipolar mode Time = 100, 400, 1000 µs Frequency = 50, 200, 450 Hz	Retention of vitamin C up to 84.3–97.1%	
	$E = 35$ kV/cm $P = 4$ µs, monopolar Time = 1000 µs Frequency = 200 Hz	Retention of vitamin C up to 84.3%	[28]

E: Electric field, E_n: Energy, *N*: Number of pulses, *P*: Pulse width.

6.5.2 *RHEOLOGICAL BEHAVIOR OF VEGETABLES*

Moens[64] studied the impact of PEF on the properties of pectin related to texture for a better understanding of changes in carrot texture through successive cooking practices. It was reported that PEF and mild heat treatment at a temperature of 60°C significantly led to the degradation of carrot tissues during cooking, which consequently affected its texture. Such degradation in the carrot matrix is found to be associated with pectin extractability, degree of methyl esterification, and Ca^{2+} crosslinking. The treatments dramatically reduced pectin extractability in water and alkali in cortical tissue but did not affect pectin or ionic cross-linkages. In vascular tissue, the PEF and combination treatments reduced pectin extractability in water and alkali, decreased the degree of methylation in pectin (by about 60 to 50%), and improved ionic cross-linkages. The addition of Ca^{2+} ions may aid in the preservation of both tissues and texture during later cooking. Fincan & Dejmek[32] investigated the effect of osmotic pretreatment and PEF on the viscoelastic properties of potato tissue. It revealed that the impact of PEF treatment on potato tissue was similar to those of hyper-osmotic treatment in terms of viscoelastic characteristics. The findings support the theory that PEF-induced mechanical property changes are driven by turgor loss. Lower treatment levels resulted in a lower residual modulus, while greater treatment levels resulted in a faster relaxation time. This could be read as more cell membrane damage, allowing cell contents to escape more quickly when the cell is stressed. The treatment results correlated roughly with the cumulative pulse on time for all field strengths; however, many brief pulses were shown to be more efficient for the same on-time.

6.6 FUTURE PERSPECTIVE

In recent years, the demand for high-quality food products has been increasing in the market. It has become important for producers to formulate food products to deliver wholesomeness products. Processed food should have better nutritional value, with improved shelf life. Such products can be processed at low temperatures strategically in combination with various nonthermal techniques. Due to this, a large number of traditional or thermal processing techniques have been replaced with nonthermal techniques for the retention of nutritional value, chemical configuration, and functional properties in processed food and to diminish the detrimental changes in processed fruits and vegetables. PEF technology is majorly used in various applications categorizing food commodities from solid to semisolid or liquid. Most of the applications of PEF in vegetables are specifically focused only on microbial and enzymatic inactivation, and few studies are reported on the nutritional value, composition, rheological properties, and sensorial attributes of PEF-processed vegetables. Rarely any report has been found on waste valorization and extraction of starch from the vegetables treated with PEF. Although it is one of the most versatile techniques, it should not be restricted to inhibiting microbial growth and enzymatic activity only. This technique was invented a long time ago but has not been appropriately explored for various commodities. Yet, detail research on various aspects of vegetables needs to be performed for better implementation of the PEF technique. On the other side, the processing conditions required in PEF for each vegetable should be studied thoroughly to facilitate the optimization conditions. Various mathematical models should also be designed to predict the efficiency of the PEF.

6.7 SUMMARY

Now a day's demand for natural and healthy food products by consumers has increased drastically. Consumers are demanding processed food products rich in minerals, vitamins, dietary fibers, etc.; therefore, it becomes crucial to treat fruits and vegetables under nonthermal processing techniques for the maximum retention of nutritional components. PEF technology is a well-known nonthermal technique. It is rather an operative technique used for the preservation and processing of a large number of commodities without altering their quality characteristics. PEF technology helps in the improvement of various quality attributes in terms of nutrition, texture, and retention of several bioactive components. This nonthermal technology is used not only for the inactivation of microorganisms or enzymes but also helps in controlling various chemical reactions which cause deterioration and affect the shelf life of the commodity. PEF also helps in the retention of color, and flavor to provide fresh and healthy food products to the consumer. The utilization of PEF is being increased and adapted by most of the food industries due to less energy consumption of operation with a greater productivity rate. PEF can also be applied to the formulation and development of various innovative food products.

KEYWORDS

- **bioactive components**
- **microbial inactivation**
- **nonthermal preservation**
- **nutritive value**
- **pulsed electric field**

REFERENCES

1. Adedeji, A. A.; Gachovska, T. K.; Ngadi, M. O.; Raghavan, G. S. V. Effect of Pretreatments on Drying Characteristics of Okra. *Dry. Technol.* **2008,** *26* (10), 1251–1256.
2. Andreou, V.; Dimopoulos, G.; Dermesonlouoglou, E.; Taoukis, P. Application of Pulsed electric Fields to Improve Product Yield and Waste Valorization in Industrial Tomato Processing. *J. Food Eng.* **2020,** *270*, e109778.
3. Andreou, V.; Kourmbeti, E.; Dimopoulos, G.; Psarianos, M.; Katsaros, G.; Taoukis, P. Optimization of Virgin Olive oil Yield and Quality Applying Nonthermal Processing. *Food Bioprocess Technol.* **2022,** *15* (4), 891–903.
4. Ade-Omowaye, B. I, Taiwo, K.; Eshtiaghi, N.; Angersbach, A.; Knorr, D. Comparative Evaluation of the Effects of Pulsed Electric Field and Freezing on Cell Membrane Permeabilization and Mass Transfer During Dehydration of Red Bell Peppers. *Innovat. Food Sci. Emerg. Technol.* **2003,** *4,* 177–188.
5. Aguilo-Aguayo, I.; Odriozola-Serrano, I.; Quintão-Teixeira, L. J.; Martin-Belloso, O. Inactivation of Tomato Juice Peroxidase by High-Intensity Pulsed Electric Fields as Affected by Process Conditions. *Food Chem.* **2008,** *107* (2), 949–955.
6. Alam, M. R.; Lyng, J. G.; Frontuto, D.; Marra, F.; Cinquanta, L. Effect of Pulsed Electric Field Pretreatment on Drying Kinetics, Color, and Texture of Parsnip and Carrot. *J. Food Sci.* **2018,** *83* (8), 2159–2166.
7. Amit, S. K.; Uddin, M. M.; Rahman, R.; Islam, S. M. R.; Khan, M. S. A Review on Mechanisms and Commercial Aspects of Food Preservation and Processing. *Agri. Food Sec.* **2017,** *6* (1), 1–22.
8. Angersbach, A.; Heinz, V.; Knorr, D. Effects of Pulsed Electric Fields on Cell Membranes in Real Food Systems. *Innovat. Food Sci. Emerg. Technol.* **2000,** *1* (2), 135–149.
9. Arshad, R. N.; Abdul-Malek, Z.; Munir, A.; Buntat, Z.; Ahmad, M. H.; Jusoh, Y. M. M.; Bekhit, A. E. D.; Roobab, U.; Manzoor, M. F.; Aadil, R. M. Electrical Systems for Pulsed Electric Field Applications in the Food Industry: An Engineering Perspective. *Trend Food Sci. Technol.* **2020,** *104,* 1–13.
10. Bai, T. G.; Zhang, L.; Qian, J. Y.; Jiang, W.; Wu, M.; Rao, S. Q.; Li, Q.; Zhang, C.; Wu, C. Pulsed Electric Field Pretreatment Modifying Digestion, Texture, Structure and Flavor of Rice. *LWT* **2021,** *138*, E-article ID 110650.

11. Barba, F. J.; Parniakov, O.; Pereira, S. A.; Wiktor, A.; Grimi, N.; Boussetta, N.; Saraiva, J. A.; Raso, J.; Martin-Belloso, O.; Witrowa-Rajchert, D.; Lebovka, N.; Vorobiev, E. Current Applications and New Opportunities for the use of Pulsed Electric Fields in Food Science and Industry. *Food Res. Int.* **2015,** *77* (4), *773–798.*
12. Barbosa–Canovas, G. V.; Gongora, M. M.; Pothakamury, U. R. In: *Preservation of Foods with Electric Fields;* Barbosa-Cánovas, G. V. (Ed.); Academic Press: San Diego, USA, 1999. pp. 1–19.
13. Barbosa-Cánovas, G. V.; Altunakar, B. Pulsed Electric Fields Processing of Foods : An Overview. In: *Pulsed Electric Fields Technology for the Food Industry;* Raso, J., Heinz, V. (Eds.); Food Engineering Series: Boston, MA: Springer, 2006, pp. 3–26.
14. Bassey, E. J.; Cheng, J. H.; Sun, D. W. Novel Nonthermal and Thermal Pretreatments for Enhancing Drying Performance and Improving Quality of Fruits and Vegetables. *Trend Food Sci. Technol.* **2021,** *112,* 137–148.
15. Bazhal, M. I.; Lebovka, N. I.; Vorobiev, E. Pulsed electric Field Treatment of Apple Tissue During Compression for Juice Extraction. *J. Food Eng.* **2001,** *50* (3), 129–139.
16. Bemiller, J. N.; Huber, K. C. Physical Modification of Food Starch Functionalities. *Ann. Rev.Food Sci. Technol.* **2015,** *6,* 19–69.
17. Ben Ammar, J.; Van Hecke, E.; Lebovka, N.; Vorobiev, E.; Lanoisellé, J. L. Freezing and Freeze-Drying of Vegetables: Benefits of a Pulsed Electric Fields Pre-Treatment. In: *6th International CIGR Technical Symposium—Towards a Sustainable Food Chain: Food Process, Bioprocessing and Food Quality Management*, France, 2011; pp. 1–6.
18. Bendicho, S., Barbosa–Canovas, G. V.; Martin O. Reduction of Protease Activity in Simulated Milk Ultrafiltrate by Continuous Flow High Intensity Pulsed Electric Field Treatments. *J. Food Sci.* **2003,** *68* (3), 952–957.
19. Beveridge, J. R.; Wall, K.; MacGregor, S. J.; Anderson, J. G.; Rowan, N. J. Pulsed Electric Field Inactivation of Spoilage Microorganisms in Alcoholic Beverages. *Proc. IEEE* **2004,** *92* (7), 1138–1143.
20. Bhat, S.; Saini, C. S.; Sharma, H. K. Changes in Total Phenolic Content and Color of Bottle Gourd (*Lagenaria siceraria*) Juice Upon Conventional and Ohmic Blanching. *Food Sc. Biotechnol.* **2017,** *26,* 29–36.
21. Cameron, M. D.; Timofeevski, S.; Aust, S. D. Enzymology of Phanerochaete Chrysosporium with Respect to the Degradation of Recalcitrant Compounds and Xenobiotics. *Appl. microbiol. biotechnol.* **2000,** *54* (6), 751–758.
22. Corrales, M.; Toepfl, S.; Butz, P.; Knorr, D.; Tauscher, B. Extraction of Anthocyanins from Grape By-Products Assisted by Ultrasonics, High Hydrostatic Pressure or Pulsed Electric Fields: A Comparison. *Innovat. Food Sci. Emerg. Technol.* **2008,** *9* (1), 85–91.
23. Da Silva, R. P.; Rocha-Santos, T. A.; Duarte, A. C. Supercritical Fluid Extraction of Bioactive Compounds. *TrAC Trend. Anal. Chem.* **2016,** *76*, 40–51.
24. Demirbas, A. Waste Management, Waste Resource Facilities and Waste Conversion Processes. *En. Conver. Manag.* **2011,** *52* (2), 1280–1287.
25. Devkota, L.; He, L.; Bittencourt, C.; Midgley, J.; Haritos, V. S. Thermal and Pulsed Electric Field (PEF) Assisted Hydration of Common Beans. *LWT-Food Sci. Technol.* **2022,** *158,* e113163.
26. El-Belghiti, K.; Rabhi, Z.; Vorobiev, E. Effect of Process Parameters on Solute Centrifugal Extraction from Electropermeabilized Carrot Gratings, *Food Bioproducts Process.* **2007,** *85,* 24–28.

27. El-Belghiti, K.; Moubarik, A.; Vorobiev, E. Aqueous Extraction of Solutes from Fennel (*Foeniculum vulgare*) Assisted by Pulsed Electric Field. *J. Food Process Eng.* **2008,** *31* (4), 548–563.
28. Elez-Martínez, P.; Soliva-Fortuny, R. C.; Martín-Belloso, O. Comparative Study on Shelf-Life of Orange Juice Processed by High Intensity Pulsed Electric Fields or Heat Treatment. *Eur. Food Res. Technol.* **2006,** *222* (3), 321–329.
29. Elez-Mart´ınez, P.; Mart´ın-Belloso, O. Effects of High Intensity Pulsed Electric Field Processing Conditions on Vitamin C and Antioxidant Capacity of Orange Juice and *"gazpacho"*, a Cold Vegetable Soup. *Food Chem.* **2007,** *102,* 201–209.
30. Espachs-Barroso, A.; Van Loey, A.; Hendrickx, M.; Martín-Belloso, O. Iactivation of Plant Pectin Methylesterase by Thermal or High Intensity Pulsed Electric Field Treatments. *Innovat. Food Sci. Emerg. Technol.* **2006,** *7* (1–2), 40–48.
31. Fachin, D.; Van Loey, A. M.; Nguyen, B. L.; Verlent, I.; Hendrickx, M. E. Comparative Study of the Inactivation Kinetics of Pectinmethylesterase in Tomato Juice and Purified form. *Biotechnol. Prog.* **2002,** *18* (4), 739–744.
32. Fincan, M.; DeVito, F.; Dejmek, P. Pulsed Electric Field Treatment for Solid-Liquid Extraction of Red Beetroot Pigment. *J. Food Eng.* **2004,** *64* (3), 381–388.
33. Gachovska, T.; Cassada, D.; Subbiah, J.; Hanna, M.; Thippareddi, H.; Snow, D. Enhanced Anthocyanin Extraction from Red Cabbage using Pulsed Electric Field Processing. *J. Food Sci.* **2010,** *75* (6), e323–e329.
34. Giner, J.; Gimeno, V.; Espachs, A.; Elez, P.; Barbosa-Cánovas, G. V.; Martín, O. Inhibition of Tomato (*Licopersicon esculentum* Mill.) Pectin Methylesterase by Pulsed Electric Fields. *Innovat. Food Sci. Emerg. Technol.* **2000,** *1* (1), 57–67.
35. Gómez, B.; Munekata, P. E. S.; Gavahian, M.; Barba, F. J.; Martí-Quijal, F. J.; Bolumar, T.; Campagnol, P. C. B.; Tomasevic, I.; Lorenzo, J. M. Application of Pulsed Electric Fields in Meat and Fish Processing Industries: An Overview. *Food Res. Int.* **2019,** *123,* 95–105.
36. González-Casado, S.; Martín-Belloso, O.; Elez-Martínez, P.; Soliva-Fortuny, R. Induced Accumulation of Individual Carotenoids and Quality Changes in Tomato Fruits Treated with Pulsed Electric Fields and Stored at Different Post-Treatments Temperatures. *Postharv. Biol. Technol.* **2018,** *146,* 117–123.
37. Guderjan, M.; Töpfl, S.; Angersbach, A.; Knorr, D. Impact of Pulsed Electric Field Treatment on the Recovery and Quality of Plant Oils. *J. Food Eng.* **2005,** *67* (3), 281–287.
38. Guerrero-Beltrán, J. A.; Welti-Chanes, J. Pulsed Electric Fields. *Encyclop. Food Health* **2015,** 561–565.
39. Gustavsson, J.; Cederberg, C.; Sonesson, U.; van Otterdijk, R.; Meybeck, A. Global Food Losses and Food Waste: Extent, Causes and Prevention. In: *Global Food Losses and Food Waste—Extent, Causes and Prevention*; Save Food Congress: Rome, Italy, 2011.
40. Han, Z.; Yu, Q.; Zeng, X. A.; Luo, D. H.; Yu, S. J.; Zhang, B. S.; Chen, X. D. Studies on the Microstructure and Thermal Properties of Pulsed Electric Fields (PEF)—Treated Maize Starch. *Int. J. Food Eng.* **2012,** *8* (1).
41. Han, Z.; Zeng, X. A.; Fu, N.; Yu, S. J.; Chen, X. D.; Kennedy, J. F. Effects of Pulsed Electric Field Treatments on some Properties of Tapioca Starch. *Carbohydr. Polym.* **2012,** *89* (4), 1012–1017.
42. Han, Z.; Zeng, X. A.; Yu, S. J.; Zhang, B. S.; Chen, X. D. Effects of Pulsed Electric Fields (PEF) Treatment on Physicochemical Properties of Potato Starch. *Innovat. Food Sci. Emerg. Technol.* **2009,** *10* (4), 481–485.

43. Hong, J.; Zeng, X. A.; Han, Z.; Brennan, C. S. Effect of Pulsed Electric Fields Treatment on the Nanostructure of Esterified Potato Starch and their Potential Glycemic Digestibility. *Innovat. Food Sci. Emerg. Technol.* **2018,** *45,* 438–446.
44. Hossain, M. B.; Aguiló-Aguayo, I.; Lyng, J. G.; Brunton, N. P.; Rai, D. K. Effect of Pulsed Electric Field and Pulsed Light Pre-Treatment on the Extraction of Steroidal Alkaloids from Potato Peels. *Innovat. Food Sci. Emerg. Technol.* **2015,** *29,* 9–14.
45. Ignat, A.; Manzocco, L.; Brunton, N. P.; Nicoli, M. C.; Lyng, J. G. The Effect of Pulsed Electric Field Pre-Treatments Prior to Deep-Fat Frying on Quality Aspects of Potato Fries. *Innovat. Food Sci. Emerg. Technol.* **2015,** *29,* 65–69.
46. Jain, D.; Pathare, P. B. Selection and Evaluation of Thin Layer Drying Models for Infrared Radiative and Convective Drying of Onion Slices. *Biosyst. Eng.* **2004,** *89* (3), 289–296.
47. Jalte, M.; Lanoisellé, J. L.; Lebovka, N. I.; Vorobiev, E. Freezing of Potato Tissue Pre-Treated by Pulsed Electric Fields. *LWT—Food Sci. Technol.* **2009,** *42* (2), 576–580.
48. Lammerskitten, A.; Mykhailyk, V.; Wiktor, A.; Toepfl, S.; Nowacka, M.; Bialik, M.; Czyżewski, J.; Witrowa-Rajchert, D.; Parniakov, O. Impact of Pulsed Electric Fields on Physical Properties of Freeze-Dried Apple tissue. *Innovat. Food Sci. Emerg. Technol.* **2019,** *57,* 102211.
49. Lebovka, N. I.; Shynkaryk, N. V.; Vorobiev, E. Pulsed Electric Field Enhanced Drying of Potato Tissue. *J. Food Eng.* **2007,** *78* (2), 606–613.
50. Leong, S. Y.; Oey, I. Effect of Pulsed Electric Field Treatment on Enzyme Kinetics and Thermostability of Endogenous Ascorbic Acid Oxidase in Carrots (*Daucus carota cv. Nantes*. *Food Chem.* **2014,** *146,* 538–547.
51. Liu, C.; Grimi, N.; Bals, O.; Lebovka, N.; Vorobiev, E. Effects of Pulsed Electric Fields and Preliminary Vacuum Drying on Freezing Assisted Processes in Potato Tissue. *Food Bioprod. Process.* **2021,** *125,* 126–133.
52. Liu, C.; Grimi, N.; Lebovka, N.; Vorobiev, E. Convective Air, Microwave, and Combined Drying of Potato Pre-Treated by Pulsed Electric Fields. *Dry. Technol.* **2019,** *37* (13), 1704–1713.
53. Liu, C.; Pirozzi, A.; Ferrari, G.; Vorobiev, E.; Grimi, N. Effects of Pulsed Electric Fields on Vacuum Drying and Quality Characteristics of Dried Carrot. *Food Bioprocess Technol.* **2020,** *13* (1), 45–52.
54. Loginova, K. V.; Lebovka, N. I.; Vorobiev, E. Pulsed Electric Field Assisted Aqueous Extraction of Colorants from Red Beet. *J. Food Eng.* **2011,** *106* (2), 127–133.
55. Lopez, N.; Puertolas, E.; Condon, S.; Raso, J.; Alvarez, I. Enhancement of the Solid-Liquid Extraction of Sucrose from Sugar Beet (Beta vulgaris) by Pulsed Electric Fields. *LWT—Food Sci. Technol.* **2009,** *42* (10), 1674–1680.
56. Maniglia, B. C.; Pataro, G.; Ferrari, G.; Augusto, P. E. D.; Le-Bail, P.; Le-Bail, A. Pulsed Electric Fields (PEF) Treatment to Enhance Starch 3 D Printing Application: Effect on Structure, Properties, and Functionality of Wheat and Cassava Starches. *Innovat. Food Sci. Emerg. Technol.* **2021,** *68,* e102602.
57. Mannozzi, C.; Rompoonpol, K.; Fauster, T.; Tylewicz, U.; Romani, S.; Dalla Rosa, M.; Jaeger, H. Influence of Pulsed Electric Field and Ohmic Heating Pretreatments on Enzyme and Antioxidant Activity of Fruit and Vegetable Juices. *Foods* **2019,** *8* (7), 247–255.
58. Manzocco, L.; Alongi, M.; Sillani, S.; Nicoli, M. C. Technological and Consumer Strategies to Tackle Food Wasting. *Food Eng. Rev.* **2016,** *8*, 457–467.
59. McHugh, T.; Toepfl, S. Pulsed Electric Field Processing for Fruits and Vegetables. *Food Technol.* **2016,** *70,* 73–74

60. Meneses, N.; Jaeger, H.; Knorr, D. Minimization of Thermal Impact by Application of Electrode Cooling in a Co-Linear PEF Treatment Chamber. *J. Food Sci.* **2011,** *76* (8), 536–543.
61. Min, S.; Zhang, Q. H. Effects of Commercial-Scale Pulsed Electric Field Processing on Flavor and Color of Tomato Juice. *J. Food Sci.* **2003,** *68* (5), 1600–1606.
62. Min, S.; Jin, Z. T.; Zhang, Q. H. Commercial Scale Pulsed Electric Field Processing of Tomato Juice. *J. Agri. Food Chem.* **2003,** *51* (11), 3338–3344.
63. Mirzaei-Baktash, H.; Hamdami, N.; Torabi, P.; Fallah-Joshaqani, S.; Dalvi-Isfahan, M. Impact of Different Pretreatments on Drying Kinetics and Quality of Button Mushroom Slices Dried by Hot-Air or Electrohydrodynamic Drying. *LWT–Food Sci. Technol.* **2022,** *155,* e112894.
64. Moens, L. G.; Huang, W.; Van Loey, A. M.; Hendrickx, M. E. G. Effect of Pulsed Electric Field and Mild Thermal Processing on Texture-Related Pectin Properties to Better Understand Carrot (*Daucus carota*) Texture Changes During Subsequent Cooking. *Innovat. Food Sci. Emerg. Technol.* **2021,** *70,* e102700.
65. Mohamed, M.; Eissa, A. Pulsed Electric Fields for Food Processing Technology. *Struct. Funct. Food Eng.* **2012,** *32,* 275–306.
66. Naderinezhad, S.; Etesami, N.; Najafabady, A. P.; Falavarjani, M. G. Mathematical Modeling of Drying of Potato Slices in a Forced Convective Dryer Based on Important Parameters. *Food Sci. Nutr.* **2016,** *4* (1), 110–118.
67. Odriozola-Serrano, I.; Soliva-Fortuny, R.; Martin-Belloso, O. Changes of Health-Related Compounds Throughout Cold Storage of Tomato Juice Stabilized by Thermal or High Intensity Pulsed Electric Field Treatments. *Innovat. Food Sci. Emerg. Technol.* **2008,** *9,* 272–279.
68. Ohshima, T.; Tamura, T.; Sato, M. Influence of Pulsed Electric Field on Various Enzyme Activities. *J. Electrostat.* **2007,** *65* (3), 156–161.
69. Ostermeier, R.; Giersemehl, P.; Siemer, C.; Töpfl, S.; Jäger, H. Influence of Pulsed Electric Field (PEF) Pre-Treatment on the Convective Drying Kinetics of Onions. *J. Food Eng.* **2018,** *237,* 110–117.
70. Pallarés, N.; Berrada, H.; Tolosa, J.; Ferrer, E. Effect of High Hydrostatic Pressure (HPP) and Pulsed Electric Field (PEF) Technologies on Reduction of Aflatoxins in Fruit Juices. *LWT–Food Sci. Technol.* **2021,** *142,* e111000.
71. Panda, S. K.; Mishra, S. S.; Kayitesi, E.; Ray, R. C. Microbial-Processing of Fruit and Vegetable Wastes for Production of Vital Enzymes and Organic Acids: Biotechnology and Scopes. *Environ. Res.* **2016,** *146,* 161–172.
72. Parniakov, O.; Barba, F. J.; Grimi, N.; Lebovka, N.; Vorobiev, E. Extraction Assisted by Pulsed Electric Energy as a Potential Tool for Green and Sustainable Recovery of Nutritionally Valuable Compounds from Mango Peels. *Food Chem.* **2016,** *192,* 842–848.
73. Phoon, P. Y.; Galindo, F. G.; Vicente, A.; Dejmek, P. Pulsed Electric Field in Combination with Vacuum Impregnation with Trehalose Improves the Freezing Tolerance of Spinach Leaves. *J. Food Eng.* **2008,** *88* (1), 144–148.
74. Plazzotta, S.; Manzocco, L.; Nicoli, M. C. Fruit and Vegetable Waste Management and the Challenge of Fresh-Cut Salad. *Trend Food Sci. Technol.* **2017,** *63,* 51–59.
75. Poojary, M. M.; Lund, M. N.; Barba, F. J. Pulsed Electric Field (PEF) as an Efficient Technology for Food Additives and Nutraceuticals Development. In *Pulsed Electric Fields to Obtain Healthier and Sustainable Food for Tomorrow;* Barba, F. J., Parniakov, O., Wiktor, A. Eds.; Academic Press: 2020. pp. 65–69.

76. Praporscic, I.; Ghnimi, S.; Vorobiev, E. Enhancement of Pressing of Sugar Beet Cuts by Combined Ohmic Heating and Pulsed Electric Field Treatment. *J. Food Process. Preserv.* **2005,** *29* (5–6), 378–389.
77. Puértolas, E.; Cregenzán, O.; Luengo, E.; Álvarez, I.; Raso, J. Pulsed-Electric-Field-Assisted Extraction of Anthocyanins from Purple-Fleshed Potato. *Food Chem.* **2013,** *136* (3–4), 1330–1336.
78. Puértolas, E.; Luengo, E.; Álvarez, I.; Raso, J. Improving Mass Transfer to Soften Tissues by Pulsed Electric Fields: Fundamentals and Applications. *Ann. Rev.Food Sci. Technol.* **2012,** *3* (1), 263–282.
79. Puértolas, E.; Saldaña, G.; Raso, J. Pulsed Electric Field Treatment for Fruit and Vegetable Processing. In: *Handbook of Electroporation;* Miklavčič, D. Ed.; Springer: Berlin, Germany, 2017. pp. 1–21.
80. Quitão-Teixeira, L. J.; Aguiló-Aguayo, I.; Ramos, A. M.; Martín-Belloso, O. Inactivation of Oxidative Enzymes by High-Intensity Pulsed Electric Field for Retention of Color in Carrot Juice. *Food Bioprocess Technol.* **2008,** 1 (4), 364–373.
81. Reineke, K.; Schottroff, F.; Meneses, N. Sterilization of Liquid Foods by Pulsed Electric Fields—an Innovative Ultra–High Temperature Process. *Front Microbiol.* **2015,** *6,* 400–408.
82. Riener, J.; Noci, F.; Cronin, D. A.; Morgan, D. J.; Lyng, J. G. Combined Effect of Temperature and Pulsed Electric Fields on Apple Juice Peroxidase and Polyphenoloxidase Inactivation. *Food Chem.* **2008,** *109* (2), 402–407.
83. Roohinejad, S.; Everett, D. W.; Oey, I. Effect of Pulsed Electric Field Processing on Carotenoid Extractability of Carrot Purée. *Int. J. Food Sci. Technol.* **2014,** *49* (9), 2120–2127.
84. Rybak, K.; Samborska, K.; Jedlinska, A.; Parniakov, O.; Nowacka, M.; Witrowa-Rajchert, D.; Wiktor, A. The Impact of Pulsed Electric Field Pretreatment of Bell Pepper on the Selected Properties of Spray Dried Juice. *Innovat. Food Sci. Emerg. Technol.* **2020,** 65, e102446.
85. Sánchez-Vega, R.; Elez-Martínez, P.; Martín-Belloso, O. Effects of High-Intensity Pulsed Electric Fields Processing Parameters on the Chlorophyll Content and its Degradation Compounds in Broccoli Juice. *Food Bioprocess Technol.* **2014,** *7* (4), 1137–1148.
86. Sánchez-Vega, R.; Rodríguez-Roque, M. J.; Elez-Martínez, P.; Martín-Belloso, O. Impact of Critical High-Intensity Pulsed Electric Field Processing Parameters on Oxidative Enzymes and Color of Broccoli Juice. *J. Food Process. Preserv.* **2020,** *44* (3), e14362.
87. Schilling, S.; Schmid, S.; Jäger, H.; Ludwig, M.; Dietrich, H.; Toepfl, S.; Carle, R. Comparative Study of Pulsed Electric Field and Thermal Processing of Apple Juice with Particular Consideration of Juice Quality and Enzyme Deactivation. *J. Agri. Food Chem.* **2008,** *56* (12), 4545–4554.
88. Segrè, A.; Falasconi, L. Eds. *Il Libro Nero Dello Spreco in Italia: Il Cibo. (The Black Book of Waste in Italy: Food);* (Vol. 12); Edizioni Ambiente: Milano, Italy, 2011.
89. Shayanfar, S.; Chauhan, O.; Toepfl, S.; Heinz, V. Pulsed Electric Field Treatment Prior to Freezing Carrot Discs Significantly Maintains Their Initial Quality Parameters after Thawing. *Int. J. Food Sci. Technol.* **2014,** *49* (4), 1224–1230.
90. Shynkaryk, M. V.; Lebovka, N. I.; Vorobiev, E. Pulsed Electric Fields and Temperature Effects on Drying and Rehydration of Red Beetroots. *Dry. Technol.* **2008,** *26* (6), 695–704.
91. Sila, D. N.; Van Buggenhout, S.; Duvetter, T.; Fraeye, I.; De Roeck, A.; Van Loey, A.; Hendrickx, M. Pectins in Processed Fruits and Vegetables: Part II—Structure–function Relationships. *Comprehen. Rev. Food Sci. Food Safe.* **2009,** *8* (2), 86–104.

92. Sun, D. W. (Ed.). Handbook of Frozen Food Processing and Packaging (2nd edn.), CRC Press: 2012.
93. Sweiggert, R. M.; Carle, R. Carotenoid Deposition in Plant and Animal Foods and its Impact on Bioavailability. *Crit. Rev. Food Sci. Nutr.* **2017,** *57,* 1807–1830.
94. Syed, Q. A.; Ishaq, A.; Rahman, U. U.; Aslam, S.; Shukat, R. Pulsed Electric Field Technology in Food Preservation: A Review. *J. Nutr. Health Food Eng.* **2017,** *6* (6), 168–172.
95. Takeoka, G. R.; Dao, L.; Flessa, S.; Gillespie, D. M.; Jewell, W. T.; Huebner, B.; Bertow, D.; Ebeler, S. E. Processing Effects on Lycopene Content and Antioxidant Activity of Tomatoes. *J. Agri. Food Chem.* **2001,** *49,* 3713–3717.
96. Teissie, J.; Golzio, M.; Rols, M. P. Mechanisms of Cell Membrane Electro Permeabilization: A Mini Review of Our Present (lack of?) Knowledge. *Biochim. Biophy. Acta (BBA)—Gen. Subject.* **2005,** *1724* (3), 270–280.
97. Toepfl, S.; Siemer, C.; Heinz, V. Effect of High-Intensity Electric Field Pulses on Solid Foods. In: *Emerging Technologies for Food Processing;* Sun, D. W. ed.:. Academic Press: San Diego: 2014. pp. 147–154.
98. Torregrosa, F.; Esteve, M. J.; Frígola, A.; Cortés, C. Ascorbic Acid Stability During Refrigerated Storage of Orange-Carrot Juice Treated by High Pulsed Electric Field and Comparison with Pasteurized Juice. *J. Food Eng.* **2006,** *73* (4), 339–345.
99. Vallverdu-Queralt, A.; Oms-Oliu, G.; Odriozola-Serrano, I.; Lamuela-Raventos, R. M.; Martin-Belloso, O.; Elez-Martinez, P. Effects of Pulsed Electric Fields on the Bioactive Compound Content and Antioxidant Capacity of Tomato Fruit. *J. Agri. Food Chem.* **2012,** *60* (12), 3126–3134.
100. Van Loey, A.; Verachtert, B.; Hendrichx, M. Effects of High Electric Field Pulses on Enzymes. *Trend Food Sci. Technol.* **2002,** *12*, 94–102.
101. Vega-Mercado, H. U. M. B. E. R. T. O.; Martin-Belloso, O. L. G. A.; Chang, F. J.; Barbosa-Canovas, G. V.; Swanson, B. G. Inactivation of *Escherichia coli* and *Bacillus subtilis* Suspended in Pea Soup Using Pulsed Electric Fields. *J. Food Process. Preserv.* **1996,** *20* (6), 501–510.
102. Wiktor, A.; Witrowa-Rajchert, D. Drying Kinetics and Quality of Carrots Subjected to Microwave-Assisted Drying Preceded by Combined Pulsed Electric Field and Ultrasound Treatment. *Dry. Technol.* **2020,** *38* (1–2), 176–188.
103. Wiktor, A.; Schulz, M.; Voigt, E.; Knorr, D.; Witrowa-Rajchert, D. Impact of Pulsed Electric Field on Kinetics of Immersion Freezing, Thawing, and on Mechanical Properties of Carrot. *ZYWN-NAUK Technol. J.* **2015,** *21,* 124–137.
104. Wiktor, A.; Schulz, M.; Voigt, E.; Witrowa-Rajchert, D.; Knorr, D. The Effect of Pulsed Electric Field Treatment on Immersion Freezing, Thawing and Selected Properties of Apple Tissue. *J. Food Eng.* **2015,** *146,* 8–16.
105. Wilhelm, C.; Winterhalter, M.; Zimmermann, U.; Benz, R. Kinetics of Pore Size During Irreversible Electrical Breakdown of Lipid Bilayer Membranes. *Biophys. J.* **1993,** *64* (1), 121–128.
106. Wouters, P. C.; Álvarez, I.; Raso, J. Critical Factors Determining Inactivation Kinetics by Pulsed Electric Field Processing. *Trend Food Sci. Technol.* **2001,** *12,* 112–121.
107. Wu, C.; Wu, Q. Y.; Wu, M.; Jiang, W.; Qian, J. Y.; Rao, S. Q.; Zhang, L.; Li, Q.; Zhang, C. Effect of Pulsed Electric Field on Properties and Multi-Scale Structure of Japonica Rice Starch. *LWT–Food Sci. Technol.* **2019,** *116,* e108515.

108. Yang, S.; Liu, G.; Qin, Z.; Munk, D.; Otte, J.; Ahrné, L. Effects of Pulsed Electric Fields on Food Constituents, Microstructure and Sensorial Attributes of Food Products. *Emerg. Process. Method Food Qual.* **2019,** 27–67.
109. Yin, Y.; Han, Y.; Liu, J. A Novel Protecting Method for Visual Green Color in Spinach Puree Treated by High Intensity Pulsed Electric Fields. *J. Food Eng.* **2007,** *79* (4), 1256–1260.
110. Zeng, F.; Gao, Q. Y.; Han, Z.; Zeng, X. A.; Yu, S. J. Structural Properties and Digestibility of Pulsed Electric Field Treated Waxy Rice Starch. *Food Chem.* **2016,** *194,* 1313–1319.
111. Zhang, Z. H.; Wang, L. H.; Zeng, X. A.; Han, Z.; Brennan, C. S. Non-Thermal Technologies and its Current and Future Application in the Food Industry: A Review. *Int. J. Food Sci. Technol.* **2019,** *54* (1), 1–13.
112. Zhang, Z. H.; Wang, L. H.; Zeng, X. A.; Han, Z.; Wang, M. S. Effect of Pulsed Electric Fields (PEFs) on the Pigments Extracted from Spinach (*Spinacia oleracea* L.. *Innovat. Food Sci. Emerg. Technol.* **2017,** *43,* 26–34.
113. Zhao, X.; Zhang, L.; Liu, D. Biomass Recalcitrance, Part I: the Chemical Compositions and Physical Structures Affecting the Enzymatic Hydrolysis of Lignocellulose. *Biofuels, Bioprod. Bioref.* **2012,** *6* (4), 465–482.
114. Zhong, K.; Hu, X.; Zhao, G.; Chen, F.; Liao, X. Inactivation and Conformational Change of Horseradish Peroxidase Induced by Pulsed Electric Field. *Food Chem.* **2005,** *92* (3), 473–479.
115. Zhong, K.; Wu, J.; Wang, Z.; Chen, F.; Liao, X.; Hu, X.; Zhang, Z. Inactivation Kinetics and Secondary Structural Change of PEF-Treated POD and PPO. *Food Chem.* **2007,** *100* (1), 115–123.
116. Zhu, F. Impact of Ultrasound on Structure, Physicochemical Properties, Modifications, and Applications of Starch. *Trend Food Sci. Technol.* **2015,** *43* (1), 1–17.
117. Zhu, F. Modifications of Starch by Electric Field—Based Techniques. *Trend Food Sci. Technol.* **2018,** *75,* 158–169.
118. Zimmermann, U.; Pilwat, G.; Beckers, F.; Riemann, F. Effects of External Electrical Fields on Cell Membranes. *Bioelectrochem. Bioenerget.* **1976,** *3* (1), 58–83.

CHAPTER 7

APPLICATIONS AND FUTURE ASPECTS OF PULSED ELECTRIC FIELD ON EGG, MEAT, AND POULTRY INDUSTRIES

AKASHDEEP KAUR and GARGI GHOSHAL

ABSTRACT

Microbes and enzymes are sources of unpleasant chemical changes in many food characteristics, such as loss in color, texture, and flavor. Various microbes and enzymes are necessary to inactivate for preservation as well as processing of food. Therefore, pulsed electric field (PEF) has high future prospects and also high potential in food processing and preservation of various foods by reduction of microorganisms, enzymatic activity, and manufacturing of many functional constituents, for example, egg, meat, poultry, milk, and fruit juices, etc. PEF is a nonheated food processing technology, uses high voltage pulses (typically 10–80 kV/cm) that are delivered to the product. PEF is examined as superior to traditional heat treatments of foods and replaced by conventional thermal methods of processing because of its short treatment time (i.e., >1 s), which minimizes the effect of heat and energy loss during food heating, high nutrient content and sensorial quality of food, along with fresh-like characteristics of food. Various studies proved the effect of PEFs to enhance the tenderness of meat in pre- and postrigor mortis at the time of aging. Although few studies suggest that physical injury is a significant cause of PEF-induced tenderness, the enzymatic nature of tenderness appears to be the most likely mechanism. The PEF is also explained to speed up ripening, ameliorate drying, and diminish both pathogens and meat spoilage organisms, but this requires intensive processing conditions. In meat processing, meat safety, softening, and fastened aging appear to

Nonthermal Pulsed Electric Field Technology for the Food Industry.
Preeti Birwal, Megh R. Goyal, Baldev Singh Kalsi, & Ankita Kataria (Eds.)

DOI: 10.1201/9781003605904-9

be areas in which PEF can offer good opportunities, but more research is needed before PEF applications become a reality in commercialization. It must deal with dead bodies, that is, carcasses—muscle, fat, and bones that differ biochemically and in composition. However, there are many technical issues that need to be addressed. A still significant objection for electrical engineers is developing equipment that reliably generates powerful electric field pulses on an industrial scale. There is a need to develop a pulse generation system included with adequate field strength, replication rate, and power at a reasonable cost. It also shortens the life of the electrode and can cause toxicity problems due to the release of particles and heavy metals from the electrode. In addition, this method is also environmentally friendly and energy efficient.

7.1 INTRODUCTION

Conventional methods for food processing, such as cooking, refrigeration, chilling, and blanching are standard technologies, and these are mostly utilized at home. Although, in the food industry, these technologies are integrated with many other processing methods using a variety of modern processing equipment, using many technological possibilities not commonly used by consumers in home processing. Contemporary food technology goes hand in hand with advances in traditional processing methods, including high-temperature short-term heating, as well as various processes such as pulsed electric field (PEF) technology, high-pressure processing, microwave heating, and extrusion. These nontraditional methods aim to preserve food while retaining the maximum possible nutritional nutrients and quality characteristics. Thus, the regularly used concept of "minimal processing" is not completely apt because, admittedly, the fundamental principle "as little as possible, but as much as necessary" is meant.

In recent years, various nonthermal treatment methods have gained popularity as tools that can replace traditional thermal food processing methods. Moreover, the nonthermal methods have various advantages over the thermal method, including efficient use of energy, preservation of food quality such as flavor, color, taste, and nutritional component maintenance, inactivation of putrefactive microorganisms and enzymes, and deterioration of quality. PEF technology is a nonheating food-sustaining technique that uses short electrical pulses to inactivate microorganisms with minimal adverse effects on food quality.[52]

PEF technology is used in food processing that involves high voltage pulses, that is, typically 10–80 kV/cm, and attached to food placed between a set of electrodes supporting the treatment space of the chamber. It is an innovative technique for efficient nonthermal processing of food. A pulse is a single obstruction through the medium from one point to another. PEF is considered better than traditional heat treatment of foods because of short processing times, typically less than 1 s, reducing energy loss during food heating and also minimizing the heating effect along with high sensorial quality as well as retention of nutrients. It is appropriate for storage of semi-liquid and liquid foods, the removal of microorganisms, and the production of functional ingredients i.e., fruit juice, milk, soup, eggs, etc.[38] Thermal processing ensures safety as well as shows improvement in the preservation of different food products. It helps to change the composition of the chemical structure of food products as well as modification in heat-sensitive ingredients and functional properties of food, which are also affected by thermal processing.

Primarily, the development of nonheated food processing techniques such as PEF helps to extend the shelf life of various foods; these methods are also evaluated for their effects on various components of food, for example, safety, processing, and quality of product. PEF can be used as an antimicrobial treatment to enhance storage stability and food safety, and also used for recovery and extraction of some beneficial components from food origin and equalization of different food products at the time of enzyme inactivation. PEF treatment is currently focused on reducing the spoilage microorganisms in different egg products. Furthermore, PEF treatment changes functional as well as structural components to the extent of extraction and there is a high level of variableness between many studies.[54] PEF is a recent nonheating technique applied to get the attention of different technologists and scientists working on meat processing due to increased mass transfer as well as modification of the membrane structure.

Various studies observed that the power of PEF improves the tenderness of meat in both pre- and postrigor muscles at the time of aging. Although it suggested that physical disturbance as a fundamental cause of PEF improves tenderization as well as the enzymatic nature of tenderization. Research suggested that the power of PEF to intervene in the tenderization process because of the alteration in membrane properties causing primitive activation of calpain proteases and also the early release of ions of calcium other than investigation do not yet allow this assumption. The latest research described that enhanced postproteolysis after PEF treatment affects muscles during aging. PEFs have been observed to accelerate ripening, enhance drying, and decrease pathogens and other microorganisms that cause meat spoilage, and the same is true when intensive processing conditions are accomplished. During meat tenderization, an area where PEF can provide attractive choices for meat processing appears with safety as well as improved curing.

A new investigation is essential before the PEF method becomes commercialized in the case of meat. This study estimated the published information related to the subject with the level of getting a better awareness of achievable implementations of PEF in the meat formulation area to give solutions to a few serious issues that are very important to be noticed as convenient choices for the meat industry.[58] This study estimated the effect of PEF on biochemical components as well as on myofibrillar proteins from light, tender, and exudative (PSE)-like chicken breast. Many PEF treatments observed electric field intensity, that is, 0–28 kV/cm, and pulse frequency, that is, 0–1000 Hz. In myofibrillar proteins, the intensity, surface hydrophobicity, sulfhydryl, and solubility of PEF have increased significantly.

Although these properties decreased as the intensity increased to 18 kV/cm, this may be related to protein aggregation due to the interaction of functional groups. Exposure to PEFs also alters the physicochemical constituents of myofibrillar proteins. It also activates the origination of less-elastic myofibrillar proteins.[13] However, electrophoresis of polyacrylamide gels using sodium dodecyl sulfate showed that the basic structure of myofibrillar proteins did not change even after PEF treatment. As a result of circular dichroism spectroscopy, the content of α-helix was improved in the sample treated with PEF, whereas the contents of β-helix and random turns were decreased.[35]

This chapter aims to investigate the effect of PEF on the meat, egg, and poultry industries.

7.2 PRINCIPLE OF PEF

The basic principle behind PEF technology is Electroporation. The application of high voltage allows the inactivation of vegetative bacteria and yeast cells in some foods by creating an electric field that can cause reversible/irreversible permeability of cell membranes. Processing time is calculated by increasing the pulse number by the effective pulse duration. PEF is applied in the form of a square wave, oscillatory pulses, bipolar, exponentially decaying at sub-ambient temperature. After treatment, the food is aseptically packaged and stored in refrigerators.[47]

7.2.1 STEPS IN THE PROCESS OF TREATING CELLS WITH PEF

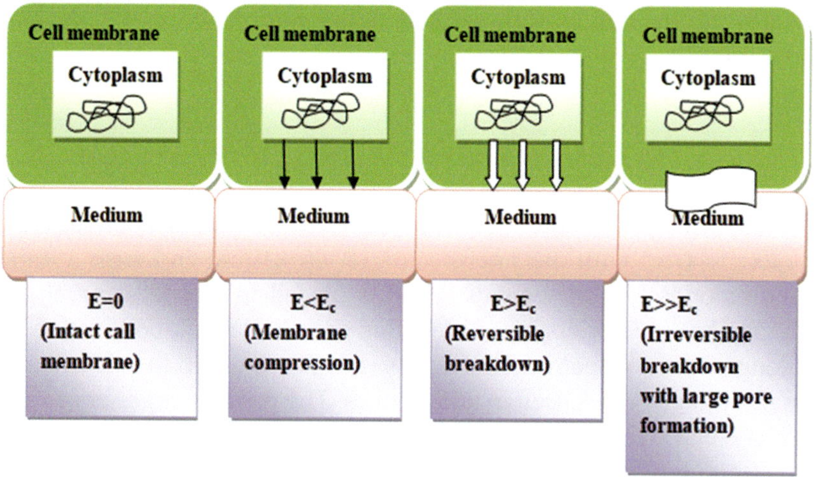

FIGURE 7.1 Potential for disruption due to stomatal development of biological cells upon PEF treatment (*E*). E_c, critical electric field.

Source: Adapted from Ref. [30]

The entire process of exposing cells to an external electric field pulse (treatment with a PEF used in food processing) can be distributed into four main steps (Figure 7.1):

- Increase in the transmembrane potential: It was observed that the power of the transmembrane increased because of the charging of cell plasma by an applied extrinsic (electric) field. This step can last from nanoseconds to milliseconds, depending upon the cell size and pulse amplitude.
- Stages of nucleation of pores lasting from nanoseconds from milliseconds: when an electric pulse is applied, then small metastable hydrophilic pores are created, which are represented by electroporation curves.
- Evolutionary stages of pore populations during electroprocessing lasting from nanoseconds to milliseconds. It was observed that at this stage, the number of pores and sizes have changed.
- Postprocessing steps lasting from milliseconds to hours or days: After the electric treatment, the entrance of extracellular substances, leakage of intracellular compounds, pore resealing, and other processes occurred.

According to observations of PEF treatment, that is, viable or dead cells, which depend upon processes that occur during all stages. In the first three stages (during the action of an electric field), more and more cells become electroporated.[30]

7.2.2 PROCESS DESIGN AND WORKING OF PEF SYSTEM

The main principle of the PEF method is short pulses application of a high electric field with the intensity of 10–80 kV/cm and with a duration of microseconds to milliseconds. Food is placed in a processing chamber, which may be static or continuous. In this chamber, a non-conductive material, along with two electrodes, is used to keep away from the electrical flow and high-voltage pulses used in this process. The product experiences a force due to an electric field that is responsible for the destruction of the genomic cell membrane of microorganisms and their interaction with charged food molecules. This occurs because of the breakdown of dielectric in microbial cell membranes and also charged molecules interaction in food. Consequently, the PEF technique has been recommended for the pasteurization of a variety of foods.[55] Different nonthermal processing techniques recommend the maintenance of food at low temperatures by use of various basic principles. This will improve the food's nutritional quality, which includes some minerals, vitamins, and some essential flavors, even as consuming less energy than thermal processing. Although PEF processing increased the temperature immediately and fast. Therefore, a combination of the type of food, processing times, electric fields, flow rate and refrigeration systems used to maintain the nutrition as well as sensory characteristics similar to fresh food.[26]

In recent years, there have been many technologies such as high hydrostatic pressure, the use of chemicals and biochemicals, oscillating magnetic fields, irradiation, and also the hurdle concept approved as emerging nonthermal techniques.[30]

7.2.2.1 PULSED ELECTRIC FIELD (PEF) SYSTEMS

Generally, PEF systems have a number of capacitors for storage of energy and a high-voltage power supply, a pulse generator, a processing chamber, that is, static/continuous, contains electrodes and a liquid food pump, and the other side, the PEF system is continuous, measurement devices such as current, voltage, temperature, and cooling and heating baths, and also a central process unit to control operations (Figure 7.2). Treatment chambers are not interconnected using insulators. Each chamber has two electrodes through which the liquid flows and indicates the electric field strength.[30]

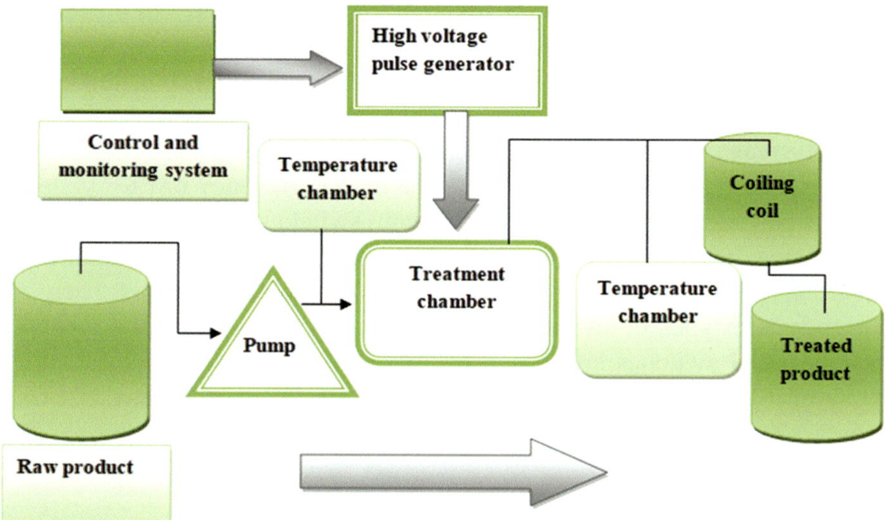

FIGURE 7.2 Illustrative diagram of PEF operation.
Source: Adapted from Ref. [30]

7.2.2.2 POWER SUPPLY

Pulses of high voltage contributed to the system through a generator of high voltage pulses of the required duration, shape, and intensity. The system's high-voltage power supply can be either a condenser charged with higher stability alternating current (AC) input or a conventional direct current (DC) power supply that provides tens of times the charge at a higher rate than a DC power supply.

Food is processed in static/continuous processing chambers by the use of a pump. Mainly, the static treatment chambers are used for preliminary laboratory research, but continuous processing chambers are advantageous for industrial-scale or pilot plant operations. Cold water for the cooling system circulated by electrodes is used to evaporate heat, which is achieved by (electric) current passing through food to avoid any undesirable thermal effects.[30]

7.2.2.3 HIGH POWER CAPACITORS AND SWITCHES

The main part of a powerful source is the on/off switch and storage capacitor due to the large consumption of resistive power, and in terms of capacitors, inductors do a small character. The power reserved in the capacitor is utilized to create magnetic and electric fields. Electric fields are used in the moving charged particles, mechanical, chemical as well as electromagnetic waves, leading to thermal destructive effects. Electromagnetic fields transmit energy in the form of electromagnetic waves. There are typical examples of electromagnetic waves, such as X-rays, microwaves, and laser beam generation.

Discharged switches play a prime role in the efficiency of PEF systems. Experts consider semiconductor (solid-state) switches to be the future (high-power) switches. Switches suitable for PEF systems for extended life include trigatron, spark gap, thyratron, ignitron, and some semiconductors. There are currently two types of switches on the market: ON switches and ON/OFF switches. ON switches are turned off automatically when the discharging is completed in the capacitor. ON switches are less costly than ON/OFF switches due to the handling of high voltages.[30]

7.2.2.4 HIGH VOLTAGE PULSE GENERATOR

High-voltage pulse generators use high-/low-value complex pulse-shaping circuits to provide a specific voltage, duration, and shape. A pulse-forming network is an electrical circuit that contains a variety of components that DC power supplies, a capacitor bank, a charging resistor, switches, pulse-shaping resistors, and also inductors. The direct current (DC) power supply charges voltage to the capacitor's bank at a particular voltage.[30]

7.2.2.5 PEF TREATMENT CHAMBER

In the processing system, the treatment chamber is one of the most important parts of this system. The idea behind the processing chamber is to support the processed sample inside during processing. Although processing chambers are grouped together for batch operation, batch systems were typically used to process fixed volumes of solid/semi-solid foods (Figure 7.3).

It consists of two electrodes and insulation that form a volume i.e., PEF treatment zone where the foods obtain pulses. Electrodes are made of carbon, stainless steel, titanium, gold, platinum, as well as metal oxides, and these are electrochemically static materials. Two electrodes joined together with a nonconductive material to keep away from electrical flow from one to another. This is needed to replace operation (at least every 100 h) because of chemical erosion caused by liquids as well as deposit configuration on the surface. A textile electrode made up of chemical-erosion-resistant material such as polyester fiber, tungsten, and titanium wire. A treatment chamber is required to guarantee the homogenous and continuous PEF treatment of food material.[30]

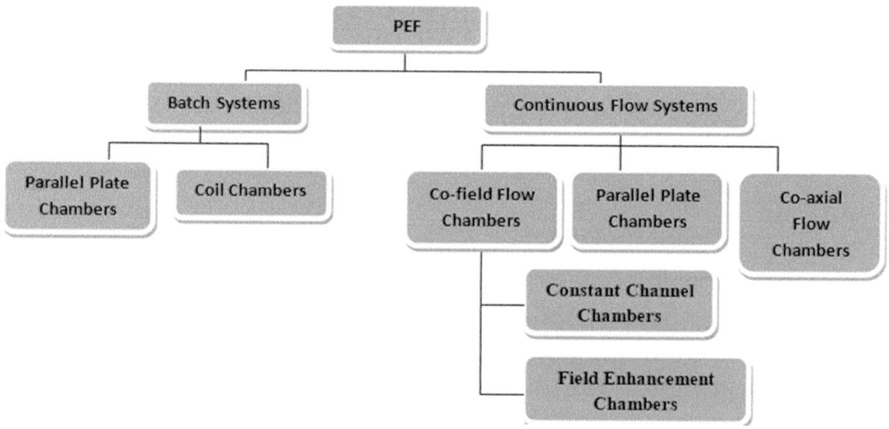

FIGURE 7.3 Different types of treatment chambers (static and continuous-flow) in PEF treatment.

Source: Adapted from Ref. [30]

7.2.2.6 CONTROLLING AND MONITORING

A PEF system is generally hooked up to a central computer, which controls the operation of pumps, electric valves, high voltage pulse generator, and voltage and pulsing. The computer also records data logs containing temperature, voltage, current, and power curves at various points, applied pulses, and product flow rates. Commercially existing software that can be used in personal computer (PC) to manage the control and data-gathering process HP V Lab® or Lab view®.[53]

7.3 FACTORS AFFECTING THE OUTCOME OF PULSED ELECTRIC FIELD TREATMENT

It is important to observe the effectiveness of the PEF technique in the process of pasteurization against pathogenic microorganisms. It is important to build up knowledge on some crucial factors affecting microbial inactivation that help in understanding the mechanism present in the inactivation of microbes by PEF and explaining PEF inactivation kinetics. There are some technical, biological, and media factors that affect the efficiency of PEF. Each factor related to equipment, process analysis, type of microorganism, and condition of the medium used.[30]

7.3.1 TECHNICAL AND BIOLOGICAL FACTORS

7.3.1.1 TECHNOLOGICAL FACTORS

There are some technological factors that affect the outcome of PEF, that is, treatment temperature, treatment time, field strength, pulse shape, and size.

a. Field strength: In the case of a PEF, the intensity must be circulated uniformly in the treatment room to perform a quality treatment. The inactivation of microbes is improving by increasing electric field strength above the critical transmembrane potential. It has low electric field strength (mainly 4–8 kV/cm), which does not affect microbial inactivation.
b. Time of treatment: It is time through several microorganisms subjected to the strength of an electric field.
c. Treatment temperature: Temperature increased in the case when PEF is implied from the flowing liquid that is positioned among the electrodes. Therefore, to reduce the inactivation of microorganisms due to the influence of PEF and prevent thermal damage, proper product cooling using a cooling water bath in the process chamber is very important.
d. Pulse shape and size: One of the most important explanatory features of PEF therapy systems is the application of electric field waveforms. The most frequent waveforms used in PEF are exponentially decaying and square waves. The square pulses are more critical and energy efficient. Square waves have peak voltages (longer) as compared with exponentially decaying pulses. Short pulses avoid too much heat or many unwanted electrolytic reactions.

7.3.1.2 BIOLOGICAL FACTORS

These factors affect the PEF on individual components of target microorganisms as well as physiological and growth states. Gram-negative plant cells are less resistant to PEFs than Gram-positive bacteria, whereas bacteria show lower sensitivity than yeast. Induction of the electric field to the cell membrane is higher when larger cells are subjected to PEF treatment.

7.3.1.3 MEDIA FACTORS

The effect of PEF on food systems is connected to the liquid portion of food. The treated environment includes several factors such as resistivity, conductivity, ionic strength, pH, dielectric properties, and composition, each of which, alone or in combination, influences PEF treatment. The effectiveness of microbial inactivation when applying a PEF is highly dependent on the physical and chemical properties of the food.[30]

7.3.2 EFFECT OF ELECTRIC FIELDS ON QUALITY OF FOOD

At the same time, a power shock is used to inactivate microorganisms and treat them with high-intensity PEF. This electric field is applied to the food placed between the two electrodes (mainly high voltage, i.e., 20–80 kV/cm) to ensure good taste, Preservation, and color retention of liquid products rather than heat treatment. Basically, high-intensity PEF treatment has been shown to produce cold-tolerant fruit juices with sensory and nutritional properties similar to freshness. Combining PEF treatment with other obstacles such as UV light, large pore microfiltration, and high-intensity light pulses, can improve the preservation of foods and preserve their organoleptic properties.[28] In the PEF method, a pulsed voltage is implemented to keep food between two electrodes for less than 1 s. *Lactobacillus acidophilus* is a very vital probiotic bacterium that is used in the manufacture of various dairy products. This research described the effect of various PEF conditions on acid tolerance in *L. acidophilus*.[43] The nutritional awareness of new methods for food processing industries has increased day by day. PEF processing is used to increase the shelf life of different foods, normally liquid foods such as yogurt, soups, milk, soups, liquid eggs, and rice pudding. This technology emphasizes excellent sensory as well as nutritional quality of the food besides extending the shelf life and ensuring safety.[29]

7.3.3 ADVANTAGES OF PEF

- Low treatment temperature, less treatment time
- Efficient inactivation of cells.
- Maintain food safety and enhance shelf life with low processing costs.
- Better retention of color, flavor, and nutrients.
- Reduction in microorganisms: 4–6 log.
- Suitable for processing heat-sensitive foods.
- It can be used to pasteurize liquids—milk, meat, juice, and soup without the use of additives.
- Contains no hazardous and environmentally friendly solvents.
- Combination possibilities with other treatment methods.

7.3.4 DISADVANTAGES OF PEF

- High cost.
- Commercial sector shortages in many parts of the world.
- Microorganisms inactivated by a PEF can survive but have spores with strong protective shells and dehydrated cells.
- Treatment does not inactivate enzymes.
- The PEF method is not appropriate for solid food products because of its nonpumpability.
- The PEF processing is controlled to food products with no air bubbles as well as less electrical conductivity.
- Limited economic and engineering studies of scalable continuous processes.
- The space region for treatment in the chamber to maintain a proper operation lesser than the highest particle size in the liquid.

7.4 APPLICATIONS OF PULSED ELECTRIC FIELD (PEF)

7.4.1 APPLICATIONS OF PEF IN EGG PROCESSING INDUSTRY

PEF changes the structural and chemical composition, modifies heat-labile components, and alters the functions of different foods. PEF is used for antimicrobial treatment, to improve extraction storability and recovery of high-value components from food matrix and also for stabilization of many food products by inactivation of enzymes. It showed that the PEF application is presently being focused on due to the prevention of egg products from several spoilage microorganisms. It reported that PEF efficiently reduces the activity of different microorganisms in egg products. There are some specific recommendations for future research directions on PEF processing which are well discussed in this review.[54]

A PEF is applied to the eggs in terms of factors under study, such as microbiome, quality, shelf life, and gelation components. Inactivation of microbes by PEF in eggs is essential; for example–3.5 decimal place reduction of the egg by PEF of *Salmonella enteritidis*, 5.5 logarithmic reduction of *Listeria innocua* due to synergistic effect of PEF, as well as nisin of liquid egg, 5.6 logarithmic reduction of *Escherichia coli* in whisked eggs by an intermittently implied PEF. The shelf life of fresh-liquid eggs treated with a PEF was preserved for 4 weeks, in terms of quality such as color, viscosity, and organoleptic properties that were not disrupted by the processing of PEF. PEF processing did not cause any appreciable changes in proteins of ovalbumin solution, and fresh egg whites dialyzed. Thus, few structural as well as functional modifications were noticed in fresh egg whites as the conclusion of processing by PEF. The microstructure of the gel, as well as the texture exposed to PEF and PEF processing conditions of egg whites, should be evaluated to reduce feasible alterations.[46] It was noticed that the effects of PEF on the inactivation of *S. Enteritidis* inoculated into liquid whole egg and the physical components and the shelf life of liquid whole egg. PEF processing conditions were 200 pps frequency, 1.2 mL/s flow rate, 25 kV/cm electric field strength, 2.12 μs pulse duration, and 250 μs total processing time. PEF treatment reduced the population in *S. enteritidis* to 1 log 10 cfu/mL in liquid whole oocytes. PEF processed samples were heated at 55C for 3.5 min for the deactivation of carcass bacteria without contaminating the liquid egg.

The amalgamation of PEF processing and heat treatment resulted in a 4.3 log 10 cfu/mL minimization in the *S. Enteritidis* population ($p < 0.05$), which caused no alteration in viscosity, color, electrical conductivity, pH, and Brix when compared to control ($p > 0.05$). Liquid whole egg samples treated with PEF at + 55°C had considerably longer shelf life at 4°C compared to control heat-processed samples ($p < 0.05$).[22] The effect of a PEF on all liquid components and microstructures of eggs, such as lipids and proteins, has been studied and compared to pasteurization. The effect of refrigeration after one week of treatment has been studied. After doing all processing treatments, only the pasteurized sample was present in the sample, even after refrigeration. It showed a significantly lower water-soluble protein value than the PEF processed and control samples. This could be related to the reinforcement of protein-protein interactions generated by the partial denaturalization of proteins as well as noticed by Cryo-SEM. In addition, a decrease in soluble protein after cooling was found in control samples and PEF-processed samples, probably due to the accumulation of egg lipoproteins during storage. In addition, slight lipolysis was observed after cooling in the control sample and PEF-processed sample. However, this effect was higher with lower as well as higher processing with PEF. Oxidation parameters studies showed intermediary lipid degradation in processed samples when differentiated from pasteurized eggs. This replicates the maximum microbiological stability of samples processed with a PEF compared to unprocessed eggs.[34] It determined the functional properties of the whole egg after PEF treatment. Liquid whole eggs are obtained from eggs from Tetra SL. They were laying hens at 27 weeks of age. Exposure Parameters to PEF were selected according to the degree of previous experience, and significant reductions in microbiota were noticed, that is, 32 and 89 kV/cm, 20, 60, and 100 pulses. After PEF treatment, it was observed that the functional properties of the liquid egg were not inferior to the control.

Moreover, the emulsifying capacity and the stability of the liquid egg froth after exposure to a PEF were significantly superior to that of the control groups. At a shear rate of 250 s^{-1}, the viscosity of the liquid egg was higher in the sample after PEF processing, that is, 112 mPas after 20 pulses and 106 mPas after 100 pulses. The noticeable viscosity for the control sample was 102 mPas. It indicated that PEFs might be useful for maintaining liquid whole eggs, mainly rheological and functional properties. Another challenge is the development of safety from microbes, which should be the main criterion for evaluating this technology.[40]

7.4.2 USE OF PEF IN THE MEAT PROCESSING INDUSTRY

Market demand for new meat products with physical and chemical modifications as well as nutritional components has generated concern in the food industry to notice various creative methods such as PEF. PEF is a new technology based on the use of electric current among two electrodes and involves the phenomena of electroporation. This study provided current information about the use of PEF processing in meat and fish for enhancing the extraction of high added-value and to improve the physicochemical as well as nutritional values and also used to enhance different processes such as preservation, tenderization, and aging. In the case of fish products and drying, PEF treatment can increase water retention properties and increase the recovery of value-added parts for meat and fish products. This research gives knowledge for helping in many purposes in the future.[18] It evaluates the potential of PEF treatment to electroporate muscle cells to produce rapidly cured meat at a temperature of 4°C. It is observed that PEF analysis, such as pulse width, electric field strength, and specific energy, had been found to observe the effect of PEF in the case of dried meat. Enhancement in the drying of pork samples at optimal conditions evaluated at laboratory scale (PW of 200 μs, strength of EF 1 kV/cm, and specific energy of 28 kJ/kg). The results investigated that the usefulness of PEF for enhancement in production is limited by the effect of the hull water barrier in the case of cured minced meat products.[45] This study allowed us to expand our knowledge of the role of PEF in the technical properties in the case of chicken meat. The results recommend PEF processing to enhance the capacity to retain water in meat without damaging protein integrity as well as functionality. Application of PEF may show a productive chance for the poultry industry to enhance consumer acceptance and meat yield by reducing the presence of unwanted liquids which is present inside meat packages. It was investigated to understand procedures involved in increasing water retention capacity after applying a PEF.[6] Various parameters of PEF were observed, including electric field strength, that is, 0–28 kV/cm, and pulse frequency, that is, 0–1000 Hz. As a result, it was shown that PEF strength, solubility, surface hydrophobicity, and sulfhydryl-group and myofibrillar proteins were notably increased. Although these properties are reduced as the strength is increased to 18 kV/cm, this may be related to protein aggregation due to the interaction of the functional group. Exposure to PEF also alters rheological properties in the case of PSE-like myofibrillar proteins and activates the formation of low-elastic myofibrillar proteins.[13] This not only improves permeability and promotes penetration of hardeners (such as spices and flavor enhancers) into the meat but also helps accelerate the curing process of the meat.

Treatment with PEF causes porcine muscles to swell and form spongy tissue.[47] It evaluated the effect of PEF processing on key technical elements of chicken meat by investigating the role of key process parameters such as pulse number (150 vs. 300 and 450 vs. 600) and strength of electric field (0.60 vs. 1.20 kV/cm). It was noticed that PEF treatment was not applied to any effect on the pH of the meat, and also, to some extent affected yellowness and lightness. Treatment with low-strength PEF significantly reduces drip loss by up to 28.5% ($p < 0.001$) after 4 days of storage in the refrigerator without compromising protein integrity as well as the function of the protein and also increases water retention capacity. Furthermore, an investigation of process parameters showed that an increase in the number of pulses is more essential for decreasing drip loss in the case of meat than twice the field strength.

From an industry perspective, the results of the study recommend the potential for PEF processing to reduce unwanted liquids in packages, thereby improving consumer acceptance.[6] It evaluated the effect of PEF processing and refrigeration on the volatile constituents and organoleptic properties of chilled and frozen lamb chops. Oxidized taste was investigated after storage for 7 days, although PEF processing of all chilled and frozen cuts was associated with signs of browning and liver taste. Partial least-square regression was applied to estimate the predictive relationship between volatile amino acids, fatty acids, and sensory attributes of PEF-processed lamb chops. The results correlate well with the juiciness and meat taste of frozen lamb cuts treated with PEFs with few volatile–compounds such as 2-nonanone, pyrrole, 2-pentylfuran, 2–ethyl–3–methylpyrazine, methylpyrazine, and thiophene. Refrigerated lamb chops processed in PEF meat have a juicy taste associated with the presence of fatty acids.[23] A PEF is used as a pretreatment of the drying process to enhance the drying rates and to obtain a higher quality product in terms of structure. The permeabilization induced by a PEF determined the improvement of mass transport during drying. Actually, a reduction of drying time was observed by approximately 25% in PEF-treated samples in comparison to control samples, and also, the permeabilization induced by the PEF on food membranes positively affected mass transfer in the juice extraction.

PEF processing to accelerate dry aging of animal protein and the treatment did not have any detrimental effects on conjugated linoleic acid content or on the total weight losses or shear force.[23] PEF application has the power to effect physical damage in case muscle tissue changes the sensory characteristics of meat in both a positive manner (enhanced tenderization) as well as in a negative manner (off-flavor production). If there is a possibility of the development of bad flavor, this requires an extensive observation on the PEF to tenderize meat and meat (turkey) chosen for principle, and it is very susceptible to oxidation. Following properties such as cooking loss, weight loss, texture, color, and lipid oxidation, are evaluated by instrumental techniques in the case of PEF. Organoleptic properties such as color, texture, and odor were performed to evaluate consumer suitability of the samples. The oxidation in lipids of all PEFs processed samples and unprocessed samples preceded at the same storage rate, and there is no difference between control and treated samples.[4]

7.4.3 EFFECT OF PEF ON MEAT TENDERIZATION

7.4.3.1 EFFECT OF PEF ON MEAT AGEING

The combination of PEF processing and aging is a good option for improving meat tenderness by increasing the rate of protein breakdown (Table 7.1).[15] Other food products such as cheese or wine and beef meat also enhance their properties, such as texture, etc., after an aging process of the pieces. Nowadays, to carry out meat tenderization, aging is one of the most important options in the industries. This aging process involves the muscle fibers breakdown by use of endogenous proteolytic enzymes such as calpains, muticatalytic proteinase complex, and lysosomal cathepsins. However, some disadvantages are that the process is slow and expensive due to long refrigeration storage and a very labor-intensive process.[1,16]

According to some authors, both aging and PEF processing affect rheological (textural) components and the cleansing of hot beef muscle loss. Previously, these processing treatments significantly reduced shear force values.[16] It investigated the effect of various PEF processing followed by aging for different time periods as well as tenderization of beef SM and LL muscles. Also, it observed the interaction between aging and PEF treatment.[7] It assessed the impact of PEF application in batch mode as well as parallel electrodes, that is, EFS (1.7–2.0 kV/cm), the energy of electric pulse (185 kJ/kg), frequency (50 Hz) and width of the pulse (20 μs) on beef muscle, its prospects are utilized to decrease meat aging and meat cost. During the aging period, it is observed that PEF processing did not affect meat color stability and cooking loss. In addition, it also observed that the PEF treatment caused some changes in the texture and the microstructure of meat. Hence, it cannot be an appropriate strategy to enhance softness, but it may also not be suitable for reducing aging time and altering functional constituents. It was observed that the utilization of the PEF technique could not be comprehensive because some important differences depend upon species (animal) and on cuts from meat. Several additional experiments and research is needed to optimize aging orders and determine their effect on meat processed with PEF.[14]

7.4.3.2 EFFECT OF PEF ON PHYSICAL AND CHEMICAL PROPERTIES OF MEAT

The PEF technique affects membranes of myocytes to determine the relation among cell membrane phospholipids, including prooxidant effect and fatty

TABLE 7.1 Applications of PEF Treatment in Fishery.

Fishery product	Production conditions	Applications	References
Fish bone	Most favorable combination of parameters: pulse number 8 and EFS 25 kV/cm.	Extraction of effective calcium.	[57]
Fish bone	Optimization of processing conditions and the best yield achieved at pulse number 9 and EFS 16.88 kV/cm.	Chondroitin sulfate extraction.	[21]
Fish bone	Combination of semi-bionic extraction method with PEF at optimum conditions at pulse number 9 and EFS 22.79 kV/cm.	Effective ingredients extraction.	[20]
Mussel	Best conditions at pulse number 9, EFS 20 kV/cm, and enzymolysis time of 2 h.	Protein extraction.	[56]
Salmon and lumpfish	Various combinations of high pressure, pulse number, and electric field processing.	PEFs applications are appropriate for pretreatments of lumpfish roes and have a better effect on salmon samples than chicken samples.	[19]
Common whelk, fillets (fresh cod and frozen), Iceland cyprine, and Pollock sirloin.	Strength of field (1.2–2.0 kV/cm), pulse width (400 μs), frequency (1–4 Hz), and number of pulses are 20, 40, 80, or 120.	Electric field strength <2.0 kV/cm. PEF was ineffective at <90 pulses and enhancement of water retention properties as well as fish drying.	[27]
Haliotis discus hannai Ino viscera	Processing time (e.g.,–100–800 μs, Intensity is 5–20 kV/cm, ratio of material solvent, i.e., (3:1–10:1).	Protein hydrolysate extraction	[31]

acids in meat.[15] Then these interactions help to originate some undesirable components, which decompose into meat by-products, which can lead to an unpleasant taste and odor in meat, which could be the reason for the reduction in sensory as well as nutritional quality of the meat. The high intensities of PEF and the number of pulse repeats can cause some bad effects on the appearance of meat due to an increase in the product temperature, which promotes the oxidation of myoglobin that contrasts with the mild PEF conditions. Furthermore, water retention capacity is also the reason that changed in the case of PEF-processed samples of meat because of the generation of pores and water movement.[1] It evaluated the enzymatic, proteolytic, and physicochemical changes of venison affected by PEF treatment. No significant effects were observed after exposure for 21 days in myofibrillar fragmentation index as well as on shear force by PEF processing. PEF conditions have significant softening effects in beef and require testing of different processing conditions to stimulate a significant response in muscles of different species.[10]

In a recent study, it demonstrated that high PEF treatment, that is, 10 kV, 200 Hz, and 20 μs, negatively affects beef quality compared to the effects assessed for low PEF treatment, that is, 2.5 kV, 200 Hz, and 20 μs. They work with a treatment chamber consisting of parallel stainless-steel plates, which present with insulation (Teflon) at a distance of 8 cm. The variables such as shear force, temperature, oxidation of lipids, color, and levels of minerals (P, Fe, and K) are badly affected by high PEF processing. It reported that color stability, cooking loss, and lipid oxidation of brisket were not affected by PEF treatment which is essential in *sous vide* treatment. But the collagen solubilization in the *sous vide* brisket significantly increased by increased the EFS.[25]

Samples of beef muscles stored at a cooling temperature (4°C) before weak PEF treatment using batch configuration as well as many other amalgamations of electric field strength (0.2–0.6 kV/cm) and frequency (1–50 Hz), there were no changes found in physical properties, such as change in the appearance of meat, as well as in color stability, pH, protein profile, and cooking losses.[16] It has also been observed to extend shelf life using samples of cooked lamb chops, and for freeze–thaw pretreatments pretreated with a PEF, there was a significant increase in volatile compounds because of protein and lipid oxidation. At the same time, all sections of the samples processed with PEF were juicy, brown in color, and associated with meat taste characteristics. Therefore, to obtain a quality meat product, it is necessary to provide an appropriate thawing step and control the temperature of the meat before applying a PEF treatment.[33]

We already know that the effect of an induced PEF is more complex for multicellular systems. This is because a multicellular system is made from different cell types with different characteristics and tissue diversity. PEF processing is used to enhance a mass transfer and also modifies the characteristics of the product in several ways. This depends on the characteristics of the product itself. It should be considered that meat is varied adipose tissue, connective tissue, nerve & vascular tissue, and multinuclear and longitudinal muscle cells.[33] A muscle fiber composition is defined significantly by species' muscle type, age, breed, diet/nutrition, environmental circumstances, and history of physical activity.[1] It studied that frozen meat was more sensitive to PEF processing from fresh samples. The main reason is that the synergistic effect of freezing and PEFs increases protein breakdown and results in a meat tenderizing effect.[2,15]

7.4.3.3 EFFECT OF PEF ON MEAT PROTEIN DIGESTIBILITY

A significant amount of un-hydrolyzed protein in the colon stimulates harmful metabolite production during gut microflora fermentation. Thus, the idea of consumption of digestible and nutritious foods is of interest to a society concerned with the quality of food and health. It concluded that beef samples processed with the PEF had greater and faster digestibility (*in vitro*) compared to other untreated samples of beef.[9] The PEF in batch mode and parallel electrodes, that is, S_1: EFS–5 kV, frequency–90 Hz & S_2: EFS–10 kV, frequency–20 Hz enhanced the protein solubility and digestibility.[9] It observed that PEF processing in periodic mode, that is, EFS (1–1.5 kV/cm), energy (48–178 kJ/kg), width of pulse (20 μs), and frequency (50 Hz), was found to enhance digestibility (in vitro) by 18% and different digestion profiles of control and PEF processed samples observed dissimilar and after 180 min of simulated digestion, in case of PEF processed samples, disruption of Z-disks and I-bands and it observed that PEF treated muscle more prone to enzymatic degradation.[12] It was reported that the rate of intracellular constituents and electroporation of muscle fibers by PEF are responsible for proteolysis processes.[3] At the same time, the beef brisket is pretreated with PEF under definite processing conditions, which significantly modifies tissue properties such as shear force, hardness, and chewiness, as well as enhanced collagen solubilization and tenderness after *sous vide* treatment without adversely affecting the protein digestibility.[2] It explained that heat generation during repeated PEF processing in continuous mode inactivates endogenous proteolytic enzymes that make meat tender.[51] It evaluated that the quality of hot-boned beef loins and topsides affected by repeated PEF processing and reduction in proteolysis also found in every treatment. The optimization of pretreatments and development to improve protein digestibility is essential to ensure better protein intake in the gut and limit the undesirable effects of eating protein-rich food.[7]

7.4.3.4 EFFECT OF PEF ON MEAT MINERALS

The modification of nutritional aspects and nutritionally important minerals was significantly observed in the case of beef and chicken. Worldwide, meat is classified as an essential origin of minerals such as Fe and Zn, and its negative commercial consequences are observed due to the alteration of different mineral concentrations. That is why, in the following years, more research studies about the effect of PEF treatment on the concentration of various minerals in meat.[24,25] In this research, it did not notice some negative alterations such as the lack of minerals, that is, Fe, P, K, Ca, Mg, and Na, from meat materials throughout in vitro gastrointestinal digestion.[9] It concluded that high PEF application, that is, EFS–10 kV, frequency–200 Hz, and pulse width–20 μs, activated fewer compositions in K, P, and Fe and negatively stimulated beef quality.[25]

7.4.3.5 EFFECT OF PEF ON RISE IN VALUE OF MEAT BY-PRODUCTS

By observing the applicability of PEFs, the potential use of PEF processing for meat products has been proposed to facilitate material extraction from waste.[19] The purpose of this study was to extract functional proteins from chicken waste using a product-based PEF. The results are protein-rich moiety, which may have antioxidant properties due to this nonthermal and nonchemical process.[17]

7.4.4 USE OF PEF IN THE FISHERIES INDUSTRY

The effect of PEF on the microstructure of chicken, salmon, and crucian carp eggs was evaluated. The results showed that salmon is more sensitive to weak PEF processing, that is, <2 kV/cm, 20–40 pulses from chicken.

However, these samples have shown a significant effect on the texture and microstructure of the chicken.[19] Many authors have noted the advantageous convenience of this method for preserving the organoleptic, physical components, and functional components of end products, for example, the introduction of minimum modifications of vitamins, flavors, and other nutrients.[39,42] The PEF is also used to improve the water retention properties of fish products and to soften the products of shellfish.[27] Unfortunately, no beneficial impact noticed on the softness of shellfish products such as common trumpets and Icelandic carp was observed. Exposing the fish's muscles to a PEF made the structure more permeable. PEF treatment suggested some efficient strategies for improving the water retention properties of fish and can also be used before the treatment of drying fish. It observed the effectiveness of protein extraction from mussels by the PEF method. In the case of PEF processing, the triangular waveform of the power pulse and pulse duration are 2 μs at the hypothesized optimal conditions, and the extraction yield of protein is maximized.[56]

7.5 FUTURE RESEARCH AND CONCLUSIONS

The application of PEF in the preservation of food provides great potential for maintaining good quality at mild heat and short residence times to preserve the nutritional properties of various products. It is also an energy-efficient and environmentally friendly substitution for the food industry. However, there are many technical issues that need to be addressed. A still significant challenge for electrical engineers is developing equipment that reliably generates powerful electric field pulses on an industrial scale. There is a need to develop a pulse generation system with sufficient field strength, power, and rate of repetition at a fair price. It also shortens the life of the electrode and can cause toxicity problems due to the release of particles and heavy metals from the electrode. So, with some challenges or problems that need further investigation, this method is expected to be used successfully in the industry as a single/combination treatment along with methods to improve the quality of the product and yield in the near future. This study provides current information about the utilization of PEF processing in the case of meat and fish for enhancement in nutritional as well as physicochemical changes (as preservation) and also to increase the recovery rate of high-value-added components. It evaluated the potential of PEF treatments for the electroporation of muscle cells dried meat faster. The applications of

PEF may be beneficial choices in the poultry industry to improve consumer acceptance and yield of meat by reducing the presence of unwanted liquids inside meat packaging. PEF helps spoil the number and microbes in food without significantly raising the meat's temperature. This helps preserve the organoleptic and nutritional composition of the meat. This technique is energy efficient, faster and contains no toxic substances as by-products. There are significant limitations to the commercialization of these methods. To solve these problems, proper research and development should be carried out first. This technique has the potential to enhance product quality and process yields. It has also been reported some significant progress has been made in the commercialization of PEF applications.[44] In the meat and fish industries, the development of convenient and inexpensive PEF systems is necessary. In recent years, a problem has been noticed: an electrochemical reaction at the interface of the electrode, which prevents the migration of the electrode and the corrosion-causing substances into the food system, and indicates the problem of replacing stainless steel electrodes with other ingredients.[24,32] Various PEF treatment variables and pre- and post-processing conditions of meat, such as meat characteristics, aging, freezing, etc., have a significant effect on food product quality.[36,41] In spite of the existence, the number of positive aspects of PEF processing for laboratory-scale fish and the processing of meat and its application in the industry is practically limited. This is due to a number of reasons (e.g., high investments in PEF equipment), the introduction of alterations in existing meat processing circuits, and the need to optimize the PEF terms of use for specific products.

7.6 SUMMARY

Based on available data, studies of individual PEF processing and its effect on many quality components in meat and fish are limited. The use of PEF in the meat and fish processing unit would be very beneficial because of the lower consumption of power and short treatment time required for PEF. Therefore, the transfer of PEF technology needs more study of the impact of PEF application on quality attributes of fish and meat by-products, such as color, lipid oxidation, tenderness, weight loss, and water retention capacity. PEFs have broad potential for preserving many foods and are also innovative technologies useful for modifying the structure of various foods. In addition, for the food processing industry, this method is an energy-efficient and environmentally friendly choice.

KEYWORDS

- **applications**
- **processing**
- **pulsed electric field**
- **meat and egg quality**
- **nonthermal processing**
- **technological properties**

REFERENCES

1. Alahakoon, A. U.; Faridnia, F.; Bremer, P. J.; Silcock, P.; Oey, I. Pulsed Electric Fields Effects on Meat Tissue Quality and Functionality. In *Handbook of Electroporation;* Miklavcic, D. Eds.; Springer, Cham: Switzerland, 2016, Vol. 4, pp. 2455–2475.
2. Alahakoon, A. U.; Oey, I. B.; Silcock, P. Process Optimization of Pulsed Electric Fields Pre-Treatment to Reduce the Sous Vide Processing Time of Beef Briskets. *Int. J. Food Sci. Technol.* **2018,** *54* (3), 823–834.
3. Alahakoon, A. U.; Oey I.; Silcock, P.; Bremer. Understanding the Effect of Pulsed Electric Fields on Thermostability of Connective Tissue Isolated from Beef Pectoralis Muscle using a Model System. *Food Res. Int.* **2017,** *100,* 261–267.
4. Arroyo, C. S.; Eslami, N. P.; Brunton, J. M.; Arimi, F. N.; Lyng, J. G. An Assessment of the Impact of Pulse Electric Fields Processing Factors on Oxidation, Color, Texture and Sensory Attributes of the Turkey Breast Meat. *Poul. Sci.* **2015,** *94* (5), 1088–1095.
5. Arroyo, C.; Lascorz, D.; O'Dowd, L.; Noci, F.; Arimi, J.; Lyng, J. G. Effect of Pulsed Electric Field Treatments at Various Stages during Conditioning on Quality Attributes of Beef *longissimus thoracis et lumborum* muscle. *Meat Sci.* **2015,** *99,* 52–59.
6. Baldi, G.; Elia, F. D.; Soglia, F.; Tappi, S.; Petracci, M.; Rocculi, P. Exploring the Effect of Pulsed Electric Fields on the Technological Properties of Chicken Meat. *Foods* **2021,** *10* (2), 1–15.
7. Bekhit, A. E.-D. A.; Suwandy, V.; Carne, A.; Vande, V. R.; Hopkins, D. L. Effect of Repeated Pulsed Electric Field Treatment on the Quality of Hot-Boned Beef Loins and Topsides. *Meat Sci.* **2016,** *111,* 136–146.
8. Bekhit, A. E.-D. A.; Vande, V. R.; Suwandy, V.; Fahri, F.; Hopkins, D. L. Effect of Pulsed Electric Field Treatment on Cold-Boned Muscles of Different Potential Tenderness. *Food and Bioprocess Technol.* **2014,** *7,* 3136–3146.
9. Bhat, Z. F.; James, D.; Morton, J. D.; Bekhit, A. E.-D. A.; Kumar, S.; Bhat, H. F. Thermal Processing Implications on the Digestibility of Meat, Fish and Seafood Proteins. *Comprehen. Rev. Food Sci. Food Safe.* **2021,** *20* (5), 4511–4548.
10. Bhat, Z. F.; Morton, J. D.; Mason, S. L.; Mungure, T. E.; Jayawardena, S. R.; Bekhit, A. E-D. A. Effect of Pulsed Electric Field on Calpain Activity and Proteolysis of Venison. *Innovat. Food Sci. Emerg. Technol.* **2019,** *52,* 131–135.

11. Boulaaba, A.; Egen, N.; Klein, G. Effect of Pulsed Electric Fields on Microbial Inactivation and Physico-Chemical Properties of Whole Porcine Blood. *Food Sci. Technol. Int.* **2014,** *20* (3), 215–225.
12. Chian, F. M.; Kaur, L.; Oey, I.; Astruc, T.; Hodgkinson, S.; Boland, M. Effect of Pulsed Electric Fields (PEF) on the Ultrastructure and *in vitro* Protein Digestibility of Bovine *longissimus thoracis*. *Food Sci. Technol.* **2019,** *103,* 253–259.
13. Dong, M.; Xu, Y.; Zhang, Y.; Han, M.; Wang, P.; Xu, X.; Zhou, G. Physicochemical and Structural Properties of Myofibrillar Proteins Isolated from Pale, Soft, Exudative (PSE)-Like Chicken Breast Meat: Effects of Pulsed Electric Field (PEF). *Innovat. Food Sci. Emerg. Technol.* 2020, 59, 1–9.
14. Faridnia, F.; Bekhit, A. E.-D. A.; Niven, B.; Oey, I. Impact of Pulsed Electric Fields and Post Mortem Vacuum Ageing on Beef *longissimus thoracis* Muscles. *Int. J. Food Sci. Technol.* **2014,** *49* (11), 2339–2347.
15. Faridnia, F.; Bremer, P.; Burritt, D. J.; Oey, I. Effects of Pulsed Electric Fieldson Selected Quality Attributes of Beef Outside Flat (*Biceps femoris*). In *1st World Congress on Electroporation and Pulsed Electric Fields in Biology, Medicine and Food & Environmental Technologies;* Jarm, T. & Kumar, P. Eds.; IFMBE Proceedings, Springer: Singapore, 2016, Vol. 53, pp. 51–54.
16. Faridnia, F.; Ma, Q. L.; Bremer, P. J.; Burritt, D. J.; Hamid, N.; Oey, I. Effect of Freezing as Pre-Treatment Prior to Pulsed Electric Field Processing on Quality Traits of Beef Muscles. *Innovat. Food Sci. Emerg. Technol.* **2015,** *29,* 31–40.
17. Ghosh, S.; Gillis, A.; Sheviryov, J.; Levkov, K.; Golberg, A. Towards Waste Meat Biorefinery: Extraction of Proteins from Waste Chicken Meat with Non-Thermal Pulsed Electric Fields and Mechanical Pressing. *J. Clean. product.* **2019,** *208* (3), 220–231.
18. Gomez, B.; Munekata, P. E.; Gavahian, M.; Barba, F. J.; Marti-Quijal, F. J.; Bolumar, T.; Campagnol, P. C. B.; Tomasevic, I.; Lorenzo, J. M. Application of Pulsed Electric Fields in Meat and Fish Processing Industries: An Overview. *Food Res. Int.* **2019,** *123,* 95–105.
19. Gudmundsson, M.; Hafsteinsson, H. Effect of Electric Field Pulses on Microstructure of Muscle Foods and Roes. *Trend Food Sci. Technol.* **2001,** *12* (3–4), 122–128.
20. He, G.; Yin, Y.; Yan, X.; Wang, Y. Semi-Bionic Extraction of Effective Ingredient from Fishbone by High Intensity Pulsed Electric Fields. *J. Food Process. Eng.* **2017,** *40* (2), 1–9.
21. He, G.; Yin, Y.; Yan, X.; Yu, Q. Optimisation Extraction of Chondroitin Sulphate from Fish Bone by High Intensity Pulsed Electric Fields. *Food Chem.* **2014,** *164,* 205–210.
22. Hermawan, N.; Evrendilek, G.; Dantzer, W.; Richter, E. R. Pulsed Electric Field Treatment of Liquid Whole Egg Inoculated with *Salmonella* Enteritidis. *J. Food Safe.* **2007,** *24,* 71–85.
23. Kantono, K.; Hamid, N.; Chadha, D.; Ma, Q.; Oey, I.; Farouk, M. M. Pulsed Electric Field (PEF) Processing of Chilled and Frozen-Thawed Lamb Meat Cuts: Relationships between Sensory Characteristics and Chemical Composition of Meat. *Foods* **2021,** *10* (5), 1–21.
24. Khan, A. A.; Randhawa, M. A.; Carne, A. M.; Ahmed, I. A.; Al-Juhaimi, F. Y.; Barr, D. & Bekhit, A. E.-D. A. Effect of Low and High Pulsed Electric Field Processing on Macro and Micro Minerals in Beef and Chicken. *Innovat. Food Sci. Emerg. Technol.* **2018,** *45,* 273–279.
25. Khan, A. A.; Randhawa, M. A.; Carne, A.; Ahmed, M. I. A.; Barr, D.; Reid, M.; Bekhit, A. E.-D. A. Effect of Low and High Pulsed Electric Field on the Quality and Nutritional

Minerals in Cold Boned Beef *M. longissimus et lumborum*. *Innovat. Food Sci. Emerg. Technol.* **2017,** *41,* 135–143.
26. Kitajima, N.; Ueda, K.; Ohshima, T.; Sato, M. Development of Textile Electrode for Microbial Inactivation with Pulsed Electric Field. *Textile Res. J.* **2007,** *77* (7), 528–534.
27. Klonowski, I.; Heinz, V.; Toepfl, S.; Gunnarsson, G.; Þorkelsson, G. Applications of Pulsed Electric Field Technology for the Food Industry. In: *Report* 06–6 *of project no. 1677,* Icelandic fishes Laboratory: Reykjavík: Iceland, 2006.
28. Koubaa, M.; Barba, F. J.; Bursać, K. D.; Putnik, P.; Santos, M. D.; Queirós, R. P.; Saraiva, J. A. Pulsed Electric Field Processing of Fruit Juices. In: *Fruit Juices Extraction, Composition, Quality and analysis;* Rajauria, G., Tiwari, B. K. Eds.; Academic Press: 2018. pp. 437–449.
29. Kumar, Y.; Kumar Patel, K.; Kumar, V. Pulsed Electric Field Processing in Food Technology. *Int. J. Eng. Studies and Tech. Approach* **2015,** *1* (2), 6–16.
30. Lasekan, O.; Ng, S.; Azeez, S.; Shittu, R.; Teoh, L.; Gholivand, S. Effect of Pulsed Electric Field Processing on Flavor and Color of Liquid Foods. *J. Food Process. Preserv.* 2016, *41* (3), 1–14.
31. Li, M.; Lin, J.; Chen, J.; Fang, T. Pulsed Electric Field-Assisted Enzymatic Extraction of Protein from Abalone (*Haliotis Discus Hannai* Ino) Viscera. *J. Food Process Eng.* **2016,** *39* (6), 702–710.
32. Loeffler, M. J. Generation and Application of High Intensity Pulsed Electric Fields. In: *Pulsed Electric Fields Technology for the Food Industry;* Raso, J.; Heinz, V.; Alvarez, I. and Toepfl, S. Eds.; Springer: Boston, MA, 2022. pp. 55–106.
33. Ma. Q.; Hamid, N.; Oey, I.; Kantono, K.; Faridnia, F.; Yoo, M.; Farouk, M. Effect of Chilled and Freezing Pre-Treatments Prior to Pulsed Electric Field Processing on Volatile Profile and Sensory Attributes of Cooked Lamb Meats. *Innovat. Food Sci. Emerg. Technol.* **2016,** *37,* 359–374.
34. Macro, M. R.; Perez, M. I.; Quiles, A.; Hernando, Effect of Pulsed Electric Fields on the Main Chemical Components of Liquid Egg and Stability at 4°C. *Czech J. Food Sci.* 2009, *27,* 109–112.
35. Mohamed, M. E. A.; Amer Eissa, H. A. Pulsed Electric Fields for Food Processing Technology. In: *Structure and Function of Food Engineering;* Amer Eissa, A. H. A. Eds.; Intech: 2012, Vol. 11, pp. 275–306.
36. Morren, J.; Roodenburg, B.; De Haan, S. W. H. Electrochemical Reactions and Electrode Corrosion in Pulsed Electric Field (PEF) Treatment Chambers. *Innovat. Food Sci. Emerg. Technol.* **2003,** *4* (3), 285–295.
37. Mungure, T. E.; Bekhit, A. E. D.; Birch, J.; Stewart, I. Effect of Rigor Temperature, Ageing and Display Time on the Meat Quality and Lipid Oxidative Stability of Hot Boned Beef *Semimembranosus* Muscle. *Meat Sci.* **2016,** *114,* 146–153.
38. Nowosad, K.; Sujka, M.; Pankiewicz, U.; Kowalski, R. The Application of PEF Technology in Food Processing and Human Nutrition. *J. Food Sci. Technol.* **2021,** *58* (2), 397–411.
39. Oziemblowski, M.; Kopec, W. Pulsed Electric Field (PEF) as an Unconventional Method of Food Preservation. *Polish J. Food Nutr. Sci.* **2005,** *14/55* (1), 31–35.
40. Oziemblowski, M.; Malicki, A.; Kopec, W.; Trziszka, T. Eds. Functional and Rheological Properties of Liquid Whole Egg after Pulsed Electric Field Treatment. In: *XIth European Symposium on the Quality of Eggs and Egg Products*; Doorwerth: The Netherlands, 2005. pp. 27–31.

41. Pataro, G.; Falcone, M.; Donsi, G.; Ferrari, G. Metal Release from Stainless Steel Electrodes of a PEF Treatment Chamber: Effects of Electrical Parameters and Food Composition. *Innovat. Food Sci. Emerg. Technol.* **2014,** *21,* 58–65.
42. Pourzaki, A.; Mirzaee, Eds. Pulsed Electric Field Generators in Food Processing. In: *18th national congress on food technology;* Mashhad: Iran, 2008. pp. 1–7.
43. Quass, D. W. Pulsed Electric Field Processing in the Food Industry. In: *A status report on PEF, Electric Power Research Institute*; Palo Alto; CA, CR-109742, 1997. pp. 2335.
44. Raso, J.; Heinz, V. Eds. Pulsed Electric Fields Technology for the Industry: *Fundamentals and applications;*Food Engineering Series: 2nd edn.; 2006. p. 467.
45. Redin, L. A.; Raso, J.; Cebrian, G.; Alvarez, I. Potential of Pulsed Electric Fields for the preparation of Spanish Dry-Cured Sausages. *Scientif. Rep.* **2019,** *9,* 1–11.
46. Sampedro, F.; Rodrigo, D.; Martinez, A.; Barbosa, C. G. V.; Rodrigo, M. Application of Pulsed Electric Field in Egg and Egg Derivatives. *Food Sci. Technol. Int.* **2006,** *12* (5), 397–405.
47. Saulis, G. Electroporation of Cell Membranes. In: The Fundamental Effects of Pulsed Electric Fields in Food Processing. *Food Eng. Rev.* **2010,** *2,* 52–73.
48. Singh, K.; Kumar, S.; Kumar, P.; Bhat, Z. F. Pulsed Light and Pulsed Electric Field-emerging Non Thermal Decontamination of Meat. *Am. J. Food Technol.* **2012,** *7, 506–516.*
49. Suwandy, V.; Carne, A.; Van de, V. R.; Bekhit, A. E.-D. A.; Hopkins, D. L. Effect of Pulsed Electric Field Treatment on Hot-Boned Muscles of Different Potential Tenderness. *Meat Sci.* **2015,** *105,* 25–31.
50. Suwandy, V.; Carne, A.; Van de, V. R.; Bekhit, A. E.-D. A.; Hopkins, D. L. Effect of Pulsed Electric Field Treatment on the Eating and Keeping Qualities of Cold-Boned Beef Loins: Impact of Initial pH and Fibre Orientation. *Food and Bioprocess Technol.* **2015,** *8* (6), 1355–1365.
51. Suwandy, V.; Carne, A.; Van de, V. R.; Bekhit, A. E.-D. A.; Hopkins, D. L. Effect of Pulsed Electric Field Treatment on the Quality of Cold-Boned Beef Loins and Topsides. *Food and Bioprocess Technol.* **2015,** *8,* 1218–1228.
52. Syed, Q. A.; Ishaq, A.; Rahman, U. U.; Aslam, S.; Shukat, R. Pulsed Electric Field Technology in Food Preservation. *J. Nutri. Health Food Eng.* **2017,** *6* (5), 168–172.
53. Toepfl, S. High Intensity Pulsed Electric Fields Applied for Food Preservation. *Chem. Eng. Process.* **2007,** *46,* 537–546.
54. Yogesh, K. Pulsed Electric Field Processing of Egg Products. *J. Food Sci. Technol.* **2015,** *53* (2), 934–945.
55. Zhang, Q. H.; Barbosa-Canovas, G. V.; Swanson, B. G. Engineering Aspects of Pulsed Electric Fields Pasteurisation, *J. Food Eng.* **1995,** *25,* 261–281.
56. Zhou, Y.; He, Q.; Zhou, D. Optimization Extraction of Protein from Mussel by High-Intensity Pulsed Electric Fields. *J. Food Process. Preserv.* **2017,** *41* (3), 1–8.
57. Zhou, Y.; Sui, S.; Huang, H.; He, G.; Wang, S.; Yin, Y.; Ma, Z. Process Optimization for Extraction of Fishbone Calcium Assisted by High Intensity Pulsed Electric Fields. *Nongye Gongcheng Xuebao/Transactions of the Chinese Society of Agricultural Engineering* **2012,** *28* (23), 265–270.
58. Zuhaib, F. B.; James, D. M.; Susan, L. M.; Alaa, El-Din, A. B. Current and Future Prospects for the use of Pulsed Electric Field in the Meat Industry. *Crit. Rev. Food Sci. Nutr.* **2019,** *59* (10), 1660–1674.

CHAPTER 8

APPLICATIONS OF PULSED ELECTRIC FIELD FOR MILK PRODUCTS

YUVRAJ K BHOSALE, NALLA BHANU PRAKASH REDDY, ADITY BANDHRAL, SAWINDER KAUR, PINTU CHOUDHARY, and PREETI BIRWAL

ABSTRACT

Milk and milk products are widely consumed worldwide throughout the age groups. However, the conventional processing of milk is known for its drawbacks, namely, slight changes in composition, nutritional degradation, and spore resistance during processing. The dairy industry is looking for novel techniques as the alternative to the thermal processing method due to increasing demand for a natural taste, rich in nutrition and safe food for consumption. Non-thermal processing has been researched and discussed, but commercial implementation plays a hurdle for industrial applications. Pulsed electric field (PEF) is emerged as the most efficient nonthermal technique alternative to traditional processing methods. This technique has less processing time at room temperature, and it has a minimum impact on components sensitive to heat and nutritional properties. It can be used for various applications in the dairy industry, such as inactivation of microbes and enzymes, enhancing the functional properties of dairy ingredients, reducing processing time, etc. Furthermore, PEF has the potential to extend the shelf life of various dairy products. It is considered as green technology, generates no waste, and is economically feasible as well as energy efficient process. This chapter is devoted to discussing the recent application of PEF for dairy products in terms of shelf life extension and value addition. The key focus will be microbial inactivation, enzyme activation, and modification of dairy ingredients such as protein and fat.

Nonthermal Pulsed Electric Field Technology for the Food Industry.
Preeti Birwal, Megh R. Goyal, Baldev Singh Kalsi, & Ankita Kataria (Eds.)

DOI: 10.1201/9781003605904-10

8.1 INTRODUCTION

In the day-to-day life of a human being, food plays a significant role in keeping one supplied with the minimum nutrition that is required for the body to conduct all activities. The food can be taken in two basic forms; firstly, solid food and liquid food. Solid food can either be vegan or meat-based food which can supply nutrition to the human body. Similarly, a wide range of liquid foods, namely, juices extracted from various fruits and vegetables, are taken as food on a daily basis. Also, milk is considered one of the nutrition-rich food products that are often milked from either buffaloes or cows, and it has an irreplaceable part in the daily diet of almost all sections of people, irrespective of age and gender. Milk and milk-based food products have multiple nutritional benefits owing to their vital nutrients, such as protein, lactose, fat, and many more required by the human body.[36] According to the Food and Agriculture Organization (FAO), global milk production stands at 852 million MT (cow milk—81%, buffalo milk—15%, goat—2%, sheep and others together contribute 2%), India leading with the production of 192 million MT (FAO, 2021). Other than the nutritional aspect, milk has a significant contribution to the economic development of farmers as most of the farmers rear cattle as a support to agriculture.

In general, cow milk has less fat percentage (3.30–5.50%) than buffalo milk (5.30–9.00%); the rest of the other components, such as lactose, total solids, and protein, are marginally rich in cow milk.[36] The richness in nutrition makes milk more susceptible to spoilage irrespective of its origin (for instance, cow, buffalo, or goat). This nutrition richness makes milk a perfect growth medium for spoilage microorganisms. Besides microbial spoilage, enzymatic spoilage is not negligible in storage, shelf life, nutrition, and consumer safety perspective. Hence, milk processing is mandatory before the initiation of spoilage by either inactivating or killing microorganisms and spores, along with inactivating undesirable enzyme activity.[31] Milk is traditionally processed by using heat as a significant source of treatment. There are several existing methods of thermal processing of milk as follows:

a. **Destruction of bacteria:** heating of milk to temperature ranges of 57–68°C for 5–20 s to destroy lipase and protease-producing psychotropic bacteria;
b. **Pasteurization:** heating of milk at various time-temperature combinations such as low-temperature long-time (62–65°C/30 min) and

high-temperature short-time (72–82°C/15–30 s.) intends to inactivate pathogenic bacteria;

c. **Sterilization:** heating of milk to ultra-high temperature (135–145°C/1–10 s) intends to kill almost all the spoilage microorganisms along with the inactivation of undesirable enzymes.

All these thermal methods involve rapid cooling of heat-treated milk to less than 4–5°C, which enables the processors to get rid of the regrowth of spoilage microbes and to extend the shelf life of milk for extended periods, sometimes up to 30 days under refrigerated storage, with suitable packaging material.[17,31,36] Also, some of the novel thermal methods, such as microwave heating,[12] ohmic heating, and radio-frequency heating,[12] were reviewed thoroughly. Several researchers opined that novel thermal technologies make milk processing beneficial due to the fact that there is less processing time, thereby less heating damages, better energy and heating efficiencies, comparatively milder loss of nutrients, and environmentally friendly technologies.[38]

Despite all these advantages, the major disadvantage of thermal or novel thermal technologies in the processing of milk or milk-based food products is the loss of nutrients and a significant amount of organoleptic/sensory properties. Also, the current consumer trend toward the consumption of minimally processed food and natural–fresh-like food materials has raised a situation that researchers had to find some way to solve this problem. They found alternatives to thermal technologies such as nonthermal technologies (broadly, a technology that does not use heat as a source of energy and heat generation during processing is insignificant).

There are several nonthermal technologies such as high-pressure processing (HPP), high-pressure homogenization, high-intensity ultrasound (HiUS), pulsed electric field (PEF), microfluidization, supercritical carbon dioxide, ultraviolet (UV), pulsed light (PL), etc.[13,15,16,19,20] These nonthermal technologies offer a wide range of benefits, such as lower processing temperature, retention of nutrients, and fresh–natural-like food products.[62]

Of these technologies, the current chapter focuses on the role of PEF in milk and milk products processing and preservation along with the advantages over other thermal, novel-thermal, and other nonthermal counterparts. Besides these, the effect of PEF on the shelf life extension by affecting spoilage microorganisms and enzymes, physicochemical, functional, and sensory properties of milk and milk products are discussed. Furthermore, the necessary information regarding the enhancement of process efficacy, and practical and commercial applicability at the industrial level is given.

8.2 APPLICATIONS OF PEF ON MILK PRODUCTS

The inactivation of harmful microorganisms and undesirable enzyme activity present in any food product using electric energy/charge started in the 1920s, called ElectroPure process for milk involves heating the milk to 70°C followed by passing through electrodes in an electric chamber to inactivate spoilage microorganisms and enzymes.[62] In PEF processing, the application of a high-intensity electric field in the range of 20–80 kV/cm is applied in short pulses ranging from nano to microseconds. The duration of treatment is generally very short. Thus, the heat generation is negligible, and nutrient loss due to heat is arrested. PEF system consists of a high-voltage pulse generator, controlling mechanism, product treatment chamber, and handling system. In PEF processing, an electric field can be applied in various waveforms such as monopolar square wave, bipolar square wave, exponential decay, etc.; liquid foods can be treated with the PEF system either in batch mode or continuous mode, depending on the chamber design. Static mode: (a) in a parallel plate system, food is placed between two electrodes (high-intensity electric voltage and ground electrode), (b) in a coaxial chamber, liquid food is placed in the concentric gap between two cylindrical-shaped electrodes.[4,20] Also, the PEF treatment and final treated food product quality depend on a variety of factors, basically process and product factors. Processing factors include intensity/voltage of electric field applied, pulse width, pulse off time, total time of treatment, cycles/second, or frequency. Product factors include electrical conductivity, moisture content, porosity, etc.[19]

PEF can be applied to milk and milk products, as tried earlier, to understand the product behavior in a better way and the effect of the processing technology on final product characteristics. Various products have been tried earlier with PEF. For instance, the effects of PEF on milk's microbial and physicochemical quality were tested,[35] as effects on fat, whey proteins, caseins, xanthine oxidase,[58] the microbial and enzymatic activity of whole milk,[44] the structure of fat globule with PEF treatment, fatty acid composition, and amino acid composition changes.[48,64]

8.2.1 SHELF-LIFE EXTENSION OF DAIRY PRODUCTS

In general, the shelf life of any food product is affected by various factors. Of them, microbial activity and enzyme activity are the two predominant factors. These two factors are the significant obstacles for producers and processors in achieving more extended storage with minimal or no nutrient spoilage.

8.2.1.1 INACTIVATION OF MICROORGANISMS

The application of high-voltage PEF makes the complex food system susceptible to chemical and structural modifications, especially electrical-sensitive components.[41] In biological cells with sensitive membranes, when exposed to PEF above the bearable transmembrane potential, pores are formed in the membrane (which helps in oozing out the cellular liquids and ultimately helps in the death of the cells) that are either reversible (temporary) or irreversible (permanent), this mechanism is called "electroporation." PEF treatment applied to liquid foods primarily offers benefits such as microbial load reduction/inactivation and structural modification.[3] Various process parameters (pulse amplitude, pulse width, pulse off time, pulse shape, number of pulses, and process temperature) and product parameters (pH and conductivity) influence microbial inactivation.[49] Some of the previous studies on the effect of PEF on milk and milk products have been listed in Table 8.1.

8.2.1.2 INACTIVATION OF ENZYMES

Enzyme inactivation is one of the major concerns after microbial spoilage in perishable liquid food products. Enzymatic compounds such as proteases, laccases, etc., affect the food product and consumer acceptability. The enzymes are basically proteins; when they are fully active, their structure is rigid. One of the ways to inactivate the enzymes is to destroy their structure and kinetically irreversible steps leading to protein aggregation, dissociation, precipitation, or degradation. This can be achieved by nonthermal technologies such as ultrasound, ozone, PEF. Of these, PEF showed some promising signs of irreversible disruption or irreversible unfolding of the structure of the enzymes, called denaturation.[34] Some of the enzyme inactivation studies were reviewed and presented in Table 8.2.

8.2.2 APPLICATION OF PEF FOR ENHANCING PHYSIOCHEMICAL PROPERTIES

8.2.2.1 SENSORY PROPERTIES

A well-known fact concerning milk is that milk proteins and milk fat globules (MFGs) are very much prone to the after-effects of heat and mechanical treatments. Still more information is required on the PEF-produced changes.

TABLE 8.1 Effect of PEF Treatment on Microbial Inactivation in Dairy-Based Food Products.

Product	Conditions	Effects on microorganisms	References
Bovine whole milk	27.5 kV/cm, 32 μs (preheated milk to 55°C), comparison with thermal pasteurization.	i. Microbiological stability was maintained for 21 days at 4 C.	[57]
	Preheating: 55 C for 24 s and PEF: continuous mode with 15.8–26.1 kV/cm for 18–100 μs.	i. Reduced inoculated *Escherichia coli* and *Listeria innocua* (8–9 log CFU/mL) to less than 2 log CFU/mL.	[57]
Milk	PEF: 34 kV/cm, 22 μs, Temp: 53°C–63°C.	i. The total microbial reduction was similar to regular pasteurization conditions (72°C for 15 s). ii. A similar shelf life was achieved with lesser temperature with PEF treatment.	[35]
Raw skimmed milk	Temperature: 20, 30, and 40°C, electric field intensity 30.77, 38.45, 46.14, and 53.86 kV/cm, and number of pulses (2 μs each) 0, 12, 24, and 30.	i. Growth of mesophilic bacteria was reduced with PEF by 6 log cycles till 25 days at 4°C.	[6]
	Electric field intensity 30.77, 38.45, 46.16, and 53.83 kV/cm, temperature 20, 30, and 40°C, and number of pulses (2 μs each) 0, 12, 21, and 30.	i. Growth of mesophilic bacteria was reduced with PEF by 7-log cycles till 25 days at 4°C but at the cost of physicochemical and organoleptic properties.	[6]
Skimmed milk	28 kV/cm, 20 μs at 15–40°C.	i. Approximately 2-log reduction of *Pseudomonas flourescens.*	[14]
Skimmed milk with 0.2% fat	24 kV/cm for 141 μs.	i. 1.8 log reduction was achieved in total microbial count.	[22]
Whole milk	PEF treatment: 21–29 kV/cm for 16–102 μs at 50°C pretreatment.	i. 5–6 log reduction of both Gram-negative and Gram-positive bacteria was achieved. ii. P*seudomonas aeruginosa* reduced when pretreated at 50°C) and *E. coli.* iii. *S. aureus* and *L. innocua* reduced below the detection limit when pretreated at 56°C.	[56]

TABLE 8.2 Effect of PEF Treatment on Enzyme Inactivation in Dairy-Based Food Products.

Product	Conditions	Effects on microorganisms	References
Bovine whole milk	27.5 kV/cm, 32 μs (preheated milk to 55°C), comparison with thermal pasteurization.	i. Alkaline phosphatase (ALP) was inactivated with and equivalent to that of thermal treatment. PEF treatment reduced plasmin (7%) and xanthine oxidase (30%) activities.	[57]
	15–35 kV/cm for treatment times of 12.5–75 μs.	i. Enzyme activities of lipase, protease, and ALP were reduced by 14, 37, and 29% at extreme PEF treatment.	[50]
	Preheating: 55°C for 24 s and PEF: continuous mode with 15.9–26.2 kV/cm for 17–101 μs.	i. Reduced ALP activity with treatment at 26.2 kV/cm for 35 μs, combined with preheating. ii. Reduced the activity of plasmin, xanthine oxidase, and lipophilic fat by 11, 33, and 81%, respectively, compared with raw milk.	[58]
	25–37 kV cm^{-1} and final PEF treatment temperatures of 15°C and 60°C.	i. 24–42% inactivation in alkaline phosphatase activity at 15°C and 29–67% inactivation at 60°C, proving the hurdle approach helps in the inactivation of enzymes.	[54]
Milk	30 kV/cm, pulse width of 2 μs with a pulse frequency of 2 Hz, and 120 pulses.	i. Inactivated proteases, but after 5 days of storage at 30°C, enzyme activity was seen growing.	[65]
Simulated milk ultrafiltrate (SMUF)	Preheating: 65–80°C, PEF treatment with an activation energy of 597.1 kJ/mol.	i. CFD models proved that only 5–12% of enzyme inactivation is due to PEF effects.	[8]
	500 kJ/l of PEF energy with different setups.	i. Continuous and coaxial modes showed better protease inactivation from *Bacillus subtilis*.	[5]

This new technology gives us the advantage of maintaining the wholeness of the milk during the treatment at low temperatures, contrary to the other traditional treatments involving high heat under pasteurization. Although it is acknowledged that there have been instances where reports have provided details regarding of PEF treatment effect on the physical & chemical characteristics of milk. Comparatively, fewer studies are available on the activity of PEF on sensory and nutritional properties. It is postulated that PEF causes microbial inactivation mainly due to the dielectric breakdown. This induces pores (electroporation) in the membrane of the cells. Although the mechanism of PEF-induced enzyme denaturation remains unknown. Some experts believe the modifications are the result of structural and configurational changes caused by high electric fields.[55] The sensory. properties of dairy-based commodities, such presence of volatile after milk processing, can indicate the impact of this technology on the sensory characteristics compared to their fresh state. Flavor is a broad term that encompasses the overall effect created by the combination of taste sensations, aromatics, and chemical sensations produced by a substance in the mouth. In contrast, taste in milk is linked to simple basic tastes. Some studies have shown that the taste and flavor sensory attributes of milk treated with PEF remained unaffected in comparison to untreated milk sensory attributes.[43]

PEF processing of milk at 30 kV/cm for 22 μs at 63°C outlet temperature attains uniform microbial stability in whole milk than that of thermal pasteurization at 72°C for 15 s having 4% fat. When milk was subjected to PEF treatment at 30 kV/cm for 22 μs, with an outlet temperature of 53°C, and stored at 4°C, its shelf life increased by approximately 3–4 days compared to untreated raw milk. The physicochemical attributes of milk did not primarily change by PEF treatments. They had a lesser impact on the volatile composition and were largely relying on the thermal factor of the treatment. It offers the advantage of preserving the subtle variations in sensory characteristics and retaining the nutritional content of milk, thereby providing a more natural and unprocessed quality to milk-based products.[59]

Sharma et al.[57] reported the sensory properties of cheddar and cottage cheese prepared from PEF treatment with cheese prepared from conventional heating of the milk. They later concluded that the cheddar cheese from PEF-treated milk is lightly varied in flavor from the other cheese prepared from conventional heating. At the same time, no flavor difference was noticed between the cottage cheese prepared from PEF-treated milk to that prepared from thermally processed milk.[57] In PEF processing of whole milk, the MFGs start showing deposition of the electrode surface.[6] On the other hand, in flash processing of whole milk, the formation of the volatile compound can be explained by other mechanisms. Zhang et al.[67] described high temperature short time (HTST) processing as the high temperature is liable for the formation of autoxidation of USFA.

PEF treatment has the least important change in lactone acids and alcohols measured compared to the raw sample. In addition, the organoleptic properties of PEF-treated milk samples have also remained equal when compared with untreated samples. PEF-treated milk samples have more retention ability of nutrients, ensuring a fresh character in milk-based products. No significant variation in taste and flavor was observed. The minimal impact of technology on the natural characteristics and nutritional attributes of milk is the reason behind its "no-difference" attribute. With PEF processing, essential nutrients, such as vitamin C, tocopherol (vitamin E), thiamine, cholecalciferol (vitamin D), and riboflavin remain unaffected, preserving their original content.[59]

8.2.2.2 PROTEIN FUNCTIONALITY

When it comes to microbial development, milk is a highly perishable item. Past research revealed that variations in milk storage are closely related to its chemical makeup (i.e., an increase in the amount of milk lipid-derived short-chain free fatty acids). These changes are caused by microbial growth, which results in microbial enzymatic activity in both raw and processed milk. Here, PEF processing at 30 kV/cm and 63°C for 22 μs can help to maintain the shelf life of whole milk to a much longer duration at chill conditions. The effects are comparable to thermal pasteurization, leaving no inimical effect on the physicochemical properties of the milk.[35]

Middendorf et al.[37] studied the effect of PEF on the coagulation of rennet enzyme in milk. They stated that the rennet coagulation of milk is decreased upon PEF processing, and ultimately, the gel strength is decreased. This also increased the coagulation time. PEF technique, when combined with thermalization, helps to develop the structure of the casein micelle in such a way that lipophilic organic compounds, such as β-carotene, can be incorporated by disintegration. Therefore, this combination of adding β-carotene was found to be 10 times higher than the classical thermalization process, with a recovery rate of 78%. On the contrary, the literature on the comprehensive mechanism of the effect of PEF treatment on Casein micelle still needs to be addressed.[37]

Plasmin is a proteolytic enzyme that has both favoring and ruinous effects on the overall quality of milk products, including texture and flavor.

This depends upon the nature of the product and degree of hydrolysis on casein micelles resulting in small peptones and some volatile components. Alkaline pH favors this enzyme to link with casein and cause changes in the milk. Vega-Mercado et al.[63] demonstrated that 90% of the plasmin could be inactivated by the PEF treatment (E = 30–46 kV/cm; t = 101 μs; T = 15°C) in the SMUF system. However, they also state that at the same PEF conditions, just lowering the temperature by 10°C, the plasmin inactivation by 60% was observed. They indicated the effect of the nonthermal technique on plasmin inactivation where the temperature ranges between 10–15°C since low temperatures were used.

Sharma et al.[57] unveiled that with the increase in PEF strength from E = 15–36 kV/cm at T = 31°C, the protease enzyme present in the raw milk and its inactivation also starts increasing. Significant inactivation during PEF processing was observed at a strength of 35 kV/cm and a duration of 75 μs, resulting in approximately a 37% reduction. Additionally, the findings indicated that the activity of protease is dependent on both the duration of processing and the strength of the PEF treatment.[1] Primarily, milk proteins remain unaffected by PEF operation, but casein micelles and whey protein structure get affected, and therefore, properties such as clumping, gelation rate, lipophilicity, gel strength, and emulsion stability start to alter.

Milk enzymes exhibit great resistance to treatment compared to vegetative forms of bacteria. As a result, PEF technology is primarily utilized to eliminate spoilage bacteria while preserving the activity of beneficial enzymes in dairy products.[59] PEF treatment presents a potential approach for improving the production yield of whey protein powder. The application of PEF on milk increases the likelihood of obtaining denatured concentrated whey protein, resulting in a higher quantity of whey protein that can be generated through this nonthermal method. This protein can then be used to prepare whey protein isolates, concentrates, and hydrolysates, which are valuable for individuals who require protein supplements, such as those with phenylketonuria. During PEF treatment, high heat was not induced up to 25-μs pulse length. The onset of denaturation of whey proteins, mainly β-lacto globulin, starts occurring at 72°C, and once the temperature reaches 116°C, the extent of denaturation is up to 90%. Variable and unstable turbidity leads to Low reproducibility, and due to this, the main issue with the whey protein nitrogen index (WPNI) emerges. Despite this, all the practical measures, such as high-intensity electric pulse, are taken to protect α-lacto globulin content in the milk sample. This means the protein is stable and less denatures at high pasteurization temperatures when compared to the β-lacto globulin.

Introducing a novel PEF technology with specific parameters, including 24 kV/cm, 20 pulses, and 25-μs pulse duration, represents an innovative advancement in the field. This technology not only reduces pasteurization and operational costs but also ensures the preservation of higher nutritional value in milk products.[51]

8.2.3 PEF FOR ENHANCING PROCESS EFFICIENCY

8.2.3.1 FERMENTATION PROCESS

PEF in fermentation works primarily by applying strong electric fields to the sample source using direct current electric pulses when positioned between two electrodes in a controlled chamber. The pulsed voltage of up to 40 kV is applied to the source in 50–1000-μs pulses. The outer layer of a microorganism cell acts as a capacitor in a system, primarily composed of a material that possesses a low dielectric constant. When these cells come into contact with electric fields, the ions within them begin to move with the field, accumulating free charges at both ends of the membrane surface. This has a profound impact on signal transduction.

Increasing the electric field intensity directly influences the potential difference across the membrane in a positive manner. Once the threshold value, typically ranging from 0.2 to 1 V, is surpassed, the cell membrane loses its semi-permeability. This is called electroporation and depends upon the electric field strength given and the generation of temporary (reversible) or permanent (irreversible) pores on the cell membrane.[27] The impact of PEF on the food substrate is influenced by the characteristics of the fermentation medium. At the same time, the electrical properties of the treated sample, including electrical conductivity and electric strength, are also crucial factors in determining the outcomes of PEF treatment. These factors contribute to the augmentation of electric field intensity and the sustenance of current flow within the system.

Studies have shown that fermentation mediums with high electrical conductivities are not desirable for PEF treatments since they result in lower electric fields being generated in the treatment chamber. Therefore, it is crucial to preserve the disparity in electrical conductivity between the microbial cytoplasm and the fermentation medium to achieve the maximum flow rate of ionic substances through the membrane. This disparity ultimately results in the disintegration of the inner structure of the microorganism.[30]

The characterization of conduction is essential for causing electroporation in the cellular membrane.[46] Immediate destruction by PEF can be appraised by estimating the cellular damage index *Z* depending on measuring the electrical conductivity in a lower frequency region (1–10 kHz).[33] One more factor that accounts for the PEF efficiency is cell density. The particular conditions of the treatment may or may not affect the inactivation of the specific microorganism during fermentation.

According to a study conducted by Zhang et al.,[66] the inactivation of *E. coli* in synthetic ultra-filtered milk was found to be consistent regardless of the initial concentration ranging from 10^2 to 10^9 CFU/ml. The treatment involved exposing the milk to 70 kV/cm with 16 pulses, each having a width of 3 μs. Another study revealed an increase in the population of *Saccharomyces cerevisiae* in apple juice led to reduced microbial inactivation during PEF treatment. This effect may be attributed to the clumping of *S. cerevisiae* cells or the presence of concealed microbial cells in regions of lower electric field strength. Moreover, the growth stage of the microbe also controls the strength of PEF treatment. It has been suggested that cells in the logarithmic phase exhibit greater susceptibility to stress compared to cells in the lag and stationary phases. This is attributed to the fact that during the logarithmic phase, a majority of the cells are actively dividing, making their cell membranes more sensitive to the applied electric field.[47]

Álvarez et al.[2] conducted a study on the cells of *Salmonella senftenberg* after giving 200 pulses at 19 kV/cm to various growth stages. They found that the log phase of the cell decreased by 1.5 log cycles when comparisons were made against lag and stationary phases. Promising results have been recovered from the fermented foods treated with PEF, such as the enhancement of some bioactive compounds. These findings are associated with the alterations occurring at microscopic and structural levels within the food system, both in terms of physical and chemical changes. Factors such as microbial characteristics, including microorganism and medium type, the composition of the medium, and processing parameters, including power, electric field intensity, frequency, treatment time, temperature, pH, and several other product parameters, determine the influence of PEF strength on fermentation.

8.2.3.2 PASTEURIZATION

ALP is the indicator to measure the effectiveness of pasteurization in milk. Its D-value is 5.4 s at 70°C means it is more heat stable than other vegetative pathogenic microorganisms seen in the milk.[29] If the ALP test comes out to be negative, this means acceptable thermal pasteurization is achieved. This is useful in maintaining safe microbial conditions within the food system.[54] To inactivate the ALP enzyme under PEF treatment, the process and medium conditions need to be controlled. It is noted that E = 20 kV/cm; t = 400 µs of PEF parameters have no such effect on ALP inactivation in milk.[61] Whereas in dairy products, up to 65% inactivation of ALP was achieved in restricted conditions of PEF (E = 22 kV/cm; t = 75 µs; T < 44°C).[11] About 24–42% inactivation of ALP in PEF-treated milk is achieved by maintaining E = 25–37 kV/cm; t = 19.5 µs and T = 15°C. In the same way, the ALP enzyme inactivation of up to 67% was achieved when the final temperature reached 60°C.[53]

Sharma et al.[54] compared the destruction of ALP enzyme in full-fat milk by both PEF and heat treatment and concluded that the activity of the enzyme had a similar effect by heat treatment and PEF at E = 25.6 kV/cm; t = 34 µs at T = 55°C when applied for 22 s. Also, they inferred that about 97% of the enzyme is directly inactivated by both. Another essential enzyme in milk is lactoperoxidase (LPO), an important indigenous milk enzyme with antimicrobial activity. It is considered an important unit of the LPO–SCN–H_2O_2 antimicrobial system that helps to control microorganisms by catalyzing H_2O_2 in fresh milk under cooling conditions.[32]

The heat sensitivity of the enzyme makes it inactivate even at a normal pasteurization temperature of 70°C in 20 min.[8] Further investigation is required to comprehensively understand the mechanism by which PEF affects the LPO enzyme in dairy products. Additionally, the impact of heat generated during PEF treatment remains to be fully acknowledged and explored. It was indicated that about 5–12% of LPO enzyme gets inactivated by the minimal temperature of PEF treatment itself, as the temperature is a deciding factor for enzyme activity.[29] Despite the challenges posed by PEF processing, LPO has shown remarkable resistance compared to other enzymes. Studies have demonstrated that LPO remains active as long as the temperature during PEF treatment stays below the typical pasteurization temperature (with parameters E = 18–36 kV/cm and t = up to 500 µs)[61]. At the same time, LPO degrades when the temperature reaches 75°C at E = 40 kV/cm.[8]

8.3 SAFETY CONSIDERATIONS IN PEF

Any new technology only succeeds when the consumer fully accepts it, and this is a crucial point regarding new technology. PEF processes have been recognized as eco-friendly and highly natural.[40,42] On the contrary, various health-related concerns arose, including high prices, a lack of knowledge about technology, and general doubting nature.[40] Moreover, the possible safety concern for applying PEF plants incorporates the production of gas bubbles and electrical arcing.

The production of gas bubbles can be reduced by operating under high pressure and high arcing generally occurs at very high voltages.[45] Concerning safety issues related to PEF processing, electrode material type and amount discharged in the operated products must fulfill healthy safety regulations before releasing that food into the market.[25] Migrated metals in processed foods are a significant concern worldwide regarding public health. European Union mainly deals with the minimum toxic level in the food to combat toxicology effects on the human race.

8.4 PEF LIMITATIONS IN THE DAIRY INDUSTRY

In the PEF system, the liquid food is placed between two metal electrodes in the processing chamber, electrically coupled to a pulse generator (or any electrolyte solution). An electrochemical cell, defined in the field of electrochemistry, is a device that enables chemical reactions to occur using an external electrical energy source. This energy source either drives an electric current through the cell or generates energy as a result of the amount of electricity from chemical reactions (electrolyte cells). When the PEF chamber is filled with an electrolyte solution, a double ion layer is created at each electrode–electrolyte interface to ensure electrically neutral conditions, this layer consists of directional dipoles or charged particles. The formed layer, comprising charged particles or aligned dipoles, exhibits electrical characteristics resembling a combination of resistors (faradic impedance) and capacitors in a parallel arrangement.[39] Most effective low-stage reactions arise at every electrode–electrolyte interface when no outside voltage is carried out to the electrodes. When competing reactions reach an equilibrium, the currents in each direction become equal. As a result, a net current of zero is achieved, with an exchange current flowing across the interface in both directions.[39]

8.4.1 CORROSION OF ELECTRODES

Electrode corrosion is a common side effect of electrochemical processes at the electrode–electrolyte interface in the PEF treatment chamber.

The concentration of dissolved metals in the treated media is used as an indicator to illustrate how the corrosion rate of the electrodes varies with different parameters. However, this equilibrium exchange current does not show the true electrode rate reactions, as some dissolved metals tend to accumulate and deposit on the surface of the electrode. For instance, the double-layer capacitors' ability to charge depends on variables such as voltage, pulse form, length, frequency, current magnitude, and chamber shape.[30]

Pataro et al.[44] studied the electrode material's chemical inertness. Similarly, the electrode material plays a vital role in estimating the corrosion rate due to its inertness toward chemical and C_{dl} value (eq. 8.1).[28] The treatment medium's composition and chemical-physical characteristics can be significant.[26]

During pulse treatment, when a potential difference is applied between the electrodes, an electrical current passes through the cell, leading to the deposition of charges across the layer.[7] With the exception of cases where the applied voltage remains below a specific threshold, it prevents a potential drop in layers from surpassing a certain level; no electrochemical processes take place.

The treatment medium's composition and chemical–physical characteristics can be significant.[25] By imposing a difference of potential across the electrodes during pulse treatment, electric current passes across the cell, leading to the accumulation of charges across the double layer.[7]

To reach the threshold voltage (U_{th}), which corresponds to the reaction potential (typically 1–2 V) of the electrode material, a charging current is necessary for charging a double-layer capacitor. The threshold time U_{th} (s) measures how quickly the double layer charges and can be calculated following equation[24,39,44]:

$$T_{th} = C'_{dl}\, U'_{th} / j /1/ \tag{8.1}$$

where j is the current density (in A/cm^2), whereas specific capacitance (capacitance per unit of area: C_{dl}, in μF/cm^2), and reaction potential (U_{th}, in V) of the two electrode–electrolyte interfaces in series (i.e., $C'_{dl} = (1/2)\, C_{dl}$ and $U'_{th} = 2\, U_{th}$).[52]

8.4.2 ELECTRODE LIFETIME AND EQUIPMENT RELIABILITY

Corrosion is a primary cause of electrode degradation; direct contact is maintained when surface roughness grows due to anode metal breakdown or partial metal oxide deposition on the cathode. The damage caused by the high-voltage electrode (anode) surface is more significant compared to the ground electrode (cathode), making the eroded electrodes more harmful or dangerous.[44]

A large electric charge can be transferred through the electrode area observed upon long-term trial. The electrode is made of no-inert material that can upsurge electrode erosion. The erosion of electrodes occurs to a greater extent, leading to potential distortions in local electric field enhancements within the treatment zone. This can contribute to the potential irregularities in PEF analysis as one of the factors affecting the process. Additionally, the temperature profile during the treatment and the flow nature throughout the treatment may affect the PEF treatment homogeneousness, reducing treatment efficiency throughout PEF operation.[44]

8.4.3 ELECTRODE FOULING

When PEF is operated on proteinaceous material, either solution, film of food particles, or any other material such as milk or material (fouling agent) starts collecting or aggregating on the surface of the electrode or the treatment chamber. Such effects account for the unpleasant consequences of the treated product. The reason is the electrochemical mechanism; when external electric field applied, charged particles begin to migrate toward the electrode with the opposite charge and accumulate near the electrode boundary. This is mainly observed in liquid-food products.[10] During electrophoresis, the material by which electrodes are made doesn't participate in any other function except generating an electric field.[9]

8.4.4 ELECTROLYSIS

When directed with PEF, liquid food products cause electrolysis of water as they contain enough water to sustain the reaction. This water electrolysis initiates many other reactions resulting from PEF operating food products. The predominant product of water electrolysis is water oxygen (O_2 gas) and hydrogen (H_2 gas) at the anode and cathode respectively. These undesirable effects can trigger the electrochemical reactions between the treatment chamber and food component, liberating gaseous molecules. It was also noticed that forming an electrical arch inside the treatment chamber causes food safety concerns and health hazards, thereby reducing overall system performance. Molecular oxygen is produced during the breakdown of water and can induce the oxidation or decomposition of several components. This includes lipids, phenolic compounds, and ascorbic acid, which produces various toxic compounds such as H_2O_2.[60] Uncurbed hydrogen release due to its high flammability and explosion property may give rise to safety issues in large PEF operation units.

More safety concerns are related to the chlorine in a treated medium. This chlorine comes from the reaction between the cathode and anode, resulting in the generation of H_2 (*g*) and Cl_2 (*g*), and if the operating chamber has a high amount of chloride ions, this will lead to incomplete processing of the treated product, thus causing food-safety issues. The primary challenge faced in PEF treatment is to minimize the occurrence of electrochemical reactions. This can be achieved by fine-tuning the pulse generator system, enhancing the treatment design, and choosing stable electrode materials. Additionally, it is crucial to consider the physicochemical properties of the food being processed in order to ensure technical stability and food safety in PEF technology.

8.5 SUMMARY

PEF technology is an alternative technique to conventional technology for processing milk and products. This technology can maintain the product quality attributes without affecting the chemical composition of milk and milk products. As compared to other nonthermal techniques, PEF has the potential to overcome most of the limitations of conventional techniques. PEF is also very efficient for enhancing the efficiency of processes such as pasteurization, sterilization, fermentation, enzyme inactivation, and extraction. It is also very efficient in improving the shelf life of products. Furthermore, it is well known for enhancing dairy ingredients' functional properties, especially proteins. PEF is a green technology that produces no waste that harms the environment. Certain limitations associated with PEF processing include electrode fouling and electrolysis. These limitations can be eradicated by making some modifications to the existing equipment.

KEYWORDS

- **dairy products**
- **microbial inactivation**
- **nonthermal process**
- **process efficiency**
- **pulsed electric field**

REFERENCES

1. Ahmad, T.; Butt, M. Z.; Aadil, R. M.; Inam-ur-Raheem, M.; Bekhit, A. E.; Guimarães, J. T.; Balthazar, C. F.; Rocha, R. S.; Esmerino, E. A.; Freitas, M. Q. Impact of Non-Thermal Processing on Different Milk Enzymes. *Int. J. Dairy Technol.* **2019,** *72* (4), 481–495.
2. Álvarez, I.; Raso, J.; Palop, A.; Sala, F. J. Influence of Different Factors on the Inactivation of *Salmonella senftenberg* by Pulsed Electric Fields. *Int. J. Food Microbiol.* **2000,** *55* (1–3), 143–146.
3. Arshad, R. N.; Abdul-Malek, Z.; Roobab, U.; Munir, M. A.; Naderipour, A.; Qureshi, M. I.; El-Din Bekhit, A.; Liu, Z. W.; Aadil, R. M. Pulsed Electric Field: A Potential Alternative Towards Sustainable Food Processing. *Trend. Food Sci. Technol.* **2021,** *111,* 43–54.
4. Bendicho, S.; Marsellés-Fontanet, A. R.; Barbosa-Cánovas, G. V.; Martín-Belloso, O. High-Intensity Pulsed Electric Fields and Heat Treatments were Applied to a Protease from *Bacillus subtilis*. A Comparison Study of Multiple Systems. *J. Food Eng.* **2005,** *69* (3), 317–323.
5. Bermúdez-Aguirre, D.; Fernández, S.; Esquivel, H.; Dunne, P. C.; Barbosa-Cánovas, G. V. Milk Processed by Pulsed Electric Fields: Evaluation of Microbial Quality, Physicochemical Characteristics, and Selected Nutrients at Different Storage Conditions. *J. Food Sci.* **2011,** *76* (5), S289–S299.
6. Bockris, J. O.; Reddy, A. K. N.; Gamboa-Aldeco, M. Eds; *Modern Electrochemistry 2A: Fundamentals of Electronics;* (2nd edn), Springer New York: 2000. pp. 771–1033.
7. Buckow, R.; Semrau, J.; Sui, Q.; Wan, J.; Knoerzer, K. Numerical Evaluation of Lactoperoxidase Inactivation During Continuous Pulsed Electric Field Processing. *Biotechnol. Progress* **2012,** *28* (5), 1363–1375.
8. Bushnell, A. H.; Clark, R. W.; Dunn, J. E.; Lloyd, S. W. Prevention of Electrode Fouling in High Electric Field Sterilization. *Trend. Food Sci. Technol.* **1995a,** *10* (6), 350.
9. Bushnell, A. H.; Clark, R. W.; Dunn, J. E.; Lloyd, S. W. *Prevention of Electrode Fouling in High Electric Field Systems for Killing Microorganisms in Food Products;* Google Patents: 1995b.
10. Castro, A. J.; Swanson, B. G.; Barbosa-Cánovas, G. V, & Zhang, Q. H. Pulsed Electric Field Modification of Milk Alkaline Phosphatase Activity. In: *Pulsed Electric Fields in Food Processing;* Barbosa-Cánovas, G. V.; Zhang, Q. H. Eds.; CRC Press: 2019. Vol 1, pp. 65–82.

11. Craven, H. M.; Swiergon, P.; Ng, S.; Midgely, J.; Versteeg, C.; Coventry, M. J.; Wan, J. Evaluation of Pulsed Electric Field and Minimal Heat Treatments for Inactivation of Pseudomonads and Enhancement of Milk Shelf-Life. *Innovat. Food Sci. Emerg. Technol.* **2008,** *9* (2), 211–216.
12. Evrendilek, G. A.; Zhang, Q. H. Effects of Pulse Polarity and Pulse Delaying Time on Pulsed Electric Fields-Induced Pasteurization of *E. coli* O157:H7. *J. Food Eng.* **2005,** *68* (2), 271–276.
13. FAO (Food and Agriculture Organization of the United Nations) *World Food and Agriculture—Statistical Yearbook 2021*. Rome, 2021.
14. Gad, A.; Jayaram, S. H. Electrode Material Migration During Pulsed Electric Field (PEF) Treatment. In: *Proceeding of 2011 ESA Annual Meeting on Electrostatics*; Case Western Reserve University, Cleveland, OH-USA: ESA(Electrostatics Society of America: 2011. pp. 1–9.
15. Gad, A.; Jayaram, S. H. Effect of Electric Pulse Parameters on Releasing Metallic Particles from Stainless Steel Electrodes During Pulsed Electric Field Processing of Milk. In: *Conference Electrostatics Joint Conference;* Cambridge: ON-Canada, 2012a. pp. 1–7.
16. Gad, A.; Jayaram, S. H. Effect of Food Composition and pH on Electrode Material Migration during PEF Application. In: *Proceedings of the International Conference Bio & Food Electrotechnologies (BFE 2012)*; Salerno: Italy, 2012b, pp. 49–52.
17. Galván-D'Alessandro, L.; Carciochi, R. A. Fermentation Assisted by Pulsed Electric Field and Ultrasound: A Review. *Fermentation* **2018,** *4* (1), 1.
18. Góngora-Nieto, M. M.; Sepúlveda, D. R.; Pedrow, P.; Barbosa-Cánovas, G. V; Swanson, B. G. Food Processing by Pulsed Electric Fields: Treatment Delivery, Inactivation Level, and Regulatory Aspects. *LWT-Food Sci. Technol.* **2002,** *35* (5), 375–388.
19. Jaeger, H.; Meneses, N.; Moritz, J.; Knorr, D. Model for the Differentiation of Temperature and Electric Field Effects During Thermal-Assisted PEF Processing. *J. Food Eng.* **2010,** *100* (1), 109–118.
20. Jayaram, S.; Castle, G. S. P.; Margaritis, A. Effects of High Electric Field Pulses on Lactobacillus Brevis at Elevated Temperatures. In: *Conference Record of the 1991 IEEE Industry Applications Society Annual Meeting;*, Victoria: B.C. 1991. pp. 674–681.
21. Kussendrager, K. D.; Van Hooijdonk, A. C. M. Lactoperoxidase: Physicochemical Properties, Occurrence, Mechanism of Action and Applications. *Br. J. Nutr.* **2000,** *84* (S1), 19–25.
22. Lebovka, N. I.; Bazhal, M. I.; Vorobiev, E. Estimation of Characteristic Damage Time of Food Materials in Pulsed-Electric Fields. *J. Food Eng.* **2002,** *54* (4), 337–346.
23. Martin-Belloso, O.; Robert Marsell's-Fontanet, A.; Elez-Martinez, P. Enzymatic Inactivation by Pulsed Electric Fields. In: *Emerging Technologies for Food Processing;* Sun, D. W. Eds.; Academic Press: 2014. pp. 155–168.
24. McAuley, C. M.; Singh, T. K.; Haro-Maza, J. F.; Williams, R.; Buckow, R. Microbiological and Physicochemical Stability of Raw, Pasteurized, or Pulsed Electric Field-Treated Milk. *Innovat. Food Sci. Emerg. Technol.* **2016,** *38,* 365–373.
25. Middendorf, D.; Bindrich, U.; Siemer, C.; Töpfl, S.; Heinz, V. Affecting Casein Micelles by Pulsed Electrical field (PEF) for Inclusion of Lipophilic Organic Compounds. *Appl. Sci.* **2021,** *11* (10), 4611.
26. Morren, J.; Roodenburg, B.; de Haan, S. W. H. Electrochemical Reactions and Electrode Corrosion in Pulsed Electric field (PEF) Treatment Chambers. *Innovat. Food Sci. Emerg. Technol.* **2003,** *4* (3), 285–295.

27. Nielsen, H. B.; Sonne, A.-M.; Grunert, K. G.; Banati, D.; Pollák-Tóth, A.; Lakner, Z.; Olsen, N. V.; Žontar, T. P.; Peterman, M. Consumer Perception of the use of High-Pressure Processing and Pulsed Electric Field Technologies in Food Production. *Appetite* **2009,** *52* (1), 115–126.
28. Niu, D.; Zeng, X. A.; Ren, E. F.; Xu, F. Y.; Li, J.; Wang, M. S.; Wang, R. Review of the Application of Pulsed Electric Fields (PEF) Technology for Food Processing in China. *Food Res. Int.* **2020,** *137,* 109715.
29. Olsen, N. V.; Grunert, K. G.; Sonne, A.-M. Consumer Acceptance of High-Pressure Processing and Pulsed-Electric Field: A Review. *Trend. Food Sci. Technol.* **2010,** *21* (9), 464–472.
30. Pankiewicz, U.; Góral, M.; Kozłowicz, K.; Góral, D. Application of Pulsed Electric Field in the Production of Ice Cream Enriched with Probiotic Bacteria (*L. rhamnosus* B 442) Containing Intracellular Calcium ions. *J. Food Eng.* **2020,** *275,* 109876.
31. Pataro, G.; Falcone, M.; Donsì, G.; Ferrari, G. Metal Release from Stainless Steel Electrodes of a PEF Treatment Chamber: Effects of Electrical Parameters and Food Composition. *Innovat. Food Sci. Emerg. Technol.* **2014,** *21,* 58–65.
32. Patel, H. A.; Carroll, T.; Kelly, A. L. Potential Applications of Non-Thermal Processing Technologies in the Dairy Industry. *Dairy Process. Quality Assur.* **2015,** 528–551.
33. Pliquett, U. Bioimpedance: A Review for Food Processing. *Food Eng. Rev.* **2010,** *2* (2), 74–94.
34. Pothakamury, U. R.; Vega, H.; Zhang, Q.; Barbosa-Canovas, G. V, & Swanson, B. G. Effect of Growth Stage and Processing Temperature on the Inactivation of *E. coli* by Pulsed Electric Fields. *J. Food Protect.* **1996,** *59* (11), 1167–1171.
35. Qin, S.; Li, P.; Liang, G.; Liu, J.; Huang, S.; Meng, J. Pulsed Electric Field Treatment of Raw Milk and Its Effect on Protein-Free Amino Acids Transition. *IEEE Trans. Plasma Sci.* **2022,** *50* (4), 911–919.
36. Rajashri, K.; Rastogi, N. K.; Negi, P. S. Non-Thermal Processing of Tender Coconut Water—A Review. *Food Rev. Int.* **2022,** *38* (sup1), 34–55.
37. Riener, J.; Noci, F.; Cronin, Denis A Morgan, Desmond J Lyng, J. G. Effect of High Intensity Pulsed Electric Fields on Enzymes and Vitamins in Bovine Raw Milk. *Int. J. Dairy Technol.* **2009,** *62* (1), 1–6.
38. Šalaševičius, A.; Uždavinytė, D.; Visockis, M.; Ruzgys, P.; Šatkauskas, S. Effect of Pulsed Electric Field (PEF) on Bacterial Viability and Whey Protein in the Processing of Raw Milk. *Appl. Sci.* **2021,** *11* (23), 11281.
39. Skromeda, N.; Katsube, T. J. Electrochemical Double-Layer Capacitance of Metals, Including some Precious metals: Preliminary Results. In: *Geological Survey of Canada, Current Research (Online);* Natural Resources Candan: Canada. 2008.
40. Shamsi, K. *Effects of Pulsed Electric Field Processing on Milk Properties: Non-Thermal Processing of Milk*; Doctoral dissertation, RMIT University: 2008
41. Shamsi, K.; Versteeg, C.; Sherkat, F.; Wan, J. Alkaline Phosphatase and Microbial Inactivation by Pulsed Electric Field in Bovine Milk. *Innovat. Food Sci. Emerg. Technol.* **2008,** *9* (2), 217–223.
42. Sharma, P.; Bremer, P.; Oey, I.; Everett, D. W. Bacterial Inactivation in Whole Milk using Pulsed Electric Field Processing. *Int. Dairy J.* **2014,** *35* (1), 49–56.
43. Sharma, P.; Oey, I.; Bremer, P.; Everett, D. W. Reduction of Bacterial Counts and Inactivation of Enzymes in Bovine Whole Milk using Pulsed Electric Fields. *Int. Dairy J.* **2014,** *39* (1), 146–156.

44. Sharma, P.; Oey, I.; Bremer, P.; Everett, D. W. Microbiological and Enzymatic Activity of Bovine Whole Milk Treated by Pulsed Electric Fields. *Int. J. Dairy Technol.* **2018,** *71* (1), 10–19.
45. Sharma, P.; Oey, I.; Everett, D. W. Thermal Properties of Milk Fat, Xanthine Oxidase, Caseins, and Whey Proteins in Pulsed Electric Field-Treated Bovine Whole Milk. *Food Chem.* **2016,** *207,* 34–42.
46. Soltanzadeh, M.; Peighambardoust, S. H.; Gullon, P.; Hesari, J.; Gullón, B.; Alirezalu, K.; Lorenzo, J. Quality Aspects and Safety of Pulsed Electric Field (PEF) Processing on Dairy Products: A Comprehensive Review. *Food Rev. Int.* **2020,** 1–22.
47. Tzedakis, T.; Basseguy, R.; Comtat, M. Voltammetric and Coulometric Techniques to Estimate the Electrochemical Reaction Rate During Ohmic Sterilization. *J. Appl. Electrochem.* **1999,** *29* (7), 819–826.
48. Van Loey, A.; Verachtert, B.; Hendrickx, M. Effects of High Electric Field Pulses on Enzymes. *Trend. Food Sci. Technol.* **2001,** *12* (3–4), 94–102.
49. Vega-Mercado, H.; Powers, J. R.; Martín-Belloso, O.; Luedecke, L.; Barbosa-Cánovas, G. V, & Swanson, B. G. Change in Susceptibility of Proteins to Proteolysis and the Inactivation of an Extracellular Protease from Pseudomonas Fluorescens M3/6 When Exposed to Pulsed Electric Fields. In: *Pulsed electric fields in food processing;* Barbosa-Cánovas, G. V.; Zhang, Q. H. Eds.; CRC Press: 2019. pp. 105–120.
50. Yang, S.; Suwal, S.; Andersen, U.; Otte, J.; Ahrné, L. Effects of Pulsed Electric Field on Fat Globule Structure, Lipase Activity, and Fatty Acid Composition in Raw Milk and Milk with Different Fat Globule Sizes. *Innovat. Food Sci. Emerg. Technol.* **2021,** *67,* 102548.
51. Yu, L. J.; Ngadi, M.; Raghavan, V. Proteolysis of Cheese Slurry Made from Pulsed Electric Field-Treated Milk. *Food and Bioprocess Technol.* **2012,** *5* (1), 47–54.
52. Zhang, Q.; QIN, B.; BARBOSA-CÁNOVAS, G. V, & Swanson, B. G. Inactivation of *E. coli* for Food Pasteurization by High-Strength Pulsed Electric Fields. *J. Food Process. Preserv.* **1995,** *19* (2), 103–118.
53. Zhang, S.; Yang, R.; Zhao, W.; Hua, X.; Zhang, W.; Zhang, Z. Influence of Pulsed Electric Field Treatments on the Volatile Compounds of Milk in Comparison with Pasteurized Processing. *J. Food Sci.* **2011,** *76* (1), C127–C132.

PART III

ROLE OF PULSED ELECTRIC FIELD IN PROCESSING AND EXTRACTION OF BY-PRODUCTS

CHAPTER 9

PULSED ELECTRIC FIELD FOR CEREALS, LEGUMES, AND OILSEEDS

REWA KULSHRESTHA, JAGBIR REHAL, and PREETI BIRWAL

ABSTRACT

The increasing demand of consumers for safe and quality products that retain their nutritional and phytochemical components pushes the processing industry to find new and innovative technologies. Nonthermal technologies are counted among novel technologies for the preservation of food. Out of these, one is the pulsed electric field (PEF) which promises to deliver products with superior sensory and quality parameters with enhanced shelf life. Cereal, oil seeds, and pulses are an integral part of the diet and need some processing interventions to increase their palatability, digestibility, and availability of nutrients. This chapter delves into the application of pulse electric field in these areas as well as its use in maximizing the extraction of bioactive compounds from cereals, oilseeds, and pulses.

9.1 INTRODUCTION

Cereals, legumes, and oilseeds (CLO) belong to Gramineae, Fabaceae, and Cruciferae family, respectively. They form the mainstay of food ingredients in food processing and consumption. These ingredients meet the challenges of food security and nutritional needs for all. They are a crucial seedbed of energy, carbohydrates, fat, protein, dietary fiber (DF), vitamins, minerals, and various phytochemicals. Being the powerhouse of nutrients, they also serve various functionalities as an ingredient. The cultivation and production cost of CLO is low; hence, they form the backbone of agricultural production

Nonthermal Pulsed Electric Field Technology for the Food Industry.
Preeti Birwal, Megh R. Goyal, Baldev Singh Kalsi, & Ankita Kataria (Eds.)

DOI: 10.1201/9781003605904-12

and the food supply chain system. Furthermore, the ability to fix atmospheric nitrogen, crop rotation, and contribute to the reduction of climate change are additional adjuncts. They are grown mainly for human nutrition, grain seeds, cattle feed production, silage production, fuel bioethanol, industrial products, and soil-improving green manure. Stable shelf-life and diverse use in food and forage are additional benefits of these crops. Major CLO crops of the world are given in Table 9.1.

TABLE 9.1 Major and Minor Cereals, Legumes, and Oilseeds of the World.

Cereals	Legumes	Oilseeds
Wheat	Common beans	Rapeseed
Rice	Chickpeas	Coconut
Maize	Lentils	Sunflower
Barley	Peas (fresh and dry)	Olive
Oats	Beans (fresh and dry)	Palm
Sorghum	Soybean	
Rye	Peanut	
Millets	Lupin seeds	Sesame

Apart from meeting dietary recommendations, CLO offer several health benefits in lowering the risk of cardiovascular diseases, type-2 diabetes, obesity, constipation, and cancer acquired from epidemiological studies and scientific research.[31,57] Legumes help balance protein quality by providing the requirement of essential amino acid (AA) lysine to cereal-based diets, and cereals fulfill their limitation of either sulfur AAs or methionine.[52] They are rich in complex carbohydrates such as DFs. High DF diets are linked to improvement in gastrointestinal health. Legumes have a low glycemic index compared to cereals, resulting in slower digestion, thus beneficial for diabetic people by assisting in maintaining healthy blood glucose and insulin levels.[34] The range of bioactive components such as phenolic compounds, tannins, saponins, peptides, and omega fatty acids offered by CLO are additional perquisites.

9.2 STRUCTURE AND CHEMICAL COMPOSITION

Structurally, CLO has three main units but with different ratios, composition, and purpose (Table 9.2). All these units, after processing, provide a range of products for consumer use, for instance, noodles, peanut butter, bread, protein isolates, oil, flour, and extruded products. CLO has been vastly explored for functional foods by introducing grain-based beverages, a range of soy milk, and baked goods.

TABLE 9.2 Basic Structural Units of Cereals, Legumes, and Oilseeds.

Cereals	Legumes	Oilseeds
Starchy endosperm	Cotyledons	Cotyledons
Hull and bran	Seed coat	Testa or seed coat
Embryo or germ	Hypocotyl (including plumule)	Embryo

9.2.1 PROCESSING

Most of the cereals are monocots whereas legumes and oilseeds are primarily dicots. Because of that, CLO undergoes different sets of primary and secondary processing based on structure, usage, and end product (Table 9.3). Cleaning, grading, soaking, size reduction or milling, dehulling, roasting, refining, polishing, extraction, etc. are part of the primary processing of CLO. The end products of CLO secondary processing range from flour, pasta, tortilla, beer brewing, bakery products, protein concentrates and isolates, oil, plant-based alternative milk, and animal feed. Conventional processing of CLO generates heat directly and indirectly within constituents. As a result, CLO encounters physio-chemical changes ranging from Maillard reaction, protein denaturation, aggregation, and lipid oxidation. Moreover, the bioavailability of nutrients present in food, being dependent on food processing technology, is compromised by these undesirable changes.

TABLE 9.3 Primary and Secondary Processing of Cereals, Legumes, and Oilseeds.

Primary processing	Secondary processing
Cleaning, grading, and drying	Fermentation and germination
Dehulling or decortication,	Blending, baking, and malting
Milling, pounding, and grinding	Puffing, flaking, and pearling
Tempering and parboiling,	Frying and extrusion
Conditioning and sieving	Roasting and steaming

9.3 PULSED ELECTRIC FIELD (PEF)

It is a cold pasteurization technique that employs electric field pulses (10–80 kV/cm) that are applied by means of electrodes for a period of a few microseconds only.[4,24] If the strength of the electric field is in the moderate-to-high range, that is, 100–300 V/cm, then PEF is applied in batch mode, whereas at 20–80 kV/cm, it is applied in ongoing mode.[49] The efficacy of PEF treatment depends upon the food matrix and operating parameters,[37] as shown in Figure 9.1.

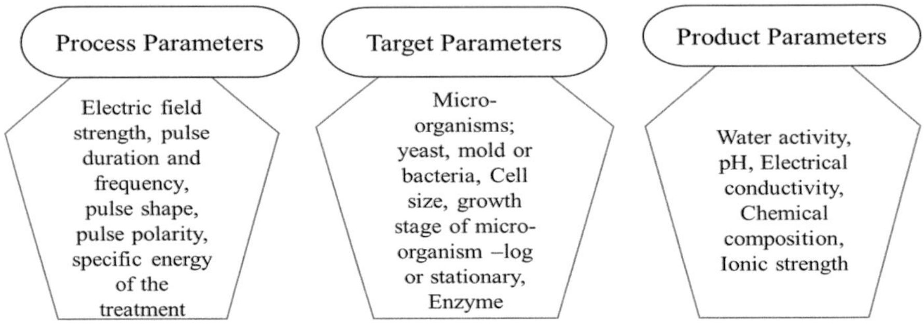

FIGURE 9.1 Factors affecting PEF treatment efficacy.

The application of PEF treatments opens the way to electroporation, that is, permeabilization of the cell membranes, which may be reversible or irreversible. Reversible electroporation occurs as the bilayer of the lipids can recover the integrity of the membrane and return to the previous state, but if the changes in the lipid layer are permanent, then the electroporation is irreversible.[50] Irreversible electroporation causes cell death, whereas reversible electroporation is gentle and finds many applications in food processing. Noteworthy advantages of PEF treatment, such as uniform treatment intensity, ongoing processing nature, short treatment time, low processing temperature, and diminished compounds degradation, (e.g., flavors,and proteins), suggest that PEF has applications at the various stages involved in food processing[39] (Table 9.4). For solid foods, PEF is a new possibility to apply fast and volumetric heating.[3]

9.4 PEF APPLICATIONS IN CEREALS, LEGUMES, AND OILSEEDS

9.4.1 *HYDRATION AND GERMINATION*

Hydration or soaking is vital for further processing in CLO. Hydration of CLO can be broadly classified into simple soaking (continuous immersion) and steeping (intermittent submersion).[30] Hydration reduces antinutritional factors such as phytic acid in legumes. Also, hydration reduces the oligosaccharides of the raffinose family, which are responsible for flatulence after pulse consumption.[69] The only drawback associated with hydration is a slow batch-wise process, which needs improvement. Cereals, millets, and legumes generally hydrate at a moderate rate, and their hydration behavior differs in native and processed states.[62] PEF-assisted hydration successfully reduced the processing time of soaking in common beans[10] by 3-fold and resulted in a better rehydration rate of 70% in dried chickpeas.[2] Promising advantages of minimized leaching losses of valuable components were also evident. The rationale proposed was the formation of tiny pores on the seed coat of legumes via PEF; thereby, water uptake increased without the need for elevated temperature.

TABLE 9.4 Different PEF Applications.

Electric field strength	Pulse duration	Application
1 kV/cm	10^{-5}–10^{-3} s	Reversible electroporation
> 3 kV/cm	10^{-5}–10^{-3} s	Irreversible electroporation
0.2–1 kV/cm	10^{-4}–10^{-1} s	Electroporation of plant cells
2–3 kV/cm	~20 μs/300 pulses	Meat processing (meat curing)
10–80 kV/cm	200–500 ms	Microbial cell lysis
10–90 kV/cm	2–500 μs	Enzyme deactivation
>30 kV/cm	0.8–4.0 ms	Maillard Reaction

Germination is an efficient technique for the breakdown of macromolecules, increase of digestibility, and enhancement of nutritional values of cereals and legumes. According to Rani and Bhardwaj,[48] the process of germination is marked by embryo development with the growth of root and shoot length with the resulting modification of endosperm. Governing factors for the germination of cereals and legumes are water, temperature, oxygen, and light.[53] Germination of barley is very important for malting and, thus, for beer making. Low-intensity PEF promotes germination (radicle emergence) in barley,[12] chickpea seeds,[29] flaxseed, and wheat.[18] The underlying mechanism for improved germination was attributed to the development of dipole in seed, which can line up with the application of an electric field. More water is absorbed due to the dipole-dipole interaction, resulting in better hydration.

9.4.2 COOKING AND SHELF-LIFE EXTENSION

Cooking renders CLO softens and palatable, destroys antinutrients, and enhances digestibility. However, thermal processing, which predominated CLO processing, has many pitfalls, such as longer duration, high energy expenses, and loss of nutrients.[14] For consumer demands, the solution lies in emerging nonthermal PEF treatment. PEF allows more homogeneous cooking of foods compared to the conventional methods of thermal cooking.[5] PEF alone or synergistically with heat cooks and extends the shelf-life of CLO products. Ratanatriwong et al.[51] were able to enhance the shelf life of reformed rice pudding significantly for 94 days when stored at 4°C on processing it by a dual treatment of mild heat up to temperatures of 55°C for 30 s, along with PEF having an EFS of 33 kV/cm, flow rate 100 L/h, pulse duration 1.47 ms and repetition rate 500 pps when compared with the control samples which were exposed to mild heat only.

The shelf-life of various CLO-based beverages is enhanced by PEF treatment; for example, pea soup[63] and enriched soy milk.[27] Esaki et al.[15] prolonged the shelf life of the bread for more than 2 days compared to the control sample by using HVEF treatment (50 kV for 220 min) of dough.

9.4.3 EXTRACTION

9.4.3.1 POLYSACCHARIDE EXTRACTION

Corn bran, known for ferulic acid content, was utilized for polysaccharide extraction using two approaches: PEF-assisted aqueous and PEF-assisted enzymatic extractions. They recovered 15.36% corn bran polysaccharide in the PEF-assisted enzyme method compared to ~9% yield in an aqueous medium of PEF.[28] Zhao et al.[71] also optimized the polysaccharides extraction, as it is rich in phytochemicals, vitamins, and minerals, from corn silk by PEF treatment. They reported the best combination to be 30 kV/cm with a 6 μs long pulse and a liquid-to-raw material ratio of 50:1, which yielded 7.3%.

Sajib[55] performed the nonstarch polysaccharide extraction from Brewer's spent grain (BSG), a residue of the brewing industry, using PEF technology (20 kV/cm, 20 pulses, and 140 Hz). PEF assists in the production of porous starch from corn for industrial requirements.[21]

9.4.3.2 PROTIEN EXTRACTION

The effectiveness of PEF as a cell disruption technique and nonthermal extraction of intracellular biomolecules is propitious. So far, PEF technology has been highly effective and preserves the activity of proteins in various microalgae.[7] However, the administration of PEF for the protein extraction from CLO is fairly unexplored. Recently, Kanthamoon and Tanongkankit[25] were able to lower the consumption of specific energy by the application of PEF for 60 min at 12.5 kV when the solid-to-liquid ratio was kept at 1:15 w/v during soy protein extraction from soybeans, which was then dried by spray drying yielding the powder with optimum functional and physicochemical properties.

9.4.3.3 OIL EXTRACTION

Oilseeds are processed either by mechanical pressing or by solvent extraction method to obtain the oil. The yield of oil by either method is not more than 80%. The rest of the oil or biomass having a significant bioactive compounds such as polyphenols and proteins is also lost by traditional methods. Before the processing, oilseeds undergo various pretreatments. Oil yield is enhanced upon the pretreatment of oilseeds using PEF for the extraction of vegetable oils[42], thus decreasing the amount of cake produced. PEF pretreatment assists in disrupting the mesocarp cells in turn releasing the oil from their lipo-vacuoles. PEF extraction technique is a green technology that generates the least waste and is energy efficient as well.[40]

TABLE 9.5 PEF Extraction of Oil from Different Oilseeds.

Oilseed	PEF treatment conditions (EFS: Frequency: Time)	% Increase in oil yield	Reference
Maize germ	0.6 kV/cm: 2 Hz: 1 min	88.4%	[20]
Soybeans	1.3 kV/cm: 50, pulses	21%	[20]
Rapeseed hulled	7.0 kV/cm: 120 pulses: 30 μs	39%	[19]
Sunflower seed	7.0 kV/cm: 15 Hz: 30 s	9.1%	[57]
Olives	2.0 kV/cm: 25 Hz: -	13.3%	[42]
Sesame seed	20 kV/cm: 0.5 Hz: 10 μs	4.9%	[56]
Sunflower seed	1.1 kV/cm	5.3%	[35]
Olive paste	2.0 kV/cm: 25 Hz: 50 μs	7.54%	[59]
Pecan Nuts	0.5 kJ/kg	21.4%	[47]

9.4.3.4 EXTRACTION OF BIOACTIVE COMPOUNDS

CLO are a source of various bioactive components, namely, dietary polyphenols. These polyphenols are part of the seed coat and impart color and are also acting as antinutritional compounds in pearl millet, legumes, and oilseeds. However, the positive part is that they act as antioxidants as they can scavenge free radicals as well as control lipid peroxidation reactions, and also chelate metal ions. The selective electroporation of PEF aids in the extraction of polyphenols from different matrices of CLO. Andreou et al.[1] explored the by-product of the virgin olive oil industry—olive pomace for extraction of polyphenols by pretreatment with PEF, where an increase in the yield of polyphenols by more than 90% was obtained on the application of 3 kV/cm for 45 ms. Similarly, obtaining polyphenols from the olive kernel and defatted canola seed cake by PEF application was studied by Roselló-Soto et al.[54] and Teh et al.,[60] respectively.

Boussetta et al.[6] investigated the hulls of flaxseed for the extraction of polyphenols. The pretreatment involved a 40 min rehydration treatment of the hulls followed first by PEF application to the tune of 20 kV/cm electric field strength for 10 ms, and second, by solid-liquid extraction using alkaline ethanol, which facilitated the polyphenol recovery to the tune of 80% from the hulls.

Corn germ meal, a by-product of the corn industry, was explored for antioxidant extraction using PEF (0.5, 2.5, and 4.5 kV/cm)-subcritical water extraction method[22] and established the use of extracted antioxidants up to 750 ppm in sesame oil for oxidative stability. The potential of PEF can be fully utilized during soy milk production to effectively modify the isoflavone profile by primarily hiking the concentration of aglycone.[33]

9.4.4 STARCH MODIFICATION

Starch comprises linear amylose (AM) and highly branched amylopectin (AP) polymers. It accounts for more than 50% of cereal weight and 20–25% of legume weight. Although the chemical activity of starch granules is inert,[23] in a food matrix, when it hydrates, its versatility expands as a thickening agent, as a bulk provider, as a gelling agent, as a stabilizing agent, etc. To overcome stability drawbacks associated with using native starches, they are altered through physical, chemical, and enzymatic methods. These methods have cascading effects on the environment and waste treatment. The alternative lies in the gentle nonthermal and selective PEF techniques.

PEF is an effective way for enhancing starch extraction and altering its crystallinity, gelatinization temperature, enthalpies, and other functional properties such as viscosity and pasting temperature.[9] The properties and various roles of starch can be modified in a homogenous, efficient, and clean manner.[46] Native glutinous rice grains were treated with 50–300 PEF pulses at the field strength of 3 kV/cm after their polishing and were examined for physicochemical properties.[44] From scanning electron microscope (SEM) images, it was observed that the treatment altered the microstructure of the surface but did not change the physical dimensions of rice grains. They also analyzed the pasting and paste properties of PEF-treated and untreated glutinous rice flour. The PEF treatment caused a fall in the peak viscosity; on the contrary, the breakdown viscosity and the final and setback viscosities showed an increase with more pulses of the PEF application, showing better susceptibility to gelatinization. They concluded that PEF-induced modifications, which will enable the processor to obtain rice-based products having a faster rate of hydration, with lower gelling temperature, due to shorter processing time, resulting in better quality.

Supporting research done on Japonica rice[65] and waxy rice starch,[67] rich in AP, shows that PEF-treated starches show elevated rapidly digestible starch levels and lesser slowly digestible starch levels than those of native starch. It implies that the sensitivity of starch toward digestive enzymes increases after PEF treatment.

PEF treatment improved the chemical modifications of starch too. During esterification, it improved the introduction of acetyl groups[22] in potato starch. However, limited such studies are available on CLO. According to Zhu,[73] the application of PEF for starch modification is at the laboratory stage only, and it is important to apply it at the industrial level in a continuous process.

9.4.5 PROTEIN MODIFICATION

PEF treatment alters the secondary and tertiary structure of protein, induces changes in the protein hydrophobicity, and decouples the noncovalent bonds, thereby hiking the protein functionality.[32,72] PEF treatment also caused aggregation of small molecules as recognized in canola protein.[70]

The impact of PEF parameters on functional, rheological, and structural properties of vital wheat gluten (VWG) was assessed by Zhang et al.[68] They observed that PEF significantly imparts stability to the network structure of VWG but does not tamper with the protein's shape at the molecular level.

PEF treatment modifies the protein structure of soy protein isolate (SPI), resulting in increased surface hydrophobicity due to partial denaturation.[66]

9.4.6 ENZYME REGULATION

The activity of enzymes in the food could be regulated during PEF processing.[41] Examples are the inactivation of lipase in brown rice[43] and lipoxygenase in soybean and Green peas.[61] Both enzymes are the reason for producing rancid off-flavor in many foods made of cereals, oilseeds, and beans.[16] Gentle PEF (12–13 kV/cm) could activate enzymes, and excess PEF results in a decrease in enzyme activity.[38]

9.4.7 AFLATOXIN REDUCTION

Mycotoxin contamination is the pandemic of CLO, and its global prevalence falls between 60 and 80%.[17] Among mycotoxins, Aflatoxins are predominant with complex structures that are stable to various chemicals and heat. Bulut et al.[8] successfully achieved a 60% reduction of *Aspergillus parasiticus* counts and decomposition of aflatoxins with reduced mutagenicity ($\approx$ 95%) at the maximum PEF energy (17.28 J) of sesame seeds with no change in the physical properties. Similar outcomes were observed in corn grains where multiple exposures at varying PEF treatments ranging from 800 to 500 Hz decreased aflatoxin production by 99.0% maximum.[13]

9.4.8 BROWNING

A pulsed electric field (PEF) at higher intensities promotes the Maillard reaction.[64] The thermal processing techniques, which are sometimes mandatory during the processing of CLO, improve the protein's nutritional quality, and it also results in the occurrence of the Maillard reaction, which adversely affects the availability of essential AA and generates toxic compounds. Maillard reaction falls under nonenzymatic reaction, which occurs between the AA group of proteins and the reducing end of carbohydrates. This is more rampant in cereals, oilseeds, and legumes than in fruits and vegetables, where enzymatic browning occurs. The browning is desirable in some cases, like baked products, owing to the color and flavor development, and it also results in the decreased nutritional value and development of harmful products such as 5-hydroxymethylfurfural (HMF) and acrylamide. PEF, since it is a nonthermal technology and operates under low temperatures, it helps in the reduction of the formation of such heat derivatives. During the production of soy milk, heat treatment is employed for the destruction of growth inhibitors and lipoxygenase inactivation, which causes denaturation and further aggregation of the proteins. Hence, the nonthermal technique such as PEF, can help in arresting these changes. Xiang et al.[65] applied a pulse electric field of 18–22 $kVcm^{-1}$ with 25–100 pulses to soy milk which resulted in an increase in the viscosity of the soy milk and exhibited significant changes in the color value with lower L* and a higher a* and b* value.

9.5 SUMMARY

Although PEF is a promising nonthermal alternative, its application in food processing poses distinct challenges. Excessive heat generation due to ohmic heating, initiation of electrolytic reactions, and presence of air bubbles leading to dielectric breakdown are the pitfalls of PEF. Also, it has been observed that greater field strengths in atmospheric conditions may cause ionization of air molecules, leading to field breakdown and arc-over conditions within the chamber. As it is in the nascent stage, consumers are skeptical about PEF products, as observed by Neilsen et al.[36] Further research in pilot and industrial products of CLO will ensure a better consumer market with renewed nutrition choices. Elaborate studies on the design and stimulation studies of PEF chamber and treatment will add an edge to an understanding of PEF treatment of CLO products.

KEYWORDS

- **cereals**
- **extraction**
- **legumes**
- **modifications**
- **oilseeds**
- **pulsed electric field**

REFERENCES

1. Andreou, V.; Psarianos, M.; Dimopoulos, G.; Tsimogiannis, D.; Taoukis, P. Effect of pulsed electric fields and high pressure on improved recovery of high-Added-Value Compounds from Olive Pomace. *J. Food Sci.* **2020,** *85* (5), 1500–1512.
2. Andreou, V.; Sigala, A.; Limnaios, A.; Dimopoulos, G.; Taoukis, P. Effect of Pulsed Electric Field Treatment on the Kinetics of Rehydration, Textural Properties, and the Extraction of Intracellular Compounds of Dried Chickpeas. *J. Food Sc.* **2021,** *86* (6), 2539–2552.
3. Astráin-Redín, L.; Moya, J.; Alejandre, M.; Beitia, E.; Raso, J.; Calvo, B.; Cebrián, G.; Álvarez, I. Improving the Microbial Inactivation Uniformity of Pulsed Electric Field Ohmic Heating Treatments of Solid Products. *LWT* **2022,** *154,* 112709.
4. Barbosa-Cánovas, G. V.; Fernandez-Molina, J. J.; Swanson, B. G. Pulsed Electric Fields: A Novel Technology for Food Preservation. *Agro Food Industry Hi-Tech.* **2001,** *12,* 9–14.
5. Blahovec, J.; Vorobiev, E.; Lebovka, N. Pulsed Electric Fields Pretreatments for the Cooking of Foods. *Food Eng. Rev.* **2017,** *9* (3), 226–236.
6. Boussetta, N.; Soichi, E.; Lanoisellé, J. L.; Vorobiev, E. Valorization of Oilseed Residues: Extraction of Polyphenols from Flaxseed Hulls by Pulsed Electric Fields. *Indust. Crop Prod.* **2014,** *52,* 347–353.
7. Buchmann, L.; Brändle, I.; Haberkorn, I.; Hiestand, M.; M athys, A. Pulsed Electric Field based Cyclic Protein Extraction of Microalgae Towards Closed-loop Biorefinery Concepts. *Biores. Techno,* **2019,** *291,* 121870.
8. Bulut, N.; Atmaca, B.; Akdemir Evrendilek, G.; Uzuner, S. Potential of Pulsed Electric Field to Control *Aspergillus parasiticus*, Aflatoxin and Mutagenicity Levels: Sesame Seed Quality. *J. Food Safe.* **2020,** *40* (6), e12855.
9. Castro, L. M.; Alexandre, E. M.; Saraiva, J. A.; Pintado, M. Starch Extraction and Modification by Pulsed Electric Fields. *Food Rev. Int.* **2021,** 1–22.
10. Devkota, L.; He, L.; Bittencourt, C.; Midgley, J.; Haritos, V. S. Thermal and Pulsed Electric Field (PEF) Assisted Hydration of Common Beans. *LWT* **2022,** *158,* 113163.
11. Duque, S. M. M.; Leong, S. Y.; Agyei, D.; Singh, J.; Larsen, N.; Oey, I. Understanding the Impact of Pulsed Electric Fields Treatment on the Thermal and Pasting Properties of Raw and Thermally Processed Oat Flours. *Food Res. Int.* **2020,** *129,* 108839.
12. Dymek, K.; Dejmek, P.; Panarese, V.; Vicente, A. A.; Wadsö, L.; Finnie, C.; Galindo, F. G. Effect of Pulsed Electric Field on the Germination of Barley Seeds. *LWT-Food Sci. Technol.* **2012,** *47* (1), 161–166.
13. Eisa, N. A.; Ali, F. M.; El-Habbaa, G. M.; Abdel-Reheem, S. K.; Abou-El-Ella, M. F. Pulsed Electric Field Technology for Checking Aflatoxin Production in Cultures and Corn Grains. *Egypt J. Phytopathol.* **2003,** *31,* 75–86.
14. Ercolini, D.; Russo, F.; Ferrocino, I.; Villani, F. Molecular Identification of Mesophilic and Psychrotrophic Bacteria from Raw Cow's Milk. *Food Microbiol.* **2009,** *26* (2), 228–231.
15. Esaki, K.; Ninomiya, F.; Hisaki, K.; Higasa, T.; Shibata, K.; Murata, K.; Aibara, S. Effects of High-Voltage Electric Field Treatment on the Water Activity of Bread. *Biosci., Biotechnol. Biochem.* **1996,** *60* (9), 1444–1449.
16. Eskin, N. A. M.; Grossman, S.; Pinsky, A. Biochemistry of Lipoxygenase in Relation to Food Quality. *Crit. Rev. Food Sci. Nutr.* **1977,** *4,* 1–40.

17. Eskola, M.; Kos, G.; Elliott, C. T.; Hajšlová, J.; Mayar, S.; Krska, R. Worldwide Contamination of Food-Crops with Mycotoxins: Validity of the Widely Cited 'FAO Estimate' of 25%. *Crit. Rev. Food Sci. Nutr.* **2020,** *60* (16), 2773–2789.
18. Evrendilek, G. A.; Atmaca, B.; Bulut, N.; Uzuner, S. Development of Pulsed Electric Fields Treatment unit to Treat Wheat Grains: Improvement of Seed Vigour and Stress Tolerance. *Comput. Elect. Agri.* **2021,** *185,* 106129.
19. Guderjan, M.; Elez-Martínez, P.; Knorr, D. Application of Pulsed Electric Fields at Oil Yield and Content of Functional Food Ingredients at the Production of Rapeseed Oil. *Innovat. Food Sci. Emerg. Technol.* **2007,** *8* (1), 55–62.
20. Guderjan, M.; Töpfl, S.; Angersbach, A.; Knorr, D. Impact of Pulsed electric Field Ttreatment on the Recovery and Quality of Plant oils. *J. Food Eng.* **2005,** *67* (3), 281–287.
21. Han, Z.; Zeng, X. A.; Zhang, B. S.; Yu, S. J. Effects of Pulsed Electric Fields (PEF) Treatment on the Properties of Corn Starch. *J. Food Eng.* **2009,** *93* (3), 318–323.
22. Hong, J.; Zeng, Xin-An.; Han, Z.; Brennan, C. S. Effect of Pulsed Electric Fields Treatment on the Nanostructure of Esterified Potato Starch and their Potential Glycemic Digestibility. *Innovat. Food Sci. Emerg. Technol.* **2018,** *45,* 438–446.
23. Hosseini, S. M.; Bojmehrani, A.; Zare, E.; Zare, Z.; Hosseini, S. M.; Bakhshabadi, H. Optimization of Antioxidant Extraction Process from Corn Meal using Pulsed Electric Field-Subcritical Water. *J. Food Process. Preserv.* **2021,** *45* (6), e15458.
24. Huang, R.; Li, C. P.; Chen, D.; Zhao, G.; Cheng, W.; Zhang, Y.; Zhao, H. Preparation of Phosphorylated Starch by Dry-Heating in the Presence of Pyrophosphate and its Ealcium-Phosphate Solubilizing Ability. *J. Food Sci. Technol.* **2013,** *50* (3), 561–566.
25. Jeyamkondan, S.; Jayas, D. S.; Holley, R. A. Pulsed Electric Field Processing of Foods: A Review. *J. Food Protect.* **1999,** *62* (9), 1088–1096.
26. Kanthamoon, K.; Tanongkankit, Y. *Pulsed Electric Field Assisted Extraction of Protein from Soybean for Production of Protein Powder.* Doctoral dissertation, Maejo University, Thailand. 2021.
27. Li, Y.; Chen, Z.; Mo, H. Effects of Pulsed Electric Fields on Physicochemical Properties of Soybean Protein Isolates. *LWT-Food Sci. Technol.* **2007,** *40* (7), 1167–1175.
28. Li, S. Q.; Zhang, Q. H. Inactivation of E. coli 8739 in Enriched Soymilk using Pulsed Electric Fields. *J. Food Sci.* **2004,** *69* (7), 169–174.
29. Liu, X.; Guo, X.; Lin, S. Improvement of Extraction rate of Corn Bran Polysaccharide by High-Voltage Pulsed Electric Field. *J. Food Safe. Quality* **2016,** *7* (6), 2419–2425.
30. Mahajan, T. S.; Pandey, O. P. Effect of Electric field (at different temperatures) on Germination of Chickpea Seed. *African J. Biotechnol.* **2014,** *13* (1), 61–67.
31. Mannam, V. Role of Hydration in Grain Processing: A Review. In *Developing Technologies in Food Science;* Meghwal, M., Goyal, M. R., Eds.; Apple Academic Press Inc.: New York, 2017. pp. 77–108.
32. Masisi, K.; Beta, T.; Moghadasian, M. H. Antioxidant Properties of Diverse Cereal Grains: A Review on *in vitro* and *in vivo* Studies. *Food Chemi.* **2016,** *196,* 90–97.
33. Melchior, S.; Calligaris, S.; Bisson, G.; Manzocco, L. Understanding the Impact of Moderate-Intensity Pulsed Electric Fields (MIPEF) on Structural and Functional Characteristics of Pea, Rice and Gluten Concentrates. *Food and Bioprocess Technol.* **2020,** *13* (12), 2145–2155.
34. Morales-de la Peña, M.; Welti-Chanes, J.; Martín-Belloso, O. Impact of Pulsed Electric Field Pre-Treatment on the Isoflavone Profile of Soymilk. *Beverages* **2022,** *8* (2), 19.

35. Mudryj, A. N.; Yu, N.; Aukema, H. M. Nutritional and Health Benefits of Pulses. *Appl. Physiol., Nutr. Metab.* **2014,** *39* (11), 1197–1204.
36. Moradi. N.; Rahimi, M. Effect of Simultaneous Ultrasound/Pulsed Electric Field Pretreatments on the Oil Extraction from Sunflower Seeds, *Sep. Sci. Technol.* **2018,** *53* (13), 2088–2099.
37. Nielsen, H. B.; Sonne, A. M.; Grunert, K. G.; Banati, D.; Pollák-Tóth, A.; Lakner, Z.; Olsen, N. V.; Žontar L.; Peterman, M. Consumer Perception of the use of High-Pressure Processing and Pulsed Electric Field Technologies in Food Production. *Appetite* **2009,** *52* (1), 115–126.
38. Nowosad, K.; Sujka, M.; Pankiewicz, U.; Kowalski, R. The Application of PEF Technology in Food Processing and Human Nutrition. *J. Food Sci. Technol.* **2021,** *58* (2), 397–411.
39. Ohshima, T.; Tamura, T.; Sato, M. Influence of Pulsed Electric Field on Various Enzyme Activities. *J. Electrostat.* **2007,** *65* (3), 156–161.
40. Park, S.; Kim, Y. R. Clean Label Starch: Production, Physicochemical Characteristics, and Industrial Applications. *Food Sci. Biotechnol.* **2021,** *30* (1), 1–17.
41. Patras, A.; Choudhary, P.; Rawson, A. Recovery of Primary and Secondary Plant Metabolites by Pulsed Electric Field Treatment. *Handbook of Electroporat. Springer, Cham.* **2017,** 1–21.
42. Poojary, M.; Roohinejad, S.; Koubaa, M.; Barba, F.; Passamonti, P.; Jambrak, A. R.; Greiner, R. Impact of Pulsed Electric Fields on Enzymes. In: Miklavčič. D. (eds.), *Handbook of Electroporation*; Springer International Publishing. 2016, pp. 1–21.
43. Puértolas, E.; de Marañón, I. M. Olive Oil Pilot-Production Assisted by Pulsed Electric Field: Impact on Extraction Yield, Chemical Parameters and Sensory Properties. *Food Chem.* **2015,** *167,* 497–502.
44. Qian, J. Y.; Gu, Y. P.; Jiang, W.; Chen, W. Inactivating Effect of Pulsed Electric Field on Lipase in Brown Rice. *Innov. Food Sci. Emerg. Technol.* **2014,** *22,* 89–94.
45. Qiu, S.; Abbaspourrad, A.; Padilla-Zakour, O. I. Changes in the Glutinous Rice Grain and Physicochemical Properties of Its Starch upon Moderate Treatment with Pulsed Electric Field. *Foods* **2021,** *10* (2), 395.
46. Qiu, S.; Li, Y.; Chen, H.; Liu, Y.; Yin, L. Effects of High-Pressure Homogenization on Thermal and Electrical Properties of Wheat Starch. *J. Food Eng.* **2014,** *128,* 53–59.
47. Quagliariello, V.; Iaffaioli, R. V.; Falcone, M.; Ferrari, G.; Pataro, G.; Donsì, F. Effect of Pulsed Electric Fields-Assisted Extraction on Anti-Inflammatory and Cytotoxic Activity of Brown Rice Bioactive Compounds. *Food Res. Int.* **2016,** *87,* 115-124.
48. Rábago-Panduro, L. M.; Morales-de la Peña, M.; Martín-Belloso, O. et al. Application of Pulsed Electric Fields PEF on Pecan Nuts *Carya illinoinensis* Wangenh. K. Koch: Oil Extraction Yield and Compositional Characteristics of the Oil and Its By-product. *Food Eng. Rev.* **2021,** *13,* 676–685.
49. Rani, H.; Bhardwaj, R. D. Quality Attributes for Barley Malt: "The backbone of beer". *J. Food Sci.* **2021,** *86* (8), 3322–3340.
50. Ranjha, M. M. A.; Kanwal, R.; Shafique, B.; Arshad, R. N.; Irfan, S.; Kieliszek, M.; Kowalczewski, P. Ł, Irfan, M.; Khalid, M. Z.; Roobab, U.; Aadil, R. M. A Critical Review on Pulsed Electric Field: A Novel Technology for the Extraction of Phytoconstituents. *Molecules* **2021,** *26* (16), 4893.
51. Raso, J.; Álvarez, I. Pulsed Electric Field Processing: Cold Pasteurization. In: *Reference Module in Food Science;* Elsevier: Amsterdam, The Netherlands, 2016, pp. 1–8.
52. Ratanatriwong, P.; Evrendilek, G. A.; Zhang, Q. H. Pulsed Electric Field Processing of Formulated Rice Pudding. *J. Food Safe.* **2008,** *28* (1), 126–141.

53. Ratnayake, W. S.; Naguleswaran, S. Utilizing Side Streams of Pulse Protein Processing: A Review. *Legume Sci.* **2021,** e120.
54. Raven, P. H. E.; Ray, F.; Eichhorn, S. E. (eds.). *Biology of Plants;* 8thedn. New York: W.H. Freeman and Company Publishers, 2005, pp. 686.
55. Roselló-Soto, E.; Barba, F. J.; Parniakov, O.; Galanakis, C. M.; Lebovka, N.; Grimi, N.; Vorobiev, E. High Voltage Electrical Discharges, Pulsed Electric Field, and Ultrasound Assisted Extraction of Protein and Phenolic Compounds from Olive Kernel. *Food Bioprocess Technol.* **2015,** *8* (4), 885–894.
56. Sajib, M. *Preparation and Evaluation of Arabinoxylan Based Prebiotics*. M.Sc. Thesis, Division of Biotechnology, LTH, Lund University, Lund, Sweden, 2017.
57. Sarkis, J. R.; Boussetta, N.; Vorobiev, E. Application of Pulsed Electric Energies for Oil and Polyphenol Extraction from Sesame Cake and Sesame Seeds. In: Miklavčič, D. (eds) *Handbook of Electroporation;* Springer, Cham. 2017.
58. Singhal, P.; Kaushik, G.; Mathur, P. Antidiabetic Potential of Commonly Consumed Legumes: A Review. *Crit. Rev. Food Sci. Nutr.* **2014,** *54* (5), 655–672.
59. Shorstkii, I.; Mirshekarloo, M. S.; Koshevoi, E. Application of Pulsed Electric Field for Oil Extraction from Sunflower Seeds: Electrical Parameter Effects on Oil Yield. *J. Food Process Eng.* **2015,** *40* (1), 1–7.
60. Tamborrino, A.; Urbani, S.; Servili, M.; Romaniello, R.; Perone, C.; Leone, A. Pulsed Electric Fields for the Treatment of Olive Pastes in the Oil Extraction Process. *Appl. Sci.* **2019,** *10* (1), 114.
61. Teh, S. S.; Niven, B. E.; Bekhit, A. E. D. A.; Carne, A.; Birch, E. J. Microwave and Pulsed Electric Field Assisted Extractions of Polyphenols from Defatted Canola Seed Cake. *Int. J. Food Sci. Technol.* **2015,** *50* (5), 1109–1115.
62. Van Loey. A.; Verachtert, B.; Hendrickx, M. Effects of High Electric Field Pulses on Ezymes. *Trends Food Sci. Technol.* **2002,** *12* (3–4), 94–102.
63. Vasudeva, S.; Vishwanathan, K. H.; Aswathanarayana, K. N.; Indhudhara Swamy, Y. M. Hydration Behaviour of Food Grains and Modelling their Moisture Pick up as per Peleg's Equation: Part I. Cereals. *J. Food Sci. Technol.* **2010,** *47* (1), 34–41.
64. Vega-Mercado, H.; Martín-Belloso, O.; Chang, F. J.; Barbosa-Ccanovas, G. V.; Swanson, B. G. Inactivation of Escherichia coli and Bacillus subtilis suspended in Pea Soup using Pulsed Electric Fields. *J. Food Process. Preserv.* **1996,** *20* (6), 501–510.
65. Wang, J.; Guan, Y.-G.; Yu, S.-J.; Zeng, X.-A.; Liu, Y.-Y.; Yuan, S.; Xu, R. Study on the Maillard Reaction Enhanced by Pulsed Electric Field in a Glycin-Glucose Model System. *Food Bioprocess Technol.* **2011,** *4* (3), 469–474.
66. Wu, C.; Wu, Q. Y.; Wu, M.; Jiang, W.; Qian, J. Y.; Rao, S. Q.; Zhang, L.; Li, Q.; Zhang, C. Effect of Pulsed Electric Field on Properties and Multi-Scale Structure of Japonica rice Starch. *LWT* **2019,** *116,* 108515.
67. Xiang, B. Y.; Simpson, M. V.; Ngadi, M. O.; Simpson, B. K. Effect of Pulsed Electric Field on the Rheological and Colour Properties of Soy Milk. *Int. J. Food Sci. Nutr.* **2011,** *62* (8), 787–793.
68. Zeng, F.; Gao, Q. Y.; Han, Z.; Zeng, X. A.; Yu, S. J. Structural Properties and Digestibility of Pulsed Electric Field Treated Waxy Rice Starch. *Food Chem.* **2016,** *194,* 1313–1319.
69. Zhang, C.; Yang, Y. H.; Zhao, X. D.; Zhang, L.; Li, Q.; Wu, C.; Ding, X.; Qian, J. Y. Assessment of Impact of Pulsed Electric Field on Functional, Rheological and Structural Properties of Vital Wheat Gluten. *LWT* **2021,** *147,* 111536.
70. Zhang, J.; Song, G.; Mei, Y.; Li, R.; Zhang, H.; Liu, Y. Present Status on Removal of Raff Inose Family Oligosaccharides—A Review. *Czech J. Food Sci.* **2019,** *37* (3), 141–154.

71. Zhang, L.; Wang, L. J.; Jiang, W.; Qian, J. Y. Effect of Pulsed Electric Field on Functional and Structural Properties of Canola Protein by Pretreating Seeds to Elevate Oil yield. *LWT* **2017,** *84*, 73–81.
72. Zhao, W, Yu, Z.; Liu, J.; Yu, Y.; Yin, Y.; Lin, S.; Chen, F. Optimized Extraction of Polysaccharides from Corn Silk by Pulsed Electric Field and Response Surface Quadratic Design. *J. Sci. Food Agri.* **2011,** *91* (12), 2201–2209.
73. Zhao, W.; Tang, Y.; Lu, L.; Chen, X.; Li, C. Pulsed Electric Fields Processing of Protein-based Foods. *Food Bioprocess Technol.* **2014,** *7* (1), 114–125.
74. Zhu, F. Modifications of Starch by Electric Field based Techniques. *Trend Food Sci. Technol.* **2018,** *75,* 158–169.

CHAPTER 10

APPLICATIONS AND FUTURE ASPECTS OF PULSED ELECTRIC FIELD IN FOOD BYPRODUCTS

ABHIMANYU SINGH, HITESH JETHANI, DIVYA AGGARWAL, and UMESH H. HEBBAR

ABSTRACT

The waste generated from food-processing industries is advancing to be a global concern. This demands an urgent need to address the issues pertaining to environmental sustainability. The processing of agricultural, livestock, and marine produce generates massive amounts of solid and liquid byproducts containing unutilized nutrients and bioactive metabolites. These biomolecules are potential nutraceuticals with numerous health benefits and should be recovered for food, feed, and pharmaceutical applications. The existing methods for the extraction of metabolites rely on the use of hazardous chemicals and thermal processing which compromises the nutritional quality and safety of the final product. In recent decades, pulsed electric field (PEF) has come up as an emerging alternative method for the valorization of byproducts. PEF-assisted extraction has several advantages over conventional techniques, such as higher yields, reduced processing time, lower requirement of harmful solvents, and avoiding extensive pretreatments. This chapter highlights the role, opportunities, and future scope of PEF technology in the recovery of high-value compounds from the byproducts originating from various food-processing industries. The pulsed electric field is a promising technology that has the capacity to add value to industrial byproducts in a sustainable manner.

Nonthermal Pulsed Electric Field Technology for the Food Industry.
Preeti Birwal, Megh R. Goyal, Baldev Singh Kalsi, & Ankita Kataria (Eds.)

DOI: 10.1201/9781003605904-13

10.1 INTRODUCTION

Due to the rapidly growing population and globalization, there is an increase in consumer demand for a wide variety of foods. Besides consuming fresh, consumers also demand various food commodities that are heavily processed, such as dried and frozen produce, juices, nectars, concentrates, jams/jellies, ketchup/sauces, chips/flakes, or crispy snacks because of the easy to go and convenient alternatives with extended shelf life. After the household sewage treatment systems, the food-processing industry attains the second position in producing environmental waste on a larger scale. These food wastes are often considered unwanted and worthless refuse, but they are potential byproducts containing high-value health beneficial compounds. The primary processing of foods triggers the generation of millions of tonnes of industrial byproducts annually. These include damaged raw materials, leftover seeds, peels/skins, cobs/hulls/husks, oilseed cakes, molasses, and spent/distillers' grains, which corresponds to around 190 million tonnes per year globally.[18] The different kinds of foods determine the production of distinct food losses. Fruits and vegetables (including roots and tubers) contribute a higher percentage of byproducts (45%), followed by fish and seafood (35%), and oilseed, dairy, and meat (20%) industry.[38, 68] The release of these byproducts into the atmosphere without any suitable pretreatment or incineration may pose a significant burden on the environment. The toxic byproducts further pollute the groundwater and cause the formation of corrosive gases, such as hydrogen sulfide and methane. Moreover, food wastes and byproducts have several indirect outcomes, such as eutrophication, global desertification, terrestrial acidification, and generation of greenhouse gas emissions.[51, 53] The economic problems associated with global food waste are yet another crucial factor.

Food scarcity and unbalanced diets are the major hurdles to the malnourished population. Certain social groups, especially the people below the poverty line, may not have appropriate access to agro-industrial crops to cater to their nutritional needs. Due to the increased risk of nutrition-related disorders, people today are becoming more aware of their consumption patterns and are shifting toward the choices of healthy foods and beverages. Here, the concept of nutraceuticals comes into play as a value-added growth possibility in both domestic and international markets. They confer immense health benefits but contribute to rising healthcare costs. Therefore, the recovery and exploitation of such compounds from various futile entities serve as natural and inexpensive alternatives for the benefit of consumers.[50, 72]

Food waste reduction requires proper waste management, disposition, or recycling procedures. The primary solution to combat the issues pertaining to food waste is by upcycling these underutilized byproducts. It involves their upgradation and recycling into essential, innovative, and functional ingredients. Moreover, several research studies have indicated that these byproducts are potential sources of pronounced components, including carbohydrates, proteins, lipids, and other bioactive compounds such as dietary fibers, alkaloids, pigments, phenolics, antioxidants, and other phytochemicals.[14, 24, 31]

In the coming decades, it is expected that agri-food-processing industries will continue to perpetuate and lead to the production of huge proportions of wastes and byproducts. Their effective valorization may contribute to regional food security and circular economy by alleviating the adverse impacts on the environment and ensuring sustainability in the entire supply chain.[8] Hence, there is a need to investigate various technological ways to recover these substrates and valorize them into value-added and high-quality foods.[7] Although various extraction techniques have been traditionally employed in the recovery of different biological compounds, they possess certain shortcomings. These conventional methods of extraction require longer processing time under stringent conditions which adversely affects the quality and yield of the extractables. These techniques may also involve the considerable use of hazardous chemicals, mechanical forces, and energy. While on the other hand, researchers are investigating techniques to be of supreme interest, which may provide excellent yields by utilizing the extensively available industrial byproducts into economical and functional substitutes.[70] One of the advanced extraction nonthermal techniques is pulsed electric field (PEF), which has gained remarkable attention in extracting beneficial substances from food byproducts. This is a greener technology as it requires less amount of solvents, produces higher quality of products with greater yields, and consumes less energy for similar output. It has been widely cited as a promising alternative to conventional technologies without adversely impacting the original characteristics of the commodities.[67]

This chapter explores the potentiality and future perspectives of the PEF technique in recovering the underutilized valuable components in the byproducts associated with different sectors of food-processing industries.

10.2 TECHNOLOGIES FOR EXTRACTION OF BIOACTIVES FROM BYPRODUCTS

An indispensable outcome during the production of processed foods is the generation of inedible waste products. While the food industry generated waste presents an economic and environmental challenge,[33] it is also a potential feedstock for various biomolecules. The food-processing waste contains both nutritional and functional compounds such as vitamins, pigments, proteins, polymers, carbohydrates, enzymes, and minerals.[78] The valuable metabolites from food waste can be separated and utilized for various applications through an extraction process. There are different extraction methods available, and usually, a pretreatment step is associated with a specific extraction method.[22]

10.2.1 CONVENTIONAL METHODS

The most commonly used methods for extracting desirable compounds from plant or animal tissues include maceration, infusion, digestion, decoction, and percolation.[94] These methods involve the use of various solvents, and efficiency of the extraction process is largely dependent on the properties of the selected solvent.

Maceration is regarded as the oldest and cheapest extraction process in which the grounded sample is placed in a selected solvent (menstruum) for a few days at ambient temperature with occasional mixing to facilitate the solubilization of the desired compound, followed by filtration. The retained solid mass (marc) is further pressed to collect the traces of residual solvent. The infusion method is somewhat similar to maceration, except for the application of hot or cold solvents to extract readily soluble components in a shorter period. The digestion process involves a high rate of extraction since continuous heating is maintained throughout the extraction process. Decoction uses heating and water as a solvent to extract water-soluble and heat-stable compounds in a shorter time of a few minutes. Percolation is a kind of continuous maceration process done in a specific apparatus called a percolator, where solvent keeps recycling and prevents the saturation of soluble compounds.[30]

These conventional methods for the extraction of metabolites have significant limitations such as extended processing time, low extraction efficiency, degradation of thermal-sensitive biomolecules, and consumption of large quantities of solvents.[77]

Certain extraction methods such as soxhlet and hydro-distillation are most commonly used in the present time. They are improved in terms of solvent consumption and efficiency of extraction.[5] Soxhlet extraction involves refluxing the solvent for continuous extraction; however, this method is not suitable for extracting heat-labile compounds. Hydro-distillation method is essentially used for extracting volatile oils. It employs water or steam or their combination along with heat treatment to hydrolyze the material and extract the required compound.[80]

10.2.2 EMERGING METHODS

In recent decades, various novel techniques have emerged which facilitate the extraction of high-value compounds from industrial byproducts with significant improvements in product quality and yield. Some of these techniques include ultrasound-assisted extraction (UAE), microwave-assisted extraction (MAE), supercritical fluid extraction (SFE), enzyme-assisted extraction (EAE), pressurized liquid extraction (PLE), and pulsed electric field-assisted extraction (PEF).[60] In general, these methods rely on the degradation of cells or tissues, enhancing the permeability and allowing effective diffusion of solvents with an improved rate of mass transfer of solutes out of the cells/tissues.[38, 91]

UAE involves the application of sound waves of frequencies more than 20 kHz to liquid and semi-liquid materials. The compression and expansion of longitudinal waves at high frequencies lead to the effect of cavitation. The resultant localized increase in the temperature and pressure facilitates cell breakdown and improves mass transfer rate during the extraction process.[61]

During MAE, the dipolar molecules and charged ions within the food material undergo vigorous vibration and rotation. This mechanism generates heat and causes the migration of desired compounds into the solvent.[17] Commercially, microwaves of the frequency 2.45 GHz can be applied to extract the byproducts.

The SFE method uses a supercritical fluid to solubilize and extract the desired bioactive compounds. A supercritical fluid is a chemical compound that possesses solubility properties like liquid and diffusivity properties that of gases above a critical pressure and temperature point. Carbon dioxide is the most commonly used solvent because it requires low critical pressure and has a critical temperature in the ambient temperature range in order to act as a supercritical fluid.[49]

The cell membrane has a complex structure made up of various kinds of polysaccharides and protein material. EAE uses different enzymes such as amylase, cellulase, proteases, and other polysaccharide degrading enzymes. These enzymes interact and degrade the cell membrane components assisting the release of intracellular metabolites.

PLE is solvent-based extraction carried out at high pressure, which increases the boiling point. Under these conditions, the solubility, diffusivity, and penetration power of the solvent increase. This method has the advantages of reduced extraction time and higher efficiency of the extraction process.

The PEF extraction method can be utilized for pretreatment before extraction or directly to extract metabolites from the byproducts of food processing. PEF-assisted extraction is discussed in detail further in this chapter.

10.3 PEF-ASSISTED EXTRACTION

Pulsed electric field treatment involves the application of a high-voltage electric field in the form of short-duration pulses to the target material. It is employed in food industries to induce microbial cell disruption and food preservation.[37] The significance and applications of PEF technology have been increasing in the food-processing industries. PEF treatment is used independently or in conjunction with different techniques to improve unit operations pertaining to drying, freezing, and extraction in a more environmentally sustainable manner.[77]

As an emerging technology, PEF treatment remains to be explored and applied commercially to recover valuable metabolites from industrial byproducts. During PEF treatment, a high-voltage electric field in the order of kV/cm is generated in the pulse form for short durations of micro to milliseconds and focused on the food components.[38] This causes the rupture of plant, animal, and microbial cell membranes due to the development of a transmembrane potential difference. The treatment of leftover processed food as byproducts with PEF aids in the extraction of intracellular components of interest by the process of cell permeabilization.[6] This application of PEF has been shown to improve the efficiency of the extraction process when compared to conventional extraction techniques in terms of reduced processing time, higher yields, preservation of heat-labile nutrients, lowering the requirement of harmful solvents, and avoiding extensive pretreatments.[52]

10.3.1 FACTORS INFLUENCING PEF-ASSISTED EXTRACTION PROCESS

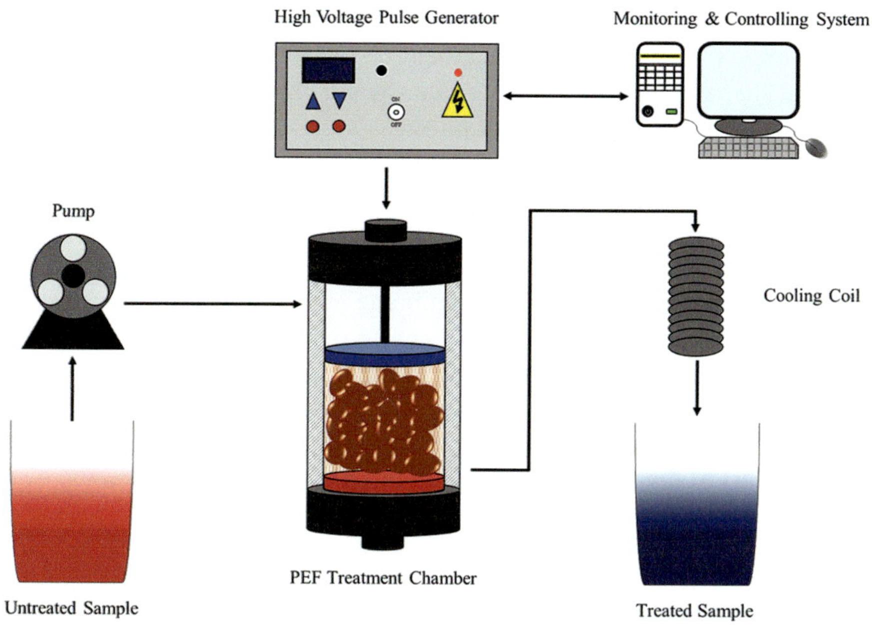

FIGURE 10.1 General components of PEF-assisted extraction unit.

Figure 10.1 shows a schematic representation of the PEF-assisted extraction system. The material from which bioactive metabolites need to be recovered is placed in an apparatus with two electrodes on the opposite ends, and the sample chamber is filled with fluid to maintain uniform electric field distribution. A high current generator supplies intense voltage pulses across these electrodes to produce an electric field through the treatment chamber. The efficiency of extraction depends on a variety of process parameters such as electric field strength, exposure time, pulse characteristics, nature of the sample, properties of the fluid medium, and the type of treatment chamber.

The electric fields low in strength could lead to reversible permeabilization of cells and tissues. The electric fields that are above a specific threshold strength can cause irreversible permeabilization of plant, animal, or microbial cells.[86] The exposure time of a sample to PEF can be calculated as the product of width and number of electric field pulses. The cells or tissues must be exposed for a minimum critical amount of time in order to establish a transmembrane potential difference beyond which PEF treatment produces an irreversible degradation of the cell membrane.[32]

The pulse wave characteristics, such as its shape and polarity, are essential to determine the effect of PEF treatment on the food material. There can be different shapes of the wave pulses like square, exponential, and oscillatory. Among these, the square shape of pulses is most efficient, while oscillatory is least effective based on the duration of continuous exposure of cell membrane to the high-intensity electric field.[64] The pulse can be monopolar or bipolar in nature, with bipolar being more effective for cell rupture. This is because the alternating polarity of electric field causes the movement of charged molecules inside the cell membrane resulting in stress development and faster breakdown.[15]

Apart from the properties of the electric field, temperature, electrical conductivity, pH, and ionic strength of the liquid medium containing the sample; consistency (solid or liquid), composition, and source (plant, animal, or microbial) of the sample; and design or type (batch or continuous) of the treatment chamber also affect the process of extraction of biomolecules.[67]

10.3.2 MECHANISM OF EXTRACTION OF BIOMOLECULES BY PEF TREATMENT

PEF treatment promotes the extraction of intracellular biologically active components from plants, animals, and microbial cells through electroporation and degeneration of the cell membrane. Figure 10.2 represents the formation of pores in the cell membrane that enhances the rate of mass transfer of desirable metabolites and the overall process of extraction. The exact mechanism of pore formation by the PEF treatment on a cellular level is poorly understood.[52]

According to Kotnik et al.,[35] a cell acts as an electrolyte due to intercellular exchange and transport of ions through the cell membrane. A cell membrane is a barrier made up of dielectric material and possesses an inherent potential difference known as resting transmembrane voltage due to the presence of free ions on both the sides. Another component of electric potential gets added under the influence of an external electric field known as induced transmembrane voltage. In order to facilitate irreversible electroporation, a threshold transmembrane potential difference needs to be generated above the critical value of 250 mV to 1 V. Zimmermann[98] reported that the increase in transmembrane potential difference due to the external electric field causes the electro-compression of the cell membrane. The pore formation in cell membranes thus initiates when the strength of an electric field capable of producing an electro-compressive force exceeds the other restoring forces in the membrane.

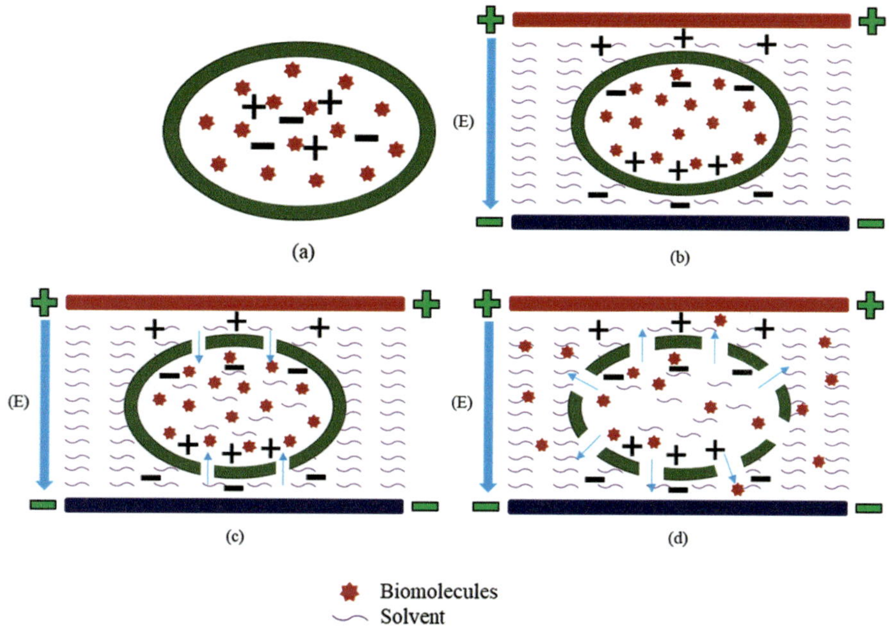

FIGURE 10.2 Schematic representation of the mechanism of extraction of biomolecules by PEF. (a) Cell containing biomolecules and randomly distributed charged ions before the application of PEF; (b) charged ions getting aligned on both sides of the cell membrane under the influence of PEF; (c) pore formation in the cell membrane and movement of solvent inside the cell; (d) enlargement of pores and mass transfer of biomolecules across cell membrane after solubilizing in the solvent.

When a high-intensity electric field in a pulsed form is applied to the cell membrane, it forms aqueous hydrophilic pores due to the rearrangement of its polar and nonpolar components.[90] Tieleman[83] simulated the process of pore formation in a lipid bilayer under the influence of an external electric field. It was observed that an electric field gradient developed due to the difference in dielectric properties of water and lipids at their interface. The water molecules act as dipoles that interact with the electric field leading to transmembrane water defects and their movement toward the core of the bilayer. The resulting local membrane defects and water-filled hydrophilic pores get further stabilized due to phospholipid lining consisting of both hydrophilic and hydrophobic ends.

The hydrophilic pores are good conductors and allow the current to flow through them, causing an increase in the local temperature. The protein channels present in the cell membrane undergo denaturation and inactivation under the influence of a high-intensity electric field and further increase the porosity.[86]

10.4 APPLICATIONS OF PEF-ASSISTED EXTRACTION

Since the emphasis on green and sustainable food processing and minimizing the carbon and water footprint is on the rise, the future of PEF technology looks promising. The application of PEF technology has thus attracted a considerable research interest in recovering the valuable compounds from processed food waste. A wide array of byproducts from various processed food groups have been studied in terms of extraction yield, processing time, and product quality. Table 10.1 highlights some of the recently conducted studies pertaining to the major kinds of byproducts obtained from food-processing industries. These individual food groups are further discussed in detail.

10.4.1 FRUITS AND VEGETABLES-PROCESSING INDUSTRY

Fruits and vegetables (FAV) contribute a significant share in food industries, as a majority of derived products are commercialized as fresh-cut slices, purees, juices, or processed products like jams and jellies. Massive amounts of waste are generated during processing along with byproducts at the field, processing industry, retail outlets, and household levels. The outer peels, seeds, and press cakes obtained after juice extraction are known to contain several bioactive constituents of chief interest. In recent decades, the application of electrotechnologies like PEF has been proven to exhibit remarkable results in terms of recovering these compounds effectively.[55]

The industrial processing of fruits and vegetables like oranges, onions, mangoes, tomatoes, and pomegranates renders their peels as the main byproducts, serving as an ideal source for the extraction of carotenoids, flavonoids, and antioxidants. In the case of oranges, PEF treatment has been shown to effectively improve the extraction of polyphenols by enhancing the levels of naringin and hesperidin from 1 to 3.1 mg/100 g and from 1.3 to 4.6 mg/100 g, respectively.[45] Similarly, PEF was demonstrated as a potential technique for selectively extracting and augmenting the recovery of ellagic

TABLE 10.1 PEF-Assisted Extraction of Bioactive Metabolites from Different Food-Processing Industry Byproducts.

Byproducts	Processing conditions	Key findings	References
		Fruits and vegetables	
Orange peels	1, 3, 5 and 7 kV/cm, 0.06–3.77 kJ/kg	Enhanced content of naringin and hesperidin, and antioxidant activity in PEF treated extract	[45]
Blueberry press cake	1, 2, and 5 kV/cm, 10 kJ/kg	Juice with higher antioxidant activity (+31%), total anthocyanin content (+60%), total phenolic content (+43%)	[10]
Grape pomace	1.2 kV/cm, 18 kJ/kg	Increased ratio (9.0) of total anthocyanins to total flavan-3-ols compared to control (7.1)	[13]
Papaya seeds	13.3 kV/cm	Higher yields (+200%) and antioxidant capacities (+20%)	[56]
Raspberry press cake	1 kV/cm, 6 kJ/kg	Higher content of total anthocyanins (26%), total phenolics (22%), and antioxidant power (24%)	[40]
Apple pomace	1, 2 and 3 kV/cm, 0.44–17.0 kJ/kg	PEF treated pomace resulted in higher concentrations of phenolic acids (chlorogenic, protocatechuic, and salicylic acid)	[44]
Mango peels	13.3 kV/cm, 1000 kJ/kg	A remarkable increase in total phenolic content (+400%)	[55]
Red prickly pear peels	20 kV/cm, 24 kJ/kg	Higher recovery of betanin (75 mg/100 g fresh material)	[36]
Sour cherry press cake	1–5 kV/cm, 10 kJ/kg	The extract obtained from PEF treated press cake exhibited higher FRAP values (16–24%)	[11]
Sweet cherry press cake	0.5–3 kV/cm, 10 kJ/kg	At 0.5 kV/cm, augmented concentration of anthocyanins (+38%) and antioxidant activity (+21%)	[58]
Thinned peach byproducts	0–5 kV/cm, 0.61–9.98 kJ/kg	Enhanced yields of coumaric acid (0.8 mg/100 g), chlorogenic acid (9.8 mg/100 g), and neocholorgenic acid (16.3 mg/100 g)	[69]
Potato peels	1 kV/cm, 5 kJ/kg	Higher antioxidant activity (9%) and total phenolics yield (10%)	[21]
Lemon peel residues	7 kV/cm	Enhanced extraction of polyphenols (300%) with 84 mg hesperidin, and 176 mg eriocitrin	[59]

TABLE 10.1 *(Continued)*

Byproducts	Processing conditions	Key findings	References
Pomegranate peels	10 kV/cm, 90–100 kJ/kg	Increased recovery of ellagic acid (≈ 740 µg/g DM)	[66]
Tomato peels	5 kV/cm, 5 kJ/kg	Increased extraction rate (27–37%) with higher lycopene yields (12–18%), and antioxidant power (18–18.2%)	[57]
Onion skin	2.5 kV/cm	Higher yield of total quercetin (33.22%) compared to untreated samples	[34]
		Cereals and legumes	
Beer waste brewing yeast	10 kV/cm	Efficient extraction of protein (2.788 ± 0.014%)	[43]
Brewer's spent grains	2.5 kV/cm	Improved recovery of total free and bound phenolic compounds by 2.7 and 1.7 times	[47]
Wheat bran	3,7 kV/cm	Increased extraction yield of arabinoxylans	[89]
		Meat, poultry, and seafood	
Eggshell powder	15 kV/cm	Eggshell derived calcium citrate with enhanced in-vitro absorption of calcium	[41]
Mussels	20 kV/cm	Improved yield of protein (>75%)	[96]
Waste chicken meat	0.2, 0.5, 1.0 kV/cm	Enhanced protein yield in extracts with higher antioxidant properties	[25]
Sea bream and sea bass byproducts	1.40 kV/cm	Enhanced yield of antioxidants:	[20]
		Sea bream—35.8% (head), 68.6% (bones), 33.8% (gills)	
		Sea bass—60.7% (head), 71.8% (bones), 22.1% (gills)	
		Plantation crops, herbs, and spices	
Borage leaves	0–5 kV/cm	Enhanced antioxidant activity with higher values of TPC	[76]
Pecan nut cake	2.0, 5.0, 7.5 kV/cm	Increase in total phenolics (17.8%) and condensed tannins (39.3%)	[65]
Rosemary and thyme byproducts	1.1 ± 0.2 kV/cm	Increased content of phenolics and antioxidants	[87]

TABLE 10.1 *(Continued)*

Byproducts	Processing conditions	Key findings	References
		Dairy	
Whey	10 kV/cm	Enhanced degree of succinylation of whey protein isolate from 88.31% to 93.45%.	[92]
		Edible oil	
Maize germ	0.6–1.3 kV/cm	Increased the recovery of phytosterols (32.4%)	[28]
Soybean	0.6–1.3 kV/cm	Increased the recovery of isoflavonoids: genistein (21%) and daidzein (20%)	[28]
Rapeseed	7 kV/cm, 84 kJ/kg	Enhanced the recovery of tocopherols, polyphenols, antioxidants, and phytosterols by solvent extraction	[29]
Flaxseed hulls	20 kV/cm, 300 kJ/kg	Efficient recovery of total polyphenols (80%)	[12]
Sesame cake	83 kJ/kg	Improved the extraction yields of polyphenols and protein	[75]
Olive kernels	18–109 kJ/kg	Improvement in the recovery of proteins, polyphenols, and antioxidants	[73]
Olive paste	2 kV/cm, 11.25 kJ/kg	Increased the yield of oil recovery (13.3%) and extraction efficiency of polyphenols (11.5%), phytosterols (9.9%), and tocopherols (15.0%)	[63]

acid from pomegranate peels,[66] lycopene from tomato peels,[57] and quercetin from onion skins.[34]

In recent years, the consumer acceptability of berry fruits has significantly increased due to their investigated health and nutritional benefits. Only 3% of red raspberries are sold fresh, while the rest are processed into juices, jams, or jellies.[88] During the production of juices, several industrial byproducts are derived, including skins, seeds, and press cakes, representing a low-cost alternative source of the key bioactive principles. Generally, mechanical pressing and solvent extraction methods are widely employed in extracting essential compounds. However, they often deteriorate the quality features of extracted products after undergoing distinct pretreatments. As a result, the application of nonthermal techniques like PEF is preferred for the optimal recovery and preservation of valuable phytonutrients. Lamanauskas et al.[40] investigated the effect of PEF pretreatment on raspberry press cake extracts and observed a considerable increase in the levels of total anthocyanins and total phenolics by 26% and 22%, respectively. Likewise, PEF treatment on blueberry press cake extracts also amplified the amounts of total anthocyanins and total phenolics by 78% and 63%, respectively.[10]

About 80% of grapes are used to manufacture wine, and these industries separate the grape pomace from the grape juice after alcoholic fermentation. The recovery of this fermented grape pomace has been carried out using PEF (1.2 kV/cm, 18 kJ/kg), which exhibited an increase in the concentrations of total polyphenols, thus paving the way toward new opportunities for the formulation of extracts with valuable biochemical compositions.[13]

In a recent study, Anwar et al.[3] reported the use of PEF in enhancing the yield of water-soluble non-starch polysaccharides from taro peels. This agro-industrial by-product can be used in the formulation of value-added food products. It may increase the commercial value of taro through the upcycling of its peel and thereby increase the economic profits for the industry.

10.4.2 CEREALS AND LEGUMES INDUSTRY

Among the diverse food categories, cereal crops generate the most waste since they belong to the group of staple food items that constitute the basis of consumers' diets worldwide. Among the cereal crops, corn, rice, and wheat grains contribute 70% higher demand, followed by barley, rye, and sorghum with comparatively lesser demand.[82]

The cereal manufacturing industry involves a range of unit operations, such as dry and wet milling, malting, and pearling, which produce discrete byproducts of differing nature and chemical compositions. During the processing of most cereal grains and pulses, debranning is the common step wherein the seed coats are often removed in order to foster digestibility and attenuate the antinutritional factors in grain products. These processing byproducts are usually undervalued and utilized as animal feed. Besides, they are a rich source of micronutrients that can be potentially recaptured in nutraceutical applications.[23, 46, 71] For example, the application of PEF treatment promotes the extraction yield of arabinoxylans from wheat bran, the major industrial by-product. These non-starch polysaccharides can be used to augment the quality of foods by improving their physicochemical and nutritional properties.[89]

Brewer's spent grain (BSG), a plentiful by-product obtained from the brewing industry, accounts for approximately 85% of the total brewing waste. Nevertheless, it contains many bioactive components, including carbohydrates, β-glucan, dietary fibers, hemicelluloses (arabinoxylans and lignin), phenolic acids, and proteins. Hence, certain sustainable technologies are employed to isolate and recover these functional ingredients. Kumari et al.[39] investigated the effect of PEF as an extraction pretreatment on the nutritional contents and bioactivities of BSG. They elucidated that PEF (2.8 kV/cm with 3000 pulses of 20 μs pulse-width) could amplify the levels of carbohydrates, reducing sugar, starch, and protein. Light BSG extracts were found to contain higher levels of free D-glucose and total free amino acids, while higher phenolic contents were reported in the dark BSG extracts.

Additionally, BSG extracts exhibited higher antimicrobial activity and confirmed the immunomodulatory activity by inducing the release of pro-inflammatory cytokines and chemokines. PEF treatment at 2.5 kV/cm, 50 Hz, for a total time of 14.5 seconds, has also been found optimal in recovering the maximum contents of flavonoids, flavan-3-ols, total free phenolics, and their derivatives from BSG extracts. The valorization of BSG samples indicates their vital role as functional ingredients in the food and pharmaceutical industry.[47] The beer waste brewing yeast (BWBY), usually discarded after a single use, is a significant source of lysine, which is generally deficient in cereals. Liu et al.[43] demonstrated the extraction of residual protein from BWBY using PEF as a satisfactory and highly efficient technique.

10.4.3 MEAT, POULTRY, AND SEAFOOD-PROCESSING INDUSTRY

Animal-based processed food waste chiefly consists of post-slaughter body parts such as skin, shells, wings, bones, tendons, gastrointestinal tract, and blood. The waste generated from meat, poultry, and seafood industries has a significant impact on greenhouse emissions, environmental sustainability, and food supply chain management.[16, 84, 93] According to the published literature, PEF treatment of livestock waste has generated a considerable amount of interest in converting nonedible byproducts into value-added products. Some examples include tenderized meat and beef,[4, 9] calcium from eggshells and fishbone,[42, 97] protein from chicken waste, molluscs, and fish waste,[25, 26, 96] and antioxidants from fish bones and gills.[20]

PEF technology is advantageous over conventional thermal methods of extraction because it retains the flavors and nutrients and does not affect the physical, organoleptic, and functional characteristics of the final product[27]. Franco et al.[20] employed the PEF process as an improved method to extract antioxidants from the heads, gills, and bones of sea bream and sea bass fishes. The treatment was given at an electric field of 1.40 kV/cm and 20 μs pulse width. The DPPH assay showed significant increase in antioxidant extraction; 35.8%, 68.6%, and 33.8% for sea bream and 60.7%, 71.8%, and 22.1% for sea bass, respectively. The eggshell powder was utilized by Lin et al.[41] to extract calcium using PEF treatment in a 2% citric acid solution. The treatment was carried out at 20 μs pulse with 15 kV/cm electric field intensity. The protein from mussels was extracted by the authors in Ref. [96] using PEF treatment at an electric field intensity of 20 kV/cm and 8 μs pulse width. The PEF treatment yielded more than 75% protein, and it also significantly reduced the off odor associated with mussel meat extract when compared with cooked samples. Similarly, Ghosh et al.[25] utilized PEF treatment to recover protein from waste chicken breast meat. The protein from meat in the form of liquid extract possessed high radical scavenging activity (ABTS, 26.93%) and total protein content of 78 mg/mL.

Collectively, meat, chicken, eggs, and seafood discards are low-value byproducts that are rich sources of muscle protein, collagen, minerals, chitosan, chitin, carotenoids, and antioxidants. Research studies have demonstrated that these byproducts can be effectively valorized with the assistance of PEF technology.

10.4.4 PLANTATION CROPS, HERBS, AND SPICES

The rising consumption of coffee and cocoa has inevitably generated crucial byproducts such as coffee silverskins and cocoa bean shells left after the roasting of beans. Typically categorized as "waste," these byproducts cause environmental concerns while also promoting diseases.[54] However, they represent an economical source of methylxanthines and various polyphenols, such as flavonols (quercetin glycosides) in cocoa and chlorogenic acids in coffee, possessing immense potential in developing value-added food ingredients or highly nutritive supplements. It has been elucidated that PEF as a novel pretreatment step has ameliorated the recovery yield of bioactives by 20% compared to the traditional extraction methods.[7] Pecan nut cake, a by-product rich in carbohydrates, dietary fiber, proteins, and phenolic compounds, is advocated as an ingredient in enhancing the functional properties (oil and water absorption capacities) of bakery products. The samples that underwent PEF treatment at 0.8 kJ/kg contained elevated concentrations of condensed tannins (30.1%) and total phenolics (9.5%) compared to untreated ones.[65]

In certain medicinal plants such as borage, only the stem is edible, while the leaves, an inedible by-product, are an inexpensive source of phytochemicals. Implementing PEF treatments (0–5 kV/cm) has profoundly improved the aqueous extraction of polyphenolic compounds and antioxidants. The studies have shown that with respect to control, the total phenolic content (TPC) and oxygen radical absorption capacity (ORAC) in PEF extracts increased significantly.[76] The efficiency of PEF treatment over ultrasound-assisted extraction was demonstrated by Tzima et al.[87] In this study, PEF treated aqueous extracts of byproducts obtained from rosemary and thyme possessed increased antioxidants and phenolic compound levels compared to the independent ultrasound-treated samples. The tamarind-processing industry generates 50–70% byproducts, including peels and seeds. These are used either as livestock feed or are renounced directly in the surroundings. The application of PEF depicts a propitious pretreatment process for the safe and selective extraction of bio-components from these materials.[48]

10.4.5 DAIRY INDUSTRY

The industrial revolution in the dairy sector has led to the increased demands for the consumption of milk and milk products, including cheese, butter, and milk powder, along with generating enough solid and liquid wastes of about 4–11 million tonnes worldwide.[2] The dairy industry often produces various wastes, such as dairy sludges, whey, and wastewater from cleaning, sanitary, and processing operations. This wastewater comprises substantial amounts of milk constituents, including inorganic salts and certain dissolved organic substances, such as casein, fat (clarified butter waste), lactose, minerals, inorganic salts, and whey, which are lost in technological cycles.

The presence of high organic contents in these dairy byproducts poses a severe threat to the external environment.[79]

The largest volume co-product obtained from the dairy industry is whey, which is generally produced in cheese manufacturing. Lactose represents the major fraction of cheese whey, contributing to 70–75% solids, followed by proteins, minerals, and non-protein nitrogen. The conventional ways to valorize whey involve the formation of powdered lactose, and dehydrated whey products, including whey powder, whey protein concentrates, and whey protein isolates.[19, 85] To alter their functional properties and widen the scope of applications, these proteins are customarily modified by either physical or chemical processes (succinylation, the most commonly used chemical method). In a study carried out by Xu et al.,[92] it was reported that when PEF was applied as a pretreatment, the degree of succinylation of whey protein isolate was enhanced from 88.31% to 93.45%. Whey is also acknowledged as a source of several bioactive ingredients, especially peptides and proteins. PEF treatment has played a vital role in enhancing their recovery and improving their immunomodulatory activity.[95]

10.4.6 EDIBLE OIL INDUSTRY

During the processing of oilseeds, a high percentage of oilseed biomass remains unutilized after the oil extraction.[62] The production of edible oils thus generates a large number of byproducts in the form of seeds, pomace, cake, and hulls. These byproducts contain various high-value phytonutrient-rich compounds, which could be recovered for better economics and value addition.[81] PEF treatment has been reported as a viable method for extracting unutilized compounds from processed oilseed biomass.

Several oilseeds have a complex structure containing hulls covering the oil-bearing cells that need to be separated before oil extraction. Boussetta et al.[12] studied the extraction of polyphenols from flaxseed hulls using PEF technology. The hulls were rehydrated in acidic and basic ethanol–water solvents before PEF treatment. It was found that PEF treatment of field strength of 20 kV/cm and a pulse duration of 10 milliseconds with an energy input of 300 kJ/kg resulted in an 80% recovery of total polyphenols. Rehydration of hulls in the alkaline ethanol (20% v/v) for 40 minutes further increased the polyphenol extraction by 37%, which was higher when compared to the treatment with acidic ethanol.

PEF can also be used as a pretreatment for oil extraction from various oilseeds. The application of PEF causes damage to the oil-bearing cells inside the seeds and significantly increases the yield when oil is further extracted by conventional methods. The higher amount of oil extraction also reduces the quantity of oilseed cake as a by-product which would contain more protein and polyphenol content. Sarkis et al.[75] employed PEF pretreatment to extract a higher yield of sesame oil from the seeds along with polyphenols, lignans, and proteins from oilseed cake. It was observed that the concentration of these target biomolecules was higher in the ethanolic extracts obtained from samples pretreated by PEF. The energy input of 83 kJ/kg was found optimum, beyond which there was no significant increase in the content of target compounds in the extract.

High-quality virgin olive oil is obtained by mechanical means, which involves converting olive fruits into a paste and then separating the oil by physical methods.[1] Olive kernels and pomace are the main byproducts of olive oil processing from which many biologically active compounds can be recovered. Roselló-Soto et al.[73] applied PEF (18–109 kJ/kg) to extract valuable compounds from olive kernels. The extracts were obtained after optimizing the pH and concentration of ethanol as a solvent. The samples pretreated by PEF showed a significant improvement in the recovery of proteins, polyphenols, and antioxidants when 109 kJ/kg energy input was applied. Puértolas and Martínez De Marañón[63] found that the treatment of olive paste by pulsed electric fields of 2 kV/cm having an energy input of 11.25 kJ/kg increased the yield of oil recovered by 13.3%. This process also resulted in a significant increase in total phenolic content (11.5%), phytosterols (9.9%), and tocopherols (15.0%) in the olive oil. This finding suggests that PEF technology can be utilized similarly to extract bioactive metabolites present in the olive pomace remaining after the extraction of olive oil.

Guderjan et al.[28] treated maize germs and soybeans with PEF (0.6–1.3 kV/cm) to enhance the recovery of functional metabolites in oil extracts. Application of PEF along with other pretreatments like tempering and drying before extraction of oil increased the recovery of phytosterols (32.4%) from maize germ oil and isoflavonoids: genistein (21%) and daidzein (20%) from soybean oil. Guderjan et al.[29] studied the effect of PEF before solvent extraction of tocopherols, polyphenols, antioxidants, and phytosterols from hulled and non-hulled rapeseed. PEF with a strength of 7 kV/cm and a pulse duration of 30 μs, when applied to non-hulled rapeseed for 120 pulses with an energy input of 84 kJ/kg, showed maximum concentrations of these metabolites in the extract.

These studies advocate that PEF can be successfully used as a pretreatment for enhancing the extraction yield of high-value compounds from byproducts of the edible oil-processing industry. The biologically active compounds extracted by this method have nutraceutical properties and can be utilized for various food applications, improving the process economics.

10.5 SUMMARY

This chapter highlights the potential of PEF as an efficient and eco-friendly nonthermal technique for extracting valuable metabolites from waste with higher yields and quality. The ever-evolving food trends have significantly shifted consumers' preferences toward more processed and convenience food. The industrial processing of fresh plants, crops, and livestock generates low-value waste which is inedible for consumption and has undesirable commercial aspects. Moreover, food waste is also becoming an increasing concern for the environment in terms of greenhouse emissions and pollution. The wastes from food-processing industries are inexpensive byproducts containing valuable nutrients, bioactive compounds, and functional biomolecules. The extraction of biologically active compounds on a commercial scale requires sustainable and green technologies. Research studies have demonstrated the application of PEF treatment in the recovery of proteins, polyphenols, antioxidants, phytonutrients, phytopigments, and minerals from a wide array of food byproducts. PEF-assisted extraction of such nutrients from byproducts having nutraceutical properties possesses significant future scope in promoting large-scale fortification for combating malnutrition.

KEYWORDS

- **bioactive metabolites**
- **byproducts**
- **extraction**
- **pulsed electric field**
- **sustainability**
- **valorization**
- **value-addition**

REFERENCES

1. Abenoza, M.; Benito, M.; Saldaña, G.; Álvarez, I.; Raso, J.; Sánchez-Gimeno, A. C. Effects of Pulsed Electric Field on Yield Extraction and Quality of Olive Oil. *Food Bioprocess Technol.* **2013,** *6* (6), 1367–1373.
2. Ahmad, T.; Aadil, R. M.; Ahmed, H.; ur Rahman, U.; Soares, B. C. V.; Souza, S. L.; Pimentel, T. C.; Scudino, H.; Guimarães, J. T.; Esmerino, E. A.; Freitas, M. Q.; Almada, R. B.; Vendramel, S. M. R.; Silva, M. C.; Cruz, A. G. Treatment and Utilization of Dairy Industrial Waste: A Review. *Trend Food Sci. Technol.* **2019,** *88,* 361–372.
3. Anwar, M.; Babu, G.; Bekhit, A. E. D. Utilization of Ultrasound and Pulse Electric Field for the Extraction of Water-Soluble Non-Starch Polysaccharide from Taro (Colocasia esculenta) Peel. *Innovat. Food Sci. Emerg. Technol.* **2021,** *70,* 102691.
4. Arroyo, C.; Lascorz, D.; O'Dowd, L.; Noci, F.; Arimi, J.; Lyng, J. G. Effect of Pulsed Electric Field Treatments at Various Stages During Conditioning on Quality Attributes of Beef Longissimus Thoracis et Lumborum Muscle. *Meat Sci.* **2014,** *99,* 52–59.
5. Azmir, J.; Zaidul, I. S. M.; Rahman, M. M.; Sharif, K. M.; Mohamed, A.; Sahena, F.; Jahurul, M. H. A.; Ghafoor, K.; Norulaini, N. A. N.; Omar, A. K. M. Techniques for Extraction of Bioactive Compounds from Plant Materials: A Review. *J. Food Eng.* **2013,** *117* (4), 426–436.
6. Barbosa-Cánovas, G. V.; Sepúlveda, D. Present Status and the Future of Pef Technology. *Novel Food Process. Technol.* **2004,** 1–44.
7. Barbosa-Pereira, L.; Guglielmetti, A.; Zeppa, G. Pulsed Electric Field Assisted Extraction of Bioactive Compounds from Cocoa Bean Shell and Coffee Silverskin. *Food Bioprocess Technol.* **2018,** *11* (4), 818–835.
8. Bhat, R. Sustainability Challenges in the Valorization of Agri-Food Wastes and Byproducts. In *Valorization of Agri-Food Wastes and Byproducts;* Bhat, R., Eds.; Academic Press. 2021, pp 1–27.
9. Bhat, Z. F.; Morton, J. D.; Mason, S. L.; Bekhit, A. E. D. A. Current and Future Prospects for the use of Pulsed Electric Field in the Meat Industry. *Crit. Rev. Food Sci. Nutr.* **2019,** *59* (10), 1660–1674.
10. Bobinaitė, R.; Pataro, G.; Lamanauskas, N.; Šatkauskas, S.; Viškelis, P.; Ferrari, G. Application of Pulsed Electric Field in the Production of Juice and Extraction of Bioactive Compounds from Blueberry Fruits and their Byproducts. *J. Food Sci. Technol.* **2015,** *52* (9), 5898–5905.
11. Bobinaitė, R.; Pataro, G.; Visockis, M.; Bobinas, Č.; Ferrari, G.; Viškelis, P. Potential Application of Pulsed Electric Fields to Improve the Recovery of Bioactive Compounds from Sour Cherries and their Byproducts. In *Proceedings of the 11th Baltic Conference on Food Science and Technology;* Jelgava, Latvia. 2017, pp 70–74.
12. Boussetta, N.; Soichi, E.; Lanoisellé, J. L.; Vorobiev, E. Valorization of Oilseed Residues: Extraction of Polyphenols from Flaxseed Hulls by Pulsed Electric Fields. *Indust. Crop Prod.* **2014,** *52,* 347–353.
13. Brianceau, S.; Turk, M.; Vitrac, X.; Vorobiev, E. Combined Densification and Pulsed Electric Field Treatment for Selective Polyphenols Recovery from Fermented Grape Pomace. *Innovat. Food Sci. Emerg. Technol.* **2015,** *29,* 2–8.
14. Capanoglu, E.; Nemli, E.; Tomas-Barberan, F. Novel Approaches in the Valorization of Agricultural Wastes and Their Applications. *J. Agri. Food Chem.* **2021,** *70* (23), 6787–6804.

15. Chang, D. C. Cell Poration and Cell Fusion using an Oscillating Electric Field. *Biophys. J.* **1989,** *56* (4), 641–652.
16. Chowdhury, P.; Viraraghavan, T.; Srinivasan, A. Biological Treatment Processes for Fish Processing Wastewater—A Review. *Biores. Technol.* **2010,** *101* (2), 439–449.
17. Destandau, E.; Michel, T.; Elfakir, C. Microwave-Assisted Extraction. In *Natural Product Extraction: Principles and Applications;* Rostagno, M.A.; Prado, J.M. Eds.; The Royal Society of Chemistry: London, UK. 2013, pp 113–156.
18. FAO (Food and Agriculture Organization). *Food Wastage Footprint: Impacts on Natural Resources: Summary Report,* Québec City, QC, Canada. 2013.
19. Figueroa Pires, A.; Garcia Marnotes, N.; Díaz Rubio, O.; Cobos Garcia, A.; Dias Pereira, C. Dairy Byproducts: A Review on the Valorization of Whey and Second Cheese Whey. *Foods* **2021,** *10,* 1067.
20. Franco, D.; Munekata, P. E. S.; Agregán, R.; Bermúdez, R.; López-Pedrouso, M.; Pateiro, M.; Lorenzo, J. M. Application of Pulsed Electric Fields for Obtaining Antioxidant Extracts from Fish Residues. *Antioxidants* **2020,** *9* (2).
21. Frontuto, D.; Carullo, D.; Harrison, S. M.; Brunton, N. P.; Ferrari, G.; Lyng, J. G.; Pataro, G. Optimization of Pulsed Electric Fields-Assisted Extraction of Polyphenols from Potato Peels Using Response Surface Methodology. *Food Bioprocess Technol.* **2019,** *12* (10), 1708–1720.
22. Galanakis, C. M. Recovery of High Added-Value Components from Food Wastes: Conventional, Emerging Technologies and Commercialized Applications. *Trend Food Sci. Technol.* **2012,** *26* (2), 68–87.
23. Galanakis, C. M. Sustainable Applications for the Valorization of Cereal Processing Byproducts. *Foods* **2022,** *11* (2), *241.*
24. Gençdağ, E.; Görgüç, A.; Yılmaz, F. M. Recent Advances in the Recovery Techniques of Plant-Based Proteins from Agro-Industrial Byproducts. *Food Rev. Int.* **2021,** 37 (4), 447–468.
25. Ghosh, S.; Gillis, A.; Sheviryov, J.; Levkov, K.; Golberg, A. Towards Waste Meat Biorefinery: Extraction of Proteins from Waste Chicken Meat with Non-Thermal Pulsed Electric Fields and Mechanical Pressing. *J. Clean. Prod.* **2019,** *208,* 220–231.
26. Golberg, A.; Sack, M.; Teissie, J.; Pataro, G.; Pliquett, U.; Saulis, G.; Stefan, T.; Miklavcic, D.; Vorobiev, E.; Frey, W. Energy-Efficient Biomass Processing with Pulsed Electric Fields for Bioeconomy and Sustainable Development. *Biotechnol. Biofuel* **2016,** *9,* 1–22.
27. Gómez, B.; Munekata, P. E. S.; Gavahian, M.; Barba, F. J.; Martí-Quijal, F. J.; Bolumar, T.; Campagnol, P. C. B.; Tomasevic, I.; Lorenzo, J. M. Application of Pulsed Electric Fields in Meat and Fish Processing Industries: An Overview. *Food Res. Int.* **2019,** *123,* 95–105.
28. Guderjan, M.; Töpfl, S.; Angersbach, A.; Knorr, D. Impact of Pulsed Electric Field Treatment on the Recovery and Quality of Plant Oils. *J. Food Eng.* **2005,** 67 (3), 281–287.
29. Guderjan, Manuela, Elez-Martínez, P.; Knorr, D. Application of Pulsed Electric Fields at Oil yield and Content of Functional Food Ingredients at the Production of Rapeseed Oil. *Innovat. Food Sci. Emerg. Technol.* **2007,** *8* (1), 55–62.
30. Handa, S. S. An Overview of Extraction Techniques for Medicinal and Aromatic Plants. *Extract. Technol. Med. Aromatic Plants* **2008,** *1* (1), 21–40.
31. Idrishi, R.; Aggarwal, D.; Sharma, V. Upcycling Technologies in the Food Industry. In *Smart and Sustainable Food Technologies;* Sehgal, S., Singh, B., Sharma, V., Eds.; Springer Nature, Singapore. 2022, pp 367–392.

32. Jeyamkondan, S.; Jayas, D. S.; Holley, R. A. Pulsed Electric Field Processing of Foods: A Review. *J. Food Protect.* **1999,** *62* (9), 1088–1096.
33. Khedkar, R.; Singh, K. Food Industry Waste: A Panacea or Pollution Hazard?. In *Paradigms in Pollution Prevention;* Jindal, T., Ed.;. Springer, Cham. 2018, pp 35–47.
34. Kim, H.; Ko, M.; Park, C.; Chung, M. Application of Pulsed Electric Field as a Pre-Treatment for Subcritical Water Extraction of Quercetin from Onion Skin. *Foods* **2022,** *11* (8), 1069.
35. Kotnik, T.; Pucihar, G.; Miklavčič, D. Induced Transmembrane Voltage and Its Correlation with Electroporation-Mediated Molecular Transport. *J. Mem. Biol.* **2010,** *236* (1), 3–13.
36. Koubaa, M.; Barba, F. J.; Grimi, N.; Mhemdi, H.; Koubaa, W.; Boussetta, N.; Vorobiev, E. Recovery of Colorants from Red Prickly Pear Peels and Pulps Enhanced by Pulsed Electric Field and Ultrasound. *Innovat. Food Sci. Emerg. Technol.* **2016,** *37,* 336–344.
37. Kumar, S.; Agarwal, N.; Raghav, P. K. Pulsed Electric Field Processing of Foods-a Review. *Int. J. Eng. Res. Modern Edu.* **2016,** *1* (1), 111–118.
38. Kumari, B.; Tiwari, B. K.; Hossain, M. B.; Brunton, N. P.; Rai, D. K. Recent Advances on Application of Ultrasound and Pulsed Electric Field Technologies in the Extraction of Bioactives from Agro-Industrial Byproducts. *Food Bioprocess Technol.* **2018,** *11* (2), 223–241.
39. Kumari, B.; Tiwari, B. K.; Walsh, D.; Griffin, T. P.; Islam, N.; Lyng, J. G.; Brunton, N. P.; Rai, D. K. Impact of Pulsed Electric Field Pre-Treatment on Nutritional and Polyphenolic Contents and Bioactivities of Light and Dark Brewer's Spent Grains. *Innovat. Food Sci. Emerg. Technol.* **2019,** *54,* 200–210.
40. Lamanauskas, N.; Pataro, G.; Bobinas, Č.; Šatkauskas, S.; Viškelis, P.; Bobinaitė, R.; Ferrari, G. Impact of Pulsed Electric Field Treatment on Juice Yield and Recovery of Bioactive Compounds from Raspberries and their Byproducts. *Zemdirbyste-Agriculture* **2016,** *103* (1), 83–90.
41. Lin, S.; Shen, T.; Ye, H. An Absorption Assessment of Eggshell Calcium Citrate Treated by PEF in Mice. *Adv. Mater. Res.* **2013,** *634* (1), 1413–1416.
42. Lin, S.; Wang, L.; Jones, G.; Trang, H.; Yin, Y.; Liu, J. Optimized Extraction of Calcium Malate from Eggshell Treated by PEF and an Absorption Assessment in vitro. *Int. J. Biol. Macromole.* **2012,** *50* (5), 1327–1333.
43. Liu, M.; Zhang, M.; Lin, S.; Liu, J.; Yang, Y.; Jin, Y. Optimization of Extraction Parameters for Protein from Beer Waste Brewing Yeast Treated by Pulsed Electric Fields (PEF). *African J. Microbiol. Res.* **2012,** *6* (22), 4739–4746.
44. Lohani, U. C.; Muthukumarappan, K. Application of the Pulsed Electric Field to Release Bound Phenolics in Sorghum Flour and Apple Pomace. *Innovat. Food Sci. Emerg. Technol.* **2016,** *35,* 29–35.
45. Luengo, E.; Álvarez, I.; Raso, J. Improving the Pressing Extraction of Polyphenols of Orange Peel by Pulsed Electric Fields. *Innovat. Food Sci. Emerg. Technol.* **2013,** *17,* 79–84.
46. Luithui, Y.; Baghya Nisha, R.; Meera, M. S. Cereal Byproducts as an Important Functional Ingredient: Effect of Processing. *J. Food Sci. Technol.* **2019,** *56* (1), 1–11.
47. Martín-García, B.; Tylewicz, U.; Verardo, V.; Pasini, F.; Gómez-Caravaca, A. M.; Caboni, M. F.; Dalla Rosa, M. Pulsed Electric Field (PEF) as Pre-Treatment to Improve the Phenolic Compounds Recovery from Brewers' Spent Grains. *Innovat. Food Sci. Emerg. Technol.* **2020,** *64,* 102402.

48. Martins, C. M.; Ferro, D. M.; de Brito, E. S.; Ferreira, S. R. S. Industrial Relevance of Tamarindus Indica L. Byproducts as Source of Valuable Active Metabolites. *Innovat. Food Sci. Emerg. Technol.* **2020,** *66,* 102518.
49. McHugh, M.; Krukonis, V. (eds.). *Supercritical Fluid Extraction: Principles and Practice* (2nd edn.). Elsevier. 2013.
50. Menon, A.; Sawant, M.; Mishra, S.; Bhatia, P.; Rathod, S. Awareness, Perception and Usage of Nutraceuticals in Indian Society. *Int. J. Scientif. Res. Sci. Technol.* **2021,** *8,* 399–414.
51. Notarnicola, B.; Hayashi, K.; Curran, M. A.; Huisingh, D. Progress in Working towards a more Sustainable Agri-Food Industry. *J. Clean. Prod.* **2012,** *28,* 1–8.
52. Nowosad, K.; Sujka, M.; Pankiewicz, U.; Kowalski, R. The Application of PEF Technology in Food Processing and Human Nutrition. *J. Food Sci. Technol.* **2021,** *58* (2), 397–411.
53. O'Connor, J.; Hoang, S. A.; Bradney, L.; Dutta, S.; Xiong, X.; Tsang, D. C. W.; Ramadass, K.; Vinu, A.; Kirkham, M. B.; Bolan, N. S. A Review on the Valorisation of Food Waste as a Nutrient Source and Soil Amendment. *Environ. Pollut.* **2021,** *272,* 115985.
54. Panak Balentić, J.; Ačkar, Đ.; Jokić, S.; Jozinović, A.; Babić, J.; Miličević, B.; Šubarić, D.; Pavlović, N. Cocoa Shell: A By-Product with Great Potential for Wide Application. *Molecules (Basel, Switzerland)* **2018,** *23* (6), 1–14.
55. Parniakov, O.; Barba, F. J.; Grimi, N.; Lebovka, N.; Vorobiev, E. Extraction Assisted by Pulsed Electric Energy as a Potential Tool for Green and Sustainable Recovery of Nutritionally Valuable Compounds from Mango Peels. *Food Chem.* **2016,** *192,* 842–848.
56. Parniakov, O.; Roselló-Soto, E.; Barba, F. J.; Grimi, N.; Lebovka, N.; Vorobiev, E. New Approaches for the Effective Valorization of Papaya Seeds: Extraction of Proteins, Phenolic Compounds, Carbohydrates, and Isothiocyanates Assisted by Pulsed Electric Energy. *Food Res. Int.* **2015,** *77,* 711–717.
57. Pataro, G.; Carullo, D.; Falcone, M.; Ferrari, G. Recovery of Lycopene from Industrially Derived Tomato Processing Byproducts by Pulsed Electric Fields-Assisted Extraction. *Innovat. Food Sci. Emerg. Technol.* **2020,** *63,* 102369.
58. Pataro, Gianpiero, Carullo, D.; Bobinaite, R.; Donsì, G.; Ferrari, G. Improving the Extraction Yield of Juice and Bioactive Compounds from Sweet Cherries and their Byproducts by Pulsed Electric Fields. *Chem. Eng. Trans.* **2017,** *57,* 1717–1722.
59. Peiró, S.; Luengo, E.; Segovia, F.; Raso, J.; Almajano, M. P. Improving Polyphenol Extraction from Lemon Residues by Pulsed Electric Fields. *Waste Biom. Valoriz.* **2019,** *10* (4).
60. Picot-Allain, C.; Mahomoodally, M. F.; Ak, G.; Zengin, G. Conventional Versus Green Extraction Techniques—a Comparative Perspective. *Curr. Opin. Food Sci.* **2021,** *40,* 144–156.
61. Pingret, D.; Fabiano-Tixier, A. S.; Chemat, F. Ultrasound-assisted extraction. In *Natural Product Extraction: Principles and Applications;* Rostagno, M. A, Prado, J. M., Eds.; The Royal Society of Chemistry: London, UK. 2013, pp 21–89.
62. Puértolas, E.; Koubaa, M.; Barba, F. J. An Overview of the Impact of Electrotechnologies for the Recovery of Oil and High-Value Compounds from Vegetable oil Industry: Energy and Economic Cost Implications. *Food Res. Int.* **2016,** *80,* 19–26.
63. Puértolas, E.; Martínez De Marañón, I. Olive Oil Pilot-Production Assisted by Pulsed Electric Field: Impact on Extraction Yield, Chemical Parameters and Sensory Properties. *Food Chem.* **2015,** *167,* 497–502.

64. Qin, B. L.; Zhang, Q.; Barbosa-C¡novas, G. V.; Swanson, B. G.; Pedrow, P. D. Inactivation of Microorganisms by Pulsed Electric Fields of Different Voltage Waveforms. *IEEE Trans. Dielect. Electric. Insulat.* **1994,** *1* (6), 1047–1057.
65. Rábago-Panduro, L. M.; Morales-de la Peña, M.; Martín-Belloso, O.; Welti-Chanes, J. Application of Pulsed Electric Fields PEF on Pecan Nuts Carya illinoinensis Wangenh. K. Koch: Oil Extraction Yield and Compositional Characteristics of the Oil and Its By-product. *Food Eng. Rev.* **2021,** *13* (3), 676–685.
66. Rajha, H. N.; Abi-Khattar, A. M.; El Kantar, S.; Boussetta, N.; Lebovka, N.; Maroun, R. G.; Louka, N.; Vorobiev, E. Comparison of Aqueous Extraction Efficiency and Biological Activities of Polyphenols from Pomegranate Peels Assisted by Infrared, Ultrasound, Pulsed Electric Fields and High-Voltage Electrical Discharges. *Innovat. Food Sci. Emerg. Technol.* **2019,** *58,* 102212.
67. Ranjha, M. M. A. N.; Kanwal, R.; Shafique, B.; Arshad, R. N.; Irfan, S.; Kieliszek, M.; Kowalczewski, P. Ł.; Irfan, M.; Khalid, M. Z.; Roobab, U.; Aadil, R. M. A Critical Review on Pulsed Electric Field: A Novel Technology for the Extraction of Phytoconstituents. *Molecules* **2021,** *26* (16), 4893.
68. Rawdkuen, S.; Kaewprachu, P. Valorization of Food Processing Byproducts as Smart Food Packaging Materials and Its Application. *Food Preserv. Waste Exploit.* **2020.**
69. Redondo, D.; Venturini, M. E.; Luengo, E.; Raso, J.; Arias, E. Pulsed Electric Fields as a Green Technology for the Extraction of Bioactive Compounds from Thinned Peach Byproducts. *Innovat. Food Sci. Emerg. Technol.* **2018,** *45,* 335–343.
70. Rifna, E. J.; Misra, N. N.; Dwivedi, M. Recent Advances in Extraction Technologies for Recovery of Bioactive Compounds Derived from Fruit and Vegetable Waste Peels: A Review. *Crit. Rev. Food Sci. Nutr.* **2023,** *63* (6), 719–752.
71. Roasa, J.; De Villa, R.; Mine, Y.; Tsao, R. Phenolics of Cereal, Pulse and Oilseed Processing Byproducts and Potential Effects of Solid-State Fermentation on their Bioaccessibility, Bioavailability and Health Benefits: A Review. *Trend Food Sci. Technol.* **2021,** *116,* 954–974.
72. Rodriguez-Lopez, A. D.; Melgar, B.; Conidi, C.; Barros, L.; Ferreira, I. C. F. R.; Cassano, A.; Garcia-Castello, E. M. Food Industry Byproducts Valorization and New Ingredients: Cases of Study. In *Sustainability of the Food System: Sovereignty, Waste, and Nutrients Bioavailability;* Betoret, N., Betoret, E., Eds.; Academic Press: Cambridge, MA, USA. 2020, pp 71–99.
73. Roselló-Soto, E.; Barba, F. J.; Parniakov, O.; Galanakis, C. M.; Lebovka, N.; Grimi, N.; Vorobiev, E. High Voltage Electrical Discharges, Pulsed Electric Field, and Ultrasound Assisted Extraction of Protein and Phenolic Compounds from Olive Kernel. *Food Bioprocess Technol.* **2015,** *8* (4), 885–894.
74. Sabater, C.; Ruiz, L.; Delgado, S.; Ruas-Madiedo, P.; Margolles, A. Valorization of Vegetable Food Waste and Byproducts Through Fermentation Processes. *Front. Microbiol.* **2020,** *11* (October), 1–11.
75. Sarkis, J. R.; Boussetta, N.; Tessaro, I. C.; Marczak, L. D. F.; Vorobiev, E. Application of Pulsed Electric Fields and High Voltage Electrical Discharges for Oil Extraction from Sesame Seeds. *J. Food Eng.* **2015,** *153,* 20–27.
76. Segovia, F. J.; Luengo, E.; Corral-Pérez, J. J.; Raso, J.; Almajano, M. P. Improvements in the Aqueous Extraction of Polyphenols from Borage (Borago officinalis L.) Leaves by Pulsed Electric Fields: Pulsed Electric Fields (PEF) Applications. *Indust. Crop Prod.* **2015,** *65,* 390–396.

77. Selvamuthukumaran, M.; Shi, J. Recent Advances in Extraction of Antioxidants from Plant Byproducts Processing Industries. *Food Quality and Safety* **2017,** *1* (1), 61–81.
78. Sharma, S. K.; Bansal, S.; Mangal, M.; Dixit, A. K.; Gupta, R. K.; Mangal, A. K. Utilization of Food Processing Byproducts as Dietary, Functional, and Novel Fiber: A Review. *Crit. Rev. Food Sci. Nutr.* **2016,** *56* (10), 1647–1661.
79. Shi, W.; Healy, M. G.; Ashekuzzaman, S. M.; Daly, K.; Leahy, J. J.; Fenton, O. Dairy Processing Sludge and Co-Products: A Review of Present and Future Re-use Pathways in Agriculture. *J. Clean. Prod.* **2021,** *314,* 128035.
80. Singh, R. A Critical Review and Opinion on Hydrodistillation. *Bio Sci. Res. Bullet.* **2017,** *33* (1), 1.
81. Smeu, I.; Dobre, A. A.; Cucu, E. M.; Mustățea, G.; Belc, N.; Ungureanu, E. L. Byproducts from the Vegetable Oil Industry: The Challenges of Safety and Sustainability. *Sustainability* **2022,** *14* (4), 2039.
82. Tapia-Hernández, J. A.; Del-Toro-Sánchez, C. L.; Cinco-Moroyoqui, F. J.; Juárez-Onofre, J. E.; Ruiz-Cruz, S.; Carvajal-Millan, E.; López-Ahumada, G. A.; Castro-Enriquez, D. D.; Barreras-Urbina, C. G.; Rodríguez-Felix, F. Prolamins from Cereal Byproducts: Classification, Extraction, Characterization and Its Applications in Micro- and Nanofabrication. *Trend Food Sci. Technol.* **2019,** *90,* 111–132.
83. Tieleman, D. P. The Molecular Basis of Electroporation. *BMC Biochem.* **2004,** *5,* 1–12.
84. Tonini, D.; Albizzati, P. F.; Astrup, T. F. Environmental Impacts of Food Waste: Learnings and Challenges from a Case study on UK. *Waste Manag.* **2018,** *76,* 744–766.
85. Tsermoula, P.; Khakimov, B.; Nielsen, J. H.; Engelsen, S. B. WHEY—The Waste-Stream that Became more Valuable than the Food Product. *Trend Food Sci. Technol.* **2021,** *118,* 230–241.
86. Tsong, T. Y. Electrically Stimulated Membrane Breakdown. In *Electrical Manipulation of Cells;* Lynch, P. T., Davey, M. R., Eds; Springer: USA. 1996, pp 15–36.
87. Tzima, K.; Brunton, N. P.; Lyng, J. G.; Frontuto, D.; Rai, D. K. The Effect of Pulsed Electric Field as a Pre-Treatment step in Ultrasound Assisted Extraction of Phenolic Compounds from Fresh Rosemary and Thyme Byproducts. *Innovat. Food Sci. Emerg. Technol.* **2021,** *69,* 102644.
88. Vladisavljević, G. T.; Vukosavljević, P.; Veljović, M. S. Clarification of Red Raspberry Juice using Microfiltration with Gas Backwashing: A Viable Strategy to Maximize Permeate Flux and Minimize a Loss of Anthocyanins. *Food and Bioprod. Process.* **2013,** *91* (4), 473–480.
89. Wahlsten, A. Pulsed Electric Field Treatment of Wheat Bran to Improve Arabinoxylan Extraction. Master thesis, Swedish University of Agricultural Sciences, Sweden. 2021.
90. Weaver, J. C. Electroporation of Biological Membranes from Multicellular to Nano Scales. *IEEE Trans. Dielect. Electric. Insulat.* **2003,** *10* (5), 754–768.
91. Wen, L.; Zhang, Z.; Sun, D. W.; Sivagnanam, S. P.; Tiwari, B. K. Combination of Emerging Technologies for the Extraction of Bioactive Compounds. *Crit. Rev. Food Sci. Nutr.* **2020,** *60* (11), 1826–1841.
92. Xu, F. Y.; Wen, Q. H.; Wang, R.; Li, J.; Chen, B. R.; Zeng, X. A. Enhanced Synthesis of Succinylated Whey Protein Isolate by Pulsed Electric Field Pretreatment. *Food Chem.* **2021,** *363,* 129892.
93. Xu, Z.; Sun, D. W.; Zeng, X. A.; Liu, D.; Pu, H. Research Developments in Methods to Reduce the Carbon Footprint of the Food System: A Review. *Crit. Rev. Food Sci. Nutr.* **2015,** *55* (9), 1270–1286.

94. Zhang, Q. W.; Lin, L. G.; Ye, W. C. Techniques for Extraction and Isolation of Natural Products: A Comprehensive Review. *Chinese Med.* **2018,** *13,* 1–26.
95. Zhang, S.; Sun, L.; Ju, H.; Bao, Z.; Zeng, X. an, & Lin, S. Research Advances and Application of Pulsed Electric Field on Proteins and Peptides in Food. *Food Res. Int.* **2021,** *139* (1), 109914.
96. Zhou, Y.; He, Q.; Zhou, D. Optimization Extraction of Protein from Mussel by High-Intensity Pulsed Electric Fields. *J. Food Process. Preserv.* **2017,** *41* (3).
97. Zhou, Y.; Sui, S.; Huang, H.; He, G.; Wang, S.; Yin, Y.; Ma, Z. Process Optimization for Extraction of Fishbone Calcium Assisted by High Intensity Pulsed Electric Fields. *Trans. Chin. Soc. Agri. Eng.* **2012,** *28* (23), 265–270.
98. Zimmermann, U. Electrical Breakdown, Electropermeabilization and Electrofusion. *Rev. Physiol., Biochem. Pharmacol.* **1986,** *105,* 176–256.

CHAPTER 11

PULSED ELECTRIC FIELD AS A GREEN TECHNOLOGY FOR THE EXTRACTION OF PHYTOCHEMICALS

RITIKA ARORA, RIZWANA HALEEM, ABHISHEK DUTT TRIPATHI, and APARNA AGARWAL

ABSTRACT

Phytochemicals have gained much attention in this modern world. Researchers are in trend by utilizing agro-waste, leaves, oilseeds, herbs, and spices to extract phytochemicals. Various novel technologies have been introduced to optimize the extraction procedure from agro-waste and other materials. A description of the pulsed electric field as a modern extraction technique of various phytochemicals present in fruits, vegetables, herbs, spices, and oilseeds is presented in this chapter. Further, the applications and impact of pulsed electric field have been discussed with different schematics and illustrations.

11.1 INTRODUCTION

Phytochemicals are a broad category of bioactive obtained from plants. This category includes carotenoids, plant sterols, glucosinolates, non-flavonoid phenolic compounds, flavonoids, and other sulfur-containing substances.[51] Soxhlet, boiling, heat reflux, maceration, water percolation, soaking, magnetic stirrer, and grinding are some of the traditional methods for obtaining phytochemicals; however, these techniques have an increased solvent expenditure, a lengthy process time, poor or minimal extraction yields, and few have thermal deterioration uncertainties. Various novel nonthermal extraction approaches

Nonthermal Pulsed Electric Field Technology for the Food Industry.
Preeti Birwal, Megh R. Goyal, Baldev Singh Kalsi, & Ankita Kataria (Eds.)

DOI: 10.1201/9781003605904-14

have been examined to address the drawbacks of these classic approaches. In current decades, the use of pulsed electric fields (PEF) to extract valuable components from food waste and by-products has drawn increased attention. The negative consequences of standard heating techniques are reduced.[52]

PEF technology is implemented to successfully separate, intensify, stabilize, and dehydrate valuable substances while maintaining their nutritional properties, making it a viable substitute for existing techniques (boiling, microwave- and ultrasound-assisted extractions). PEF was recently proposed as a method for stressing plant cells in addition to increasing extraction, promoting bioactive constituents' production. PEF's ability to electroporate cell membranes makes it a helpful pretreatment for recovering bioactive chemicals, accompanied by a conventional or novel extraction procedure. Figure 11.1 illustrates the processing of food using PEF. In the batch process, the PEF approach requires an electric field strength (EFS) of between 100 and 300 V/cm and between 20 and 80 kV/cm in continuously operated extraction. Regarding the probable PEF mechanism, several hypotheses share two points of view. One involves accelerating chemical processes involving numerous substances in the biological cell membrane to increase the solvent's solubility, and the other is the electroporation process. An external electrical force is used in electro-permeabilization or electroporation to make cell membranes more permeable.[31]

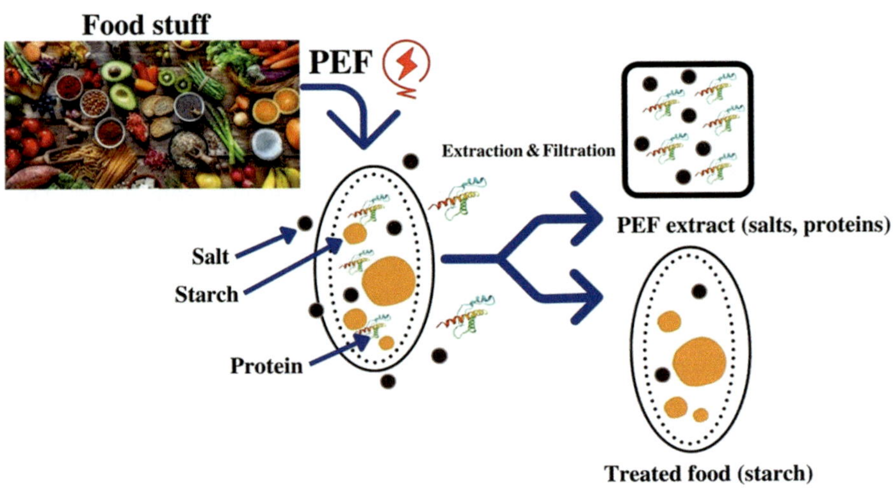

FIGURE 11.1 Processing via pulsed electric field.

Source: Adapted from Ref. [53]

This chapter highlights the pulsed electric field to extract phytochemicals from agroindustrial waste, fruits and veggies, leaves including oilseeds. Further, the factors affecting the extraction and its application have also been discussed.

11.2 PHYTOCHEMICALS DERIVED FROM AGROINDUSTRIAL WASTE

A massive proportion of non-nutritive sanative elements and nutrients for growth and development may be found in agricultural products, including nuts, herbs and spices, cereal grains, fruits and vegetables, and many more. Numerous such compounds are either primary or secondary metabolites (phytochemicals) produced by the survival systems of plants, which have health-protective properties in humans and their nutritional value. Plant food processing in the industrial sector produces massive quantities of by-products yearly. Defective raw materials, seeds, skins, husks, brans, oilseed cakes, molasses, discarded grains, and other processing by-products are common.

Numerous agribusiness by-products can be converted into a variety of biologically active substances, including polyphenolic compounds, carotenoids, vitamins, and dietary fiber. The following figure gives the schematic representation of extraction process (Figure 11.2). Natural ingredients and components are increasingly in demand for functional foods, and nutraceuticals might be met with the efficient use of by-products, a cheap raw material. The shortcomings of extraction and purification procedures, such as laborious, tedious, and environmentally unfriendly processes to extract and purify desired compounds, appear to be a significant factor in the underutilization of by-products. Implementing energy-efficient, quick, affordable, and environmentally friendly extraction processes might increase the viability of achieving the potential economic advantages of agricultural by-products.

PEF has received much research attention as a nonthermal method for food manufacturing, increased shelf-life, and microbial inactivation. On the other hand, PEF is used to recover beneficial components from byproducts. PEF technology has a significant prospect for enhanced intracellular valuable chemical extraction.[2] It helps to permeabilize cell membranes, boost intracellular chemical diffusivity, and enhance extraction effectiveness through mass transfer rate increment.

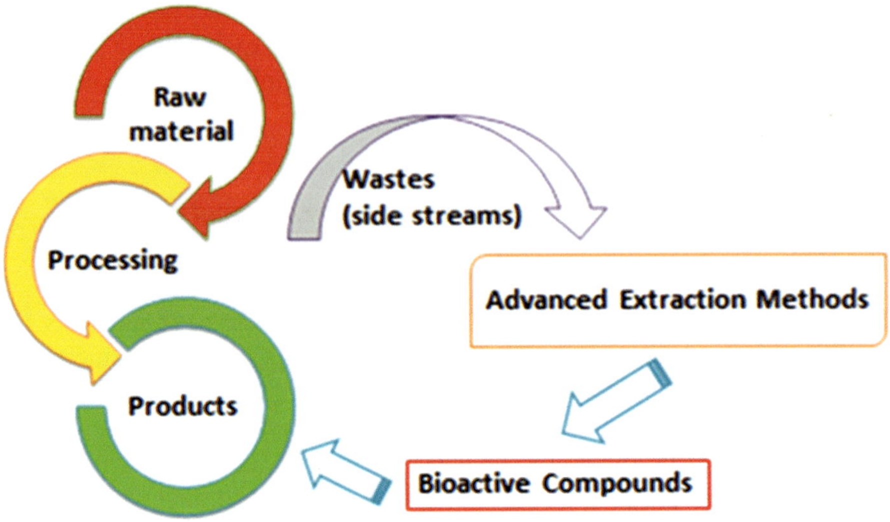

FIGURE 11.2 Schematic of extraction method.
Source: Adapted from Ref. [19]

11.3 EXTRACTION OF PHYTOCHEMICALS

Extraction utilizes certain solvents to separate the phytochemicals inside plant components. It may alternatively be described as the utilization of a solvent to treat plant components in which the elements of the phytochemicals are dissolved, but most of the inert material is left undisturbed. To separate the soluble plant metabolites from the insoluble residue, extraction is the process of doing this. The result is a rather complicated combination of metabolites that may be used internally or externally. It can be liquid, semisolid, or dry powder after removing the water.[6] The solubility differential among the solute, other chemicals in the matrix, and the stabilizing solvent serve as the basis for extraction.

When removing active compounds from plants, proper extraction procedures and techniques must be used to produce extracts and fractions rich in phytochemicals. Thus, the methods of extraction are crucial to the yield and the kind of phytochemical composition. Plant materials can be extracted using a variety of traditional and novel methods. The degree to which extraction methods vary often relies on essential variables such as extraction duration, temperature, tissue particle size, solvent-to-sample ratio, and pH. Maceration, hydro-distillation, decoction, soxhlet, and percolation are traditional methods, whereas supercritical fluid extraction, ultrasound-assisted extraction, pulsed electric field extraction, microwave-assisted extraction, and enzyme-assisted extraction are examples of innovative methods.[46]

11.3.1 PHYTOCHEMICALS FROM FRUITS AND VEGETABLES

Skins of citrus fruit, leftovers of pineapple, sugarcane bagasse, and numerous fruit leftovers (mainly peels and seeds) are only a few examples of the many fruit wastes created in huge volumes in urban settings as a consequence of intensive industrial processing of fruit's edible parts. Finding a resource and value-added application for fruit wastes is urgently needed. Recovering the bioactive components, particularly the phenolic compounds, and entirely using them in the food, pharmaceutical, and cosmetics industries is one of the most advantageous techniques. Utilizing fruit wastes as reservoirs of phytochemicals may thus have significant economic advantages and is growing in popularity.[10] Tables 11.1 and 11.2 show several phytochemicals found in fruits and veggies, respectively. Utilizing most manufacturing by-products for the generation, extraction, synthesis, or preparation of several useful substances would be one strategy to deal with the problem of fruit and vegetable wastes and assist in creating valuable substances. Peels, typically discarded, contain significant amounts of beneficial phytoconstituents and bioactive substances, such as dietary fiber, enzymes, carotenoids, organic acids, antioxidants, phenolic compounds, and many others.[37]

11.3.2 PHYTOCHEMICALS FROM HERBS AND SPICES

Plant materials like spices and herbs offer a variety of physiologically bioactive substances. For generations, people have used spices and herbs as medicines and health and well-being to better serve as sources of scent, flavor, and color, as well as preservatives. The term "biologically active" (from which the word "bioactive" is derived) refers to a substance that positively affects living things, such as tissues, cells, and organisms.[4] Bioactive molecules are substances with certain health advantages and significant effects on human development and improvement. By relieving illness circumstances, they also alter disease risks. They impact physiological or cellular functions with positive health benefits, change disease risk instead of attempting to prevent deficiency diseases, and stimulate and inhibit enzyme activity, receptor activity, and gene expression. Table 11.3 describes the different phytochemicals present in herbs and spices. They carry out crucial tasks even though they are not necessary

TABLE 11.1 Phytochemicals Obtained from Fruits.

Fruit	Phytochemical	Properties	Application
Mango (peels, seed kernel)	Flavonol glycosides, Gallic acid, ellagic acid, gallates, gallotannins, condensed tannins	Water-holding capacity, oil-holding capacity Antioxidant activity Nutraceutical potential	Prebiotic products
Apple (peel, pomace)	Epicatechin, catechins, anthocyanins, quercitin glycosides, chlorogenic acid, hydroxycinnamates, phloretin glycosides, procyanidins	Antihypertensive, antioxidative, antiproliferative, anti-inflammatory, antidiabetic, anti-carcinogenic	Biomedical and pharmaceutical application
Grapes (pomace, skin, seed)	Resveratrol, quercetin, Coumaric acid, quercetin 3-O-gluuronide, caffeic acid, ferulic acid, p-hydroxybenzoic acid, chlorogenic acid, cinnamic acid, neochlorogenic acid, protocatechuic acid, gallic acid, vanillic acid, proanthocyanidins	Antidiabetic, antioxidant, anti-platelet, anti-cholesterol, anti-inflammation, antiaging, antimicrobial, antitumor	Food additive
Banana (peel, bract)	Gallocatechin, anthocyanins, delphindin, cyaniding, catecholamine	Antimicrobial, antioxidant, anticancer, pharmacological activity	Food packaging, food products, food ingredients
Citrus fruits (peel)	Hesperidin, naringin, eriocitrin, narirutin	Anti-inflammatory, antitumor, antidiabetes, neuroprotective agent anti-atherosclerosis antiatherogenic, anticoagulant antioxidant activity, cardiovascular protective	Polystyrene production

TABLE 11.2 Phytochemicals Obtained from Vegetables.

Vegetable	Phytochemical	Properties	Application
Potato (peel)	Gallic acid, caffeic acid vanillic acid, Chlorogenic, protocatechuic acid, chlorogenic acid isomer II	Bactericidal and Bacteriostatic effect Antioxidant Antiradical scavenging activity, Antimicrobial	Edible films, food grade additives, enzyme production
Tomato	Carotenoids, Lycopene	Antioxidative, antiproliferative, anti-carcino-genic, anti-tumorigenic, anti-inflammatory, anti-mutagenic and anti-atherogenic properties	Additive for cheese, bread, and other foods
Onion	Quercetin 40-O-monoglucoside and Quercetin 3,40-O-diglucoside	Antioxidant activity, antimicrobial activity, Wound healing and anti-scar, blood sugar control, bone health Cancer risk reduction, neuroprotective effect, immunosuppression	Valuable natural resource for the development of functional foods or nutraceuticals for the prevention and control of illnesses including obesity, diabetes, cancer, cardiovascular disease, and neurodegenerative disease
Garlic	Di-ferulic acid, coumaric acid-O-glucoside, hydroxy-benzoic acid, caffeic acid-*O*-glucoside, *p*-coumaric acid, caffeoylputrescine	Antioxidant activity, anti-inflammatory activity, antimicrobial activity, immuno-modulatory activity, anti-hyperlipidemic activity, antihypertensive, cardiovascular protection	Powder, extract, oil, and capsules
Red beet	ltryptophane, *p*-coumaric and ferulic acids, cyclodopa glucoside derivatives	Antioxidant, anticancer, hepatoprotective, cardioprotective, anti-inflammatory, antimicrobial	Beetroot-based supplements, beetroot can be incorporated into a variety of bakery products, including bread, cookies, cakes, muffins, pasta, noodles, and extruded goods, as well as milk, in the form of juice, powder, pulp, puree, or extract. Functional nutraceuticals and natural coloring components in foods, including flavored milk, curd, yoghurt, beverages, films and coatings, meat products, sausages, functional gels, and many more, are available.

TABLE 11.3 Phytochemicals Obtained from Herbs and Spices.

Herbs and spices	Phytochemical	Properties	Application
Black pepper	Kaempherol, rhamnetin, quercetin, dipiperamide, Piperine, wisanine, piptigrine, alkamides	Piperine shows an antioxidant effect. In vitro, piperine helps prevent oxidative degradation by repressing or suppressing free radicals and reactive oxygen species, Lowers lipid oxidation, anti-inflammatory, and analgesic effect. Helps in digestion Weight management Increases nutrient bioavailability	It is advantageous for people to alter their metabolism to lower their likelihood of developing degenerative diseases like cancer, diabetes, obesity, and cardiovascular disease. Employed as antioxidants in food to prolong the storage life of oils and cosmetics and inhibit the onset of rancidity.
Cinnamon	Gallic acid, *p*-hydroxybenzoic acid, procatechuic acid, vanillic acid, syringic acid, ellagic acid, gallagic acid, punicalagin, gentisic acid, Cinnamaldehyde	Antioxidant Control of blood glucose Antibacterial and antifungal activity Anti-inflammatory and anti-oxidant effects Cardiovascular health Hepatoprotection Neuroprotection	Halts the development of harmful chemicals while cooking at high temperatures and prevents food oxidation.
Rosemary	Catechin, epicatechin, epigallocatechin, epigallocatechin gallate, proanthocyanidins, rosmarinic acid, rosmanol, rosmadial, rosmaridipehnol, diterpenes epirosmanol, Carnosol, carnosic acid, isorosmanol, rosmariquinone	Antioxidant and anti-inflammatory effects Cognition, mental health, and neuroprotection Vascular health Blood glucose control	Added to foods and drinks to improve flavor, color, and aroma.
Turmeric	Curcumins, 4-hydroxycinnamoyl methane, *p*-coumaric, caffeic, ferulic, chlorogenic acid, curcuminoids, curcumin	Antioxidative and anti-inflammatory effects Cardiovascular health Gastrointestinal health Brain health and cognitive function	

TABLE 11.3 *(Continued)*

Herbs and spices	Phytochemical	Properties	Application
Ginger	Gingerol, shogaol, zingerone, Quercetin, myricetin, kaempferol, isorhamnetin	Cardiovascular health Nausea and vomiting Antioxidant Joint and muscle health Anti-inflammatory effects Antiglycation and antiglycemic effects Potential weight management Neuroprotective effect	

for existence. Plants create bioactive compounds to protect themselves from diverse environmental factors for survivability and adaptability. Phenolics and flavonoids serve as structural supports for or protectors of tissues, participating in defense mechanisms, luring pollinators and animals that disperse seeds, and serving as UV ray protectors, allelopathic substances, and signal molecules in plant–environment connection.[55] By altering human metabolism in a way that is advantageous for preventing or decreasing the risk of degenerative illnesses, including cancer, diabetes, obesity, and cardiovascular diseases, these phytochemicals offer similar priceless advantages to humans. Herbs are utilized in food as antioxidants to prevent rancidity and lengthen the duration of oils' and cosmetics' storage. It significantly boosts the stability of oils high in omega-3 fatty acids.[49]

11.3.3 PHYTOCHEMICALS FROM PLANT LEAVES

The leaves retain their nutritional value whether they are eaten raw, roasted, or dried and kept for a long time. Due to inherent phytochemicals, including phenolic acids, flavonoids, carotenoids, and glucosinolates, the leaves of the plant may also be consumed as foodstuff and livestock feed, in addition to having potential uses as functional foods and nutraceuticals. For example, the main phytochemicals found in moringa leaves are crypto-chlorogenic acid, isoquercetin, and astragalin, which are known for their antioxidant, antihypertension, and anti-inflammatory properties. The phytochemicals that can be obtained from various leaves are given in Table 11.4. Several in-vitro experiments have primarily supported this plant's extract's therapeutic activities and biological activity based on the bioactive components and their antioxidant activity.[16] Figure 11.3 shows the phytochemical extract from the leaves. Various leaves are under research for their valuable phytochemical compound, which possesses a therapeutic effect.

11.3.4 PHYTOCHEMICALS FROM OILSEEDS

Owing to the diminishing availability of natural resources and the necessity for environmental preservation, the utilization of agricultural wastes has attracted much interest in recent years. The following table presents the phytochemicals derived from oilseeds (Table 11.5). Numerous studies have been done to turn waste materials into biofuels, food components, and other applications with additional value.

TABLE 11.4 Phytochemicals Obtained from Leaves.

Leaves	Phytochemical	Property	Application
Stevia rebaudiana	Polyphenols Chlorophyll Carotenoids Non-nutritice substances—oxalic acid and tannins	Highly nutritive Antioxidant activity Excellent sensory and functional property Thermal and pH stable	Nutraceuticals and functional foods, alternative to sucrose
Morus nigra L., or black mulberry	Phenolic compounds— caffeoylquinic acid, chlorogenic acid, ferulic acid, isoquercitrin, rutin and kaempferol-3-O-(6–malonyl)-glucoside	Anti-inflammatory Stops bleeding	Conventional hormone therapy for women, also used for snake bites
*Ocimum basilicum L.*or basil leaves	Phenolic acids Flavonoids—Quercetin and rutin rosmarinic acid, aromatic compounds Essential oils—eugenol, chavicol, linalool, and α-terpineol	Therapeutic potential Preservation effects Health promoting Antimicrobial activity Radical scavenging Anti-inflammatory Antioxidant potential Antidiabetic Cardioprotective Immunoprotective neuroprotective Antistress and antitussive properties	Culinary ingredient-soup seasoning, tomato-based recipes, spinach seasoning, sandwich seasoning, dip seasoning, pasta seasoning, and squash seasoning. Food preservation Medicinal uses—cancer, tremors, dysentery, mental disorders, inflammation, biliousness, tooth decay, and bronchitis
Sea buckthorn (*Hippophaë rhamnoides*)	Sugars, sugar derivatives, organic acids, phenolic compounds, and lipophilic compounds (mainly carotenoids and tocopherols	Anti-inflammatory antioxidant functions metabolic health.	Value-added products to enhance human health
Olive leaves	Oleuropein; phenols and flavonoids	Strong antioxidant potency and inhibited cancer and endothelial cell proliferation	Pharmaceuticals

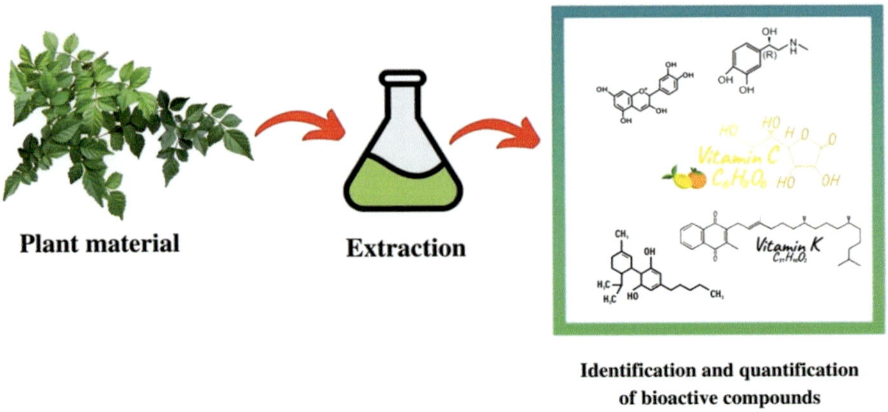

FIGURE 11.3 Phytochemicals from leaves.

Source: Adapted from Ref. [34]

The fundamental idea behind PEF-assisted extraction is that a dielectric break will cause the cell membrane to become electroporated.[40] As seen in Figure 11.4, the trans-membrane potential rises in the presence of an external electric field due to the accumulation of charges across the membrane. Further contact with an electric field raises the potential even more, which causes electrostatic interaction through the membrane between opposing charges and causes membrane thinning. The membrane will break down if the external field strength is increased beyond the critical breakdown voltage, which creates transmembrane pores. Depending on the strength of the field, the duration of the pulse, and the quantity of pulses, permeabilization can be reversible or irreversible.[35]

Reversible breakdown becomes irreversible once the dimension and amount of holes increase due to prolonged exposures exceeding critical field strengths. When the pores in relation to the overall membrane surface are small, the breakdown is reversible.

11.4 FACTORS AFFECTING EXTRACTION

1. Electric field intensity: One critical element that influences cell inactivation is the electric field's strength. When the applied field strength surpasses a threshold value known as the critical field intensity, the cell inactivation rises. Pulse width also affects this parameter. More strong electric fields generally trigger more significant amounts of extraction. It is difficult to continuously raise

TABLE 11.5 Phytochemicals Obtained from Oilseeds.

Oilseed	Phytochemical	Properties	Application
Peanuts (*Arachis hypogae L.*)	Proanthocyanidins and flavonoids, phenolic acids, and stilbenoids	Antioxidant capability	It has several positive and preventative effects on human health due to its oxidative stability and preservation in the commercial product.
Flaxseed (*Linum usitatissimum*)	Lignan (ecoisolariciresinol diglucoside), hydroxytyrosol		
Rapeseed/Canola Oil (*Brassica napus subsp. Napus*)	Sinapic acid derivatives and canolol, Glucosinolates, phytic acid	Antineoplastic, antioxidant, biocidal, and bioherbicidal	
Sunflower Seed (*Helianthus annuus*)	Phytic acid	Antioxidant capacity, hypocholesterolemic capacity, anticancerous activities inhibit the metastasis of tumor	
Sesame Seeds (*Sesamum indicum*)	Oil-soluble lignans (sesamolin, sesamolinol, sesamin, pinoresinol, and sesaminol), phenolic acids (ferulic acid, gallic acid), flavonols (quercetin), and glycosylated water-soluble lignin (sesaminol monoglucoside, pinoresinol triglucoside, pinoresinol monoglucoside, and ssesaminol triglucoside), Phytosterols	Anticarcinogenic and hypolipidemic activities also reduce the risk of atherosclerosis and CVDs', immunomodulatory functions, neuroprotective function, antihypertensive, and prevents hypoxia and brain damage, cholesterol absorption, increase its excretion, and decrease its gut absorption, which results in lower blood cholesterol levels	

the electric field intensity due to system stability concerns and the need to prevent irreparable harm. A significant increase in extraction yield occurs at the front end of a rise in electric field intensity, and a marginal rise or drop occurs when electric field intensity exceeds a certain threshold. Undoubtedly, the best values should be chosen after considering all the factors.

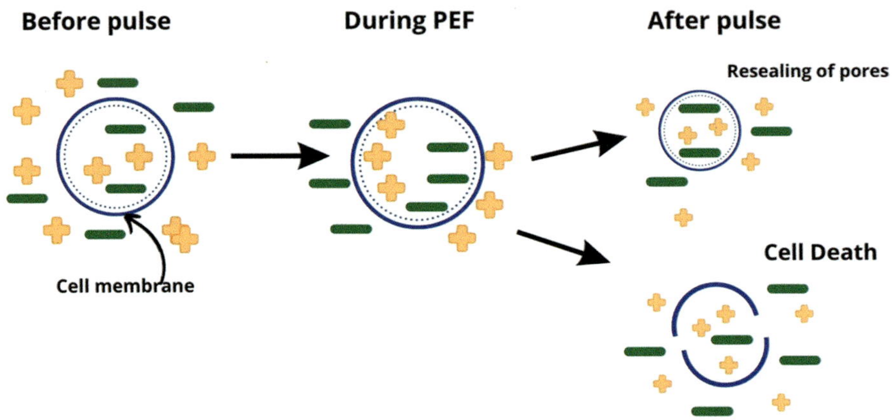

FIGURE 11.4 PEF-assisted electroporation of cell.
Source: Adapted from Ref. [48]

2. The shape of pulsed wave: Square, oscillatory, bipolar, or instantaneous reverse charges with exponential decay are some types of electric field pulses that can be used.
3. Solvent type, concentration, and pH: The liquid's conductivity, the solvent's polarity, and the solubility of a particular extract in the solvent are essential considerations when choosing a solvent. As a solvent's conductivity rises, the cell membrane's electroporation is improved, likely resulting in a higher extraction rate. As a result, using a solvent with a more excellent conductivity can improve extraction while using the same amount of input energy. On the other hand, a substance's solubility is also significant. A high mass transfer rate is represented by high extract solubility in a solvent.
4. Solvent ratio to raw material: Effective extraction depends on the solvent to raw material ratio. The final product's proportion may dilute when the proportion of solvent to raw material rises. According to He et al., with a particular amount of raw material, the proportion of bioactive compounds around the sample particles decreased when

the quantity of solvent grew, making the bioactive more straightforward to dissolve. Nevertheless, as the amount of solvent increased, the likelihood of a molecular collision decreased. Consequently, a particular variation of the solvent-to-material ratio might include an excellent value.[5]

5. Pulse period: One critical parameter to assess the effectiveness of the PEF process is the pulse period, which is determined by adding the pulse numbers and the pulse width. Cell inactivation is enhanced by a rise in pulse counts or pulse width.[28] Due to an unfavorable temperature increase, longer pulse duration may not always contribute to the best extraction. Therefore, based on experimental findings, the best processing parameters must be developed to achieve the highest inactivation rate with the most negligible heat.
6. Treatment temperature: Higher temperatures often cause liquid solvents to become less viscous, strengthening the matrix particles' penetration and improving extraction. The degradation temperature of the product's bioactive components regulates maximum temperature.

11.5 APPLICATIONS OF PULSED ELECTRIC FIELD

A nonthermal technology, the pulsed electric field, has stepped into the food-processing industry and has found its various applications, including the extraction of phytochemical constituents. Since some phytochemical tends to degrade at high temperatures, PEF has proved to be a beneficial technique. Figure 11.5 illustrates the basic flow chart of PEF. In addition to electroporating the cell membrane, enabling the passage of molecules without transport channels, several electrochemical processes occur at the electrode–solution interface once PEF is given to biomass (or any electrolyte solution) and brought in clear interaction with the procedure chamber's metal electrodes. Pulsed electric field (PEF)-assisted extraction is the application of short pulses (μs to ms) of average electric voltage (often 0.5–20 kV/cm) to the desired material placed across two electrodes. By applying high electric voltage (5–50 kV/cm), the technique can be employed for preserving materials, inactivating enzymes, and killing microorganisms.[19] In cell cultures and plant structures, PEF exposure strengths ranging from low to medium are frequently regarded as an effective pretreatment approach for boosting the concentrations of secondary metabolites during extraction.[7]

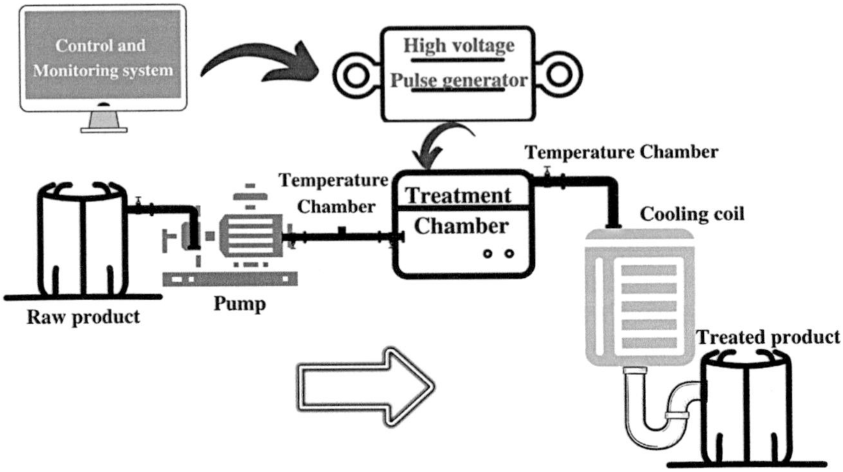

FIGURE 11.5 Flow chart of a PEF food-processing system with the essential component.
Source: Adapted from Ref. [45]

Researchers have noted the positive impacts of PEF on the juice sector, including apple, strawberry, watermelon, and citrus juices, on several enzymes' inactivation in fruit juices. Nevertheless, PEF operational parameters, including pulse shape, width, frequency, and treatment chamber dimensions, must be tuned to achieve high outcomes that affect the inactivation rate.[38]

Due to the high concentration of chemicals that are beneficial to health and have high value-added, microalgae are significant resources. Pulsed electric fields (PEFs) can potentially be employed in microalgal biorefinery applications to release compounds, including polyphenols, polyunsaturated fatty acids, and pigments. PEFs can be seen as a "moderate" cell wall rupture approach. PEFs' energy requirements and extraction efficiency rely on several variables, including the kind of microalgae present, the medium's conductivity, the number of microalgae present, and the strength of the pulse.[20]

A study was done by Pappas et al.[32] to assess the pulsed electric field's efficiency method for extracting polyphenols from olive leaves. With a pulse length of 10 s, the aqueous ethanol concentration of 25% v/v showed the most significant PEF impact. While the metabolites increased by 265.67%, the total amount of polyphenols increased by 31.85%. It was discovered that the solvent, the length of the pulse throughout the PEF procedure, and the structure of the extracted metabolites all affected the recovery of polyphenols.

Research on the *Asparagus officinalis* root (AR) was done by Symes et al.[47] Utilizing a pulsed electric field (PEF), this research sought to increase the amount of polyphenols and flavonoids extracted from green AR. Compared to traditional solvent extraction, the PEF parameters (PEF intensity of 1.6 kV/cm, frequency around 200 Hz, and pulse width of 20 s) produced more effective extraction output but reduced antioxidant activity.

Research by López-Gámez et al.[21] sought to determine the primary factors contributing to the rise in carotenoids in carrots with PEF treatment and to look at how such treatments affect the pertinent qualitative characteristics. Carrots were subjected to PEF treatments with varying electric field (E) intensities (0.8, 2, and 3.5 kV/cm) and pulse counts (5, 12, and 30). Following E $\geq$2 kV/cm application, the total and individual carotenoid contents (phytoene and β-carotene) rose; however, the concentration of lutein declined, and that of α-carotene remained close to that of untreated carrots. Better extractability is probably correlated with higher overall carotenoid content, but variations in the level of specific components show that PEF may also work by altering their production or triggering their breakdown.

Mung bean germination was the subject of research by Minh et al.[24] This study examined the effects of some PEF technical characteristics on the radicle elongation, isothiocyanate, and total phenolic of mung bean sprouts. These characteristics included pulse electric field strength (PEF) (300–1500 kV/cm), number of the pulse (50–250), and pulse width (1.0–3.0 s) at a regular frequency of 1 Hz at room temperature. Results indicated that isothiocyanate and total phenolic content were considerably raised by pulse electric field intensity and pulse number, but root length extension was retarded. While expanding root length expansion, pulse width had a detrimental effect on total phenolic and isothiocyanate.

11.6 EFFECTS OF PEF EXTRACTION PROCEDURES ON PHYTOCHEMICAL STABILITY

Pulsed electric field (PEF) technology has gained much attention in the food-processing industry over the past few decades, as a novel nonthermal preservation method and for various other practical uses.[44] Nowacka et al.[30] carried out research to assess the effects of the pulsed electric field (PEF) for use as a pretreatment technique when extracting betalains from beets at varied electric field intensities (4.38 and 6.25 kV/cm), 10–30 pulses, and 0–12.5 kJ/kg of energy input. The outcomes demonstrated that using PEF pretreatment had a substantial ($p < 0.05$) impact on the effectiveness of beetroot's bioactive components extraction. The red beet extract showed the most significant increase in betalain component concentration after 20 pulses of an electric field at the strength of 4.38 kV/cm (betanin by 329%, vulgaxanthin by 244%), in comparison to the control specimen because of the increased permeability of the cell membrane caused by leaking materials capable of conducting electricity, such as mineral salts, to the intercellular gap and using a PEF to treat the plant material improved electrical conductivity.

Vallverdu-Queralt et al.[50] investigated how the biologically active components (vitamin C content, lycopene, and total polyphenol) and tomato fruit antioxidant capacity were influenced by moderate-intensity pulsed electric fields (MIPEF). To obtain more phytochemical constituents in tomato fruit, the MIPEF treatment parameters were optimized. Various electric field intensities (from 0.4 to 2.0 kV/cm) and pulse counts (from 5 to 30) were applied to tomato fruits before they were instantly chilled at 4°C for 24 hours. The tomatoes treated with MIPEF had a greater concentration of bioactive constituents than the untreated tomatoes. At 30 or fewer pulses at 1.2 kV/cm, a 44% increase in the content of all polyphenols was achieved. Eighteen pulses at 1.2 kV/cm increased the hydrophilic antioxidant capacity by 44%, while five pulses at 1.2 kV/cm increased the lipophilic antioxidant capacity by 37%. The treated tomatoes' total bioactive compounds and antioxidant capacity were reached within sixteen pulses at 1 kV/cm. MIPEF treatments might thus be a successful way to increase tomatoes' bioactive component content and antioxidant capacity.

According to Morales-de la Peña et al.,[25] research to enhance the isoflavone profile of soymilk, pulsed electric fields (PEFs) was examined as an extraction-assisting method. In two phases of the soymilk extraction process, hydrated soybeans (HSB) and soybean slurry (SBS), low-intensity PEFs were used under various operating parameters, with the soymilk obtained using the traditional method serving as the control (CSM). Compared to CSM, the final soymilk samples from PEF-HSB and PEF-SBS had higher aglycone content and lower glucoside isoflavone concentrations.

11.7 SUMMARY

The pulsed electric field has shown itself as a promising technology for extracting phytochemicals. PEF processing allows for the higher valorization of plant cell constituents since it is a nonthermal process and uses less energy than traditional thermal extraction and dehydration processes. Agro-industrial by-products, oilseeds, herbs, and spices are a valuable reservoir of natural phytochemicals, and in order to fully use the potential of inexpensive by-products and other plant materials, efficient and cost-effective extraction procedures must be available. Utilizing cutting-edge extraction techniques may be a wise decision. PEF-assisted extractions can apply energy-efficient procedures to improve phytochemical recovery. A noteworthy feature of the mild PEF procedure for developing valorized products appears to be the targeted extraction of intracellular molecules. In the literature, the sugar beet for sucrose extraction, grapes, and red beet for colorants, green biomass for polyphenols, microalgae for lipids, and a plethora of other substances are only a few examples of targeted extraction through the electroporated cellular matrix. For potential industrial exploitation, the benefits of PEF-assisted procedures can be assessed from an economic and ecological perspective by conducting trial work on the products of relevance using the PEF process at the industrial and pilot plant levels.

KEYWORDS

- **agro-industrial waste**
- **extraction**
- **phytochemicals**
- **pulsed electric field processing**

REFERENCES

1. Abou Baker, D. H.; Ibrahim, E. A.; Salama, Z. Citrus Peels as a Source of Bioactive Compounds with Industrial and Therapeutic Applications. In *Phenolic Compounds—Chemistry, Synthesis, Diversity, Non-Conventional Industrial, Pharmaceutical and Therapeutic Applications;* Badria, F. A., Ed.; IntechOpen: London, UK. 2021, pp 1–13.
2. Ade-Omowaye, B. I. O.; Angersbach, A.; Taiwo, K. A.; Knorr, D. Use of Pulsed Electric Field Pre-Treatment to Improve Dehydration Characteristics of Plant Based Foods. *Trend Food Sci. Technol.* **2001,** *12* (8), 285–295.
3. Al-Weshahy, A.; Rao, V. A. Potato Peel as a Source of Important Phytochemical Antioxidant Nutraceuticals and their Role in Human Health-A Review: In *Phytochemicals as Nutraceuticals—Global Approaches to Their Role in Nutrition and Health;* Rao, V., Ed.; IntechOpen: London, UK, 2012.
4. Amalraj, A.; Varma, K.; Jacob, J.; Divya, C.; Kunnumakkara, A. B.; Stohs, S. J.; Gopi, S. A Novel highly Bioavailable Curcumin Formulation Improves Symptoms and Diagnostic Indicators in Rheumatoid Arthritis Patients: A Randomized, Double-Blind, Placebo-Controlled, Two-Dose, Three-Arm, and Parallel-Group Study. *J. Med. Food* **2017,** *20* (10), 1022–1030.

5. Aoude, C.; Lammerskitten, A.; Parniakov, O.; Zhang, R.; Grimi, N.; El Zakhem, H.; Vorobiev, E. Equipment and Recent Advances in Pulsed Electric Fields. *Innovat. Emerg. Technol. Bio-Marine Food Sector* **2022,** 149–172.
6. Azwanida, N. N. A Review on the Extraction Methods use in Medicinal Plants, Principle, Strength and Limitation. *Med. Aromat. Plants* **2015,** *4* (196), 2167–0412.
7. Barbosa-Pereira, L.; Guglielmetti, A.; Zeppa, G. Pulsed Electric Field Assisted Extraction of Bioactive Compounds from Cocoa Bean Shell and Coffee Silverskin. *Food and Bioprocess Technol.* **2018,** *11* (4), 818–835.
8. Chaudhary, P.; Sharma, A.; Singh, B.; Nagpal, A. K. Bioactivities of Phytochemicals Present in Tomato. *J. Food Sci. Technol.* **2018,** *55* (8), 2833–2849.
9. Coman, V.; Teleky, B. E.; Mitrea, L.; Martău, G. A.; Szabo, K.; Călinoiu, L. F.; Vodnar, D. C. Bioactive potential of Fruit and Vegetable Wastes. *Adv. Food Nutr. Res.* **2020,** *91,* 157–225.
10. Deng, G. F.; Shen, C.; Xu, X. R.; Kuang, R. D.; Guo, Y. J.; Zeng, L. S.; Gao, L.L.; Lin, X.; Xie, J.F.; Xia, E.Q. & Li, H. B. Potential of Fruit Wastes as Natural Resources of Bioactive Compounds. *Int. J. Mole. Sci.* **2012,** *13* (7), 8308–8323.
11. Embuscado, M. E. Bioactives from Culinary Spices and Herbs: A Review. *J. Food Bioactiv.* **2019,** *6.*
12. Ezeorba, T. P. C.; Chukwudozie, K. I.; Ezema, C. A.; Anaduaka, E. G.; Nweze, E. J.; Okeke, E. S. Potentials for Health and Therapeutic Benefits of Garlic Essential Oils: Recent Findings and Future Prospects. *Pharmacol. Res. Modern Chin. Med.* **2022,** *3,* 100075.
13. Fierascu, R. C.; Fierascu, I.; Avramescu, S. M.; Sieniawska, E. Recovery of Natural Antioxidants from Agro-Industrial Side Streams through Advanced Extraction Techniques. *Molecules* **2019,** *24* (23), 4212.
14. Goulas, V.; Exarchou, V.; Troganis, A. N.; Psomiadou, E.; Fotsis, T.; Briasoulis, E.; Gerothanassis, I. P. Phytochemicals in Olive-Leaf Extracts and their Antiproliferative Activity against Cancer and Endothelial Cells. *Mole. Nutr. Food Res.* **2009,** *53* (5), 600–608.
15. Javed, M.; Ahmed, W.; Mian, R.; Ahmad, A. M. R. Garlic as a Potential Nominee in Functional Food Industry. In *Herbs and Spices-New Processing Technologies*; Ahmad, R.S., Ed.; IntechOpen: London, UK. 2021.
16. Kashyap, P.; Kumar, S.; Riar, C. S.; Jindal, N.; Baniwal, P.; Guiné, R. P.; Correia, P. M.; Mehra, R.; Kumar, H. Recent Advances in Drumstick (Moringa oleifera) Leaves Bioactive Compounds: Composition, Health Benefits, Bioaccessibility, and Dietary Applications. *Antioxidants* **2022,** *11* (2), 402.
17. Kovačević, D. B.; Maras, M.; Barba, F. J.; Granato, D.; Roohinejad, S.; Mallikarjunan, K.; Montesano, D.; Lorenzo, J. M.; Putnik, P. Innovative Technologies for the Recovery of Phytochemicals from Stevia Rebaudiana Bertoni Leaves: A Review. *Food Chem.* **2018,** *268,* 513–521.
18. Kumar, M.; Chandran, D.; Tomar, M.; Bhuyan, D. J.; Grasso, S.; Sá, A. G. A.; Carciofi, B. A. M.; Radha, Dhumal, S.; Singh, S.; Mekhemar, M. Valorization Potential of Tomato (*Solanum lycopersicum* L.) Seed: Nutraceutical Quality, Food Properties, Safety Aspects, and Application as a Health-Promoting Ingredient in Foods. *Horticulturae* **2022,** *8* (3), 265.
19. Kumari, B.; Tiwari, B. K.; Hossain, M. B.; Brunton, N. P.; Rai, D. K. Recent Advances on Application of Ultrasound and Pulsed Electric Field Technologies in the Extraction

of Bioactives from Agro-Industrial By-Products. *Food Bioprocess Technol.* **2018,** *11* (2), 223–241.

20. Lafarga, T.; Aguiló-Aguayo, I. Pulsed Electric Fields for the Extraction of Lipids, Pigments, and Polyphenols from Cultured Microalgae. In: *Innovative and Emerging Technologies in the Bio-marine Food Sector;* Garcia-Vaquero, M., Rajauria, G., Eds.; Academic Press, 2022, pp 197–221.
21. López-Gámez, G.; Elez-Martínez, P.; Martín-Belloso, O.; Soliva-Fortuny, R. Changes of Carotenoid Content in Carrots after Application of Pulsed Electric Field Treatments. *LWT* **2021,** *147,* 111408.
22. Ma, X.; Yang, W.; Kallio, H.; Yang, B. Health Promoting Properties and Sensory Characteristics of Phytochemicals in Berries and Leaves of Sea Buckthorn (*Hippophaë rhamnoides*). *Crit. Rev. Food Sci. Nutr.* **2022,** *62* (14), 3798–3816.
23. Ma, Z. F.; Zhang, H. Phytochemical Constituents, Health Benefits, and Industrial Applications of Grape seeds: A Mini-Review. *Antioxidants* **2017,** *6* (3), 71.
24. Minh, N. P. Total Phenolic, Isothiocyanate and Radicle Elongation Determination of Mung Bean Sprout During Germination Affected by Different Variables of Pulsed Electric Field Treatment. *Plant Sci. Today* **2022,** *9* (2), 504–508.
25. Morales-de la Peña, M.; Welti-Chanes, J.; Martín-Belloso, O. Impact of Pulsed Electric Field Pre-Treatment on the Isoflavone Profile of Soymilk. *Beverages* **2022,** *8* (2), 19.
26. Morya, S.; Menaa, F.; Jiménez-López, C.; Lourenço-Lopes, C.; BinMowyna, M. N.; Alqahtani, A. Nutraceutical and Pharmaceutical Behavior of Bioactive Compounds of Miracle Oilseeds: An Overview. *Foods* **2022,** *11* (13), 1824.
27. Nadeem, H. R.; Akhtar, S.; Sestili, P.; Ismail, T.; Neugart, S.; Qamar, M.; Esatbeyoglu, T. Toxicity, Antioxidant Activity, and Phytochemicals of Basil (*Ocimum basilicum* L.) Leaves Cultivated in Southern Punjab, Pakistan. *Foods* **2022,** *11* (9), 1239.
28. Naliyadhara, N.; Kumar, A.; Girisa, S.; Daimary, U. D.; Hegde, M.; Kunnumakkara, A. B. Pulsed Electric Field (PEF): Avant-Garde Extraction Escalation Technology in Food Industry. *Trend Food Sci. Technol.* **2022,** *122,* 238–255.
29. Nastić, N.; Borrás-Linares, I.; Lozano-Sánchez, J.; Švarc-Gajić, J.; Segura-Carretero, A. Optimization of the Extraction of Phytochemicals from Black Mulberry (*Morus nigra* L.) leaves. *J. Indust. Eng. Chem.* **2018,** *68,* 282–292.
30. Nowacka, M.; Tappi, S.; Wiktor, A.; Rybak, K.; Miszczykowska, A.; Czyzewski, J.; Drozdzal, K.; Witrowa-Rajchert, D.; Tylewicz, U. The Impact of Pulsed Electric field on the Extraction of Bioactive Compounds from Beetroot. *Foods* **2019,** *8* (7), 244.
31. Panja, P. Green Extraction Methods of Food Polyphenols from Vegetable Materials. *Curr. Opin. Food Sci.* **2018,** *23,* 173–182.
32. Pappas, V. M.; Lakka, A.; Palaiogiannis, D.; Bozinou, E.; Ntourtoglou, G.; Batra, G.; Lalas, S. I. Use of Pulsed Electric Field as a Low-Temperature and High-Performance "green" Extraction Technique for the Recovery of High Added Value Compounds from Olive Leaves. *Beverages* **2021,** *7* (3), 45.
33. Punia Bangar, S.; Singh, A.; Chaudhary, V.; Sharma, N.; Lorenzo, J. M. Beetroot as a Novel Ingredient for its Versatile Food Applications. *Crit. Rev. Food Sci. Nutr.* **2022,** 1–25.
34. Putnik, P.; Lorenzo, J. M.; Barba, F. J.; Roohinejad, S.; Režek Jambrak, A.; Granato, D.; Montesano, D.; Bursać Kovačević, D. Novel Food Processing and Extraction Technologies of High-Added Value Compounds from Plant Materials. *Foods* **2018,** *7* (7), 106.

35. Ranjha, M. M. A.; Kanwal, R.; Shafique, B.; Arshad, R. N.; Irfan, S.; Kieliszek, M.; Kowalczewski, P.Ł.; Irfan, M.; Khalid, M.Z.; Roobab, U.; Aadil, R. M. A Critical Review on Pulsed Electric Field: A Novel Technology for the Extraction of Phytoconstituents. *Molecules* **2021,** *26* (16), 4893.
36. Ranjith, S. Bioactive Composition of Onion (*Allium cepa*) and its Health Benefits: A Review. *Pharma Innovat. J.* **2022,** *11* (6), 1148–1153
37. Rifna, E. J.; Misra, N. N.; Dwivedi, M. Recent Advances in Extraction Technologies for Recovery of Bioactive Compounds Derived from Fruit and Vegetable Waste Peels: A Review. *Crit. Rev. Food Sci. Nutr.* **2023,** *63* (6), 719–752.
38. Roobab, U.; Abida, A.; Chacha, J. S.; Athar, A.; Madni, G. M.; Ranjha, M. M. A. N.; Rusu, A.V.; Zeng, X.A.; Aadil, R.M.; Trif, M. Applications of Innovative Non-Thermal Pulsed Electric Field Technology in Developing Safer and Healthier Fruit Juices. *Molecules* **2022,** *27* (13), 4031.
39. Sagar, N. A.; Pareek, S.; Sharma, S.; Yahia, E. M.; Lobo, M. G. Fruit and Vegetable Waste: Bioactive Compounds, their Extraction, and Possible Utilization. *Comprehensive Rev. Food Sci. Food Safe.* **2018,** *17* (3), 512–531.
40. Salee, N.; Chaiyana, W.; Yawootti, A.; Naruenartwongsakul, S.; Klangpetch, W.; Walter, P.; Utama-Ang, N. Optimization of the Pulse Electric Field Assisted Extraction of Black Rice Grain for Antioxidant and Sirtuin1 Enzyme Stimulation Activities. *Scientif. Rep.* **2022,** *12* (1), 1–11.
41. Sepelev, I.; Galoburda, R. Industrial Potato Peel Waste Application in Food Production: A Review. *Res. Rural Develop.* **2015,** *1,* 130–136.
42. Serna-Cock, L.; García-Gonzales, E.; Torres-León, C. Agro-Industrial Potential of the Mango Peel Based on its Nutritional and Functional Properties. *Food Rev. Int.* **2016,** *32* (4), 364–376.
43. Shehzadi, K.; Rubab, Q.; Asad, L.; Ishfaq, M.; Shafique, B.; Ali Nawaz Ranjha, M. M.; & Sabtain, B. A Critical Review on Presence of Polyphenols in Commercial Varieties of Apple Peel, their Extraction and Health Benefits. *Open Access J. Biogener. Sci. Res.* **2020,** *6,* 18.
44. Soliva-Fortuny, R.; Balasa, A.; Knorr, D.; Martín-Belloso, O. Effects of Pulsed Electric Fields on Bioactive Compounds in Foods: A Review. *Trend Food Sci. Technol.* **2009,** *20* (11–12), 544–556.
45. Soni, A.; Brightwell, G. Effect of Hurdle Approaches Using Conventional and Moderate Thermal Processing Technologies for Microbial Inactivation in Fruit and Vegetable Products. *Foods* **2022,** *11* (12), 1811.
46. Stéphane, F. F. Y.; Jules, B. K. J.; Batiha, G. E.; Ali, I.; Bruno, L. N. Extraction of Bioactive Compounds from Medicinal Plants and Herbs. In *Natural Medicinal Plants*; El-Shemy, H., Ed.: IntechOpen: London, UK, 2021.
47. Symes, A.; Shavandi, A.; Bekhit, A. E. D. A. Effects of Ionic Liquids and Pulsed Electric Fields on the Extraction of Antioxidants from Green Asparagus Roots. *Int. J. Food Sci. Technol.* **2022,** *58* (7), 3935–3945.
48. Thulasidas, J. S.; Varadarajan, G. S.; Sundararajan, R. Pulsed Electric Field for Enhanced Extraction of Intracellular Bioactive Compounds from Plant Products: An Overview. *Nov. Approach. Drug. Des. Dev.* **2019,** *5* (2), 1–6.
49. Tome-Carneiro, J.; Visioli, F. Polyphenol-Based Nutraceuticals for the Prevention and Treatment of Cardiovascular Disease: Review of Human Evidence. *Phytomedicine* **2016,** *23* (11), 1145–1174.

50. Vallverdu-Queralt, A.; Oms-Oliu, G.; Odriozola-Serrano, I.; Lamuela-Raventos, R. M.; Martin-Belloso, O.; Elez-Martinez, P. Effects of Pulsed Electric Fields on the Bioactive Compound Content and Antioxidant Capacity of Tomato Fruit. *J. Agr. Food Chem.* **2012,** *60* (12), 3126–3134.
51. Wink, M. Current Understanding of Modes of Action of Multicomponent Bioactive Phytochemicals: Potential for Nutraceuticals and Antimicrobials. *Ann. Rev. Food Sci. Technol.* **2022,** *13*, 337–359.
52. Xiao, J.; Bai, W. Bioactive Phytochemicals. *Crit. Rev. Food Sci. Nutr.* **2019,** *59* (6), 827–829.
53. Zaini, H. M.; Roslan, J.; Saallah, S.; Munsu, E.; Sulaiman, N. S.; Pindi, W. Banana Peels as a Bioactive Ingredient and its Potential Application in the Food Industry. *J. Funct. Foods* **2022,** *92,* 105054.
54. Zhang, S.; Sun, L.; Ju, H.; Bao, Z.; Zeng, X. A.; Lin, S. Research Advances and Application of Pulsed Electric Field on Proteins and Peptides in Food. *Food Res. Int.* **2021,** *139,* 109914.
55. Zheng, J.; Zhou, Y.; Li, Y.; Xu, D. P.; Li, S.; Li, H. B. Spices for Prevention and Treatment of Cancers. *Nutrients* **2016,** *8* (8), 495.

INDEX

I

J

K

L

M

N

O

P

Q

S